Privacy in Dynamical Systems

Farhad Farokhi

Editor

Privacy in Dynamical Systems

 Springer

Editor
Farhad Farokhi
Department of Electrical and Electronic
Engineering
The University of Melbourne
and CSIRO's Data61
Melbourne, VIC, Australia

ISBN 978-981-15-0495-2   ISBN 978-981-15-0493-8   (eBook)
https://doi.org/10.1007/978-981-15-0493-8

This Springer imprint is published by the registered company Springer Nature Singapore Pte Ltd.
The registered company address is: 152 Beach Road, #21-01/04 Gateway East, Singapore 189721,
Singapore

To Jodie and my parents

Preface

Privacy literature within the computer-science community has been mostly focused on responding to queries on large static datasets kept securely by a data curator. In practice, however, the underlying data in possession of the curator might change over time, sometimes according to a dynamical system. This has motivated a body of research on privacy in dynamical systems. This book attempts at presenting the state-of-the-art in this relatively new field of research.

The contributions of this book focus on two distinct approaches for providing privacy. One approach is based on statistics. The authors in this case use randomness to ensure privacy and rely on information-theoretic measures of privacy or differential privacy. The other approach is based on using multi-party secure computation and homomorphic encryption for restricting access to private data. In what follows, I provide a brief summary of the contributions in this book.

In Chap. 1, the authors propose a measure of privacy using the Fisher information. They particularly focus on smart meter privacy and use heating, ventilation, and air conditioning units for masking private information that can be extracted from the energy consumption of the households. They then compute optimal privacy-preserving policy by minimizing a weighted sum of the Fisher information and the operation cost of the unit under constraints on the temperatures of the house.

In Chap. 2, the authors also consider smart-meter privacy. They use information-theoretic measures for quantifying private information leakage. This chapter considers privacy-enabling techniques based on both data manipulation and demand shaping (using renewable energy sources and rechargeable batteries). The authors particularly use tools from the Markov decision processes and rate–distortion theory to find the optimal energy management strategies.

In Chap. 3, the authors consider hypothesis-testing adversaries for modeling privacy risks. They use fundamental information-theoretic bounds for hypothesis testing to measure the private information leakage and to devise privacy-preserving methods. This chapter also uses smart-meter privacy for demonstration of theoretic results and privacy-enhancing technologies.

In Chap. 4, the authors use Bayesian statistics and information theory to develop privacy measures in the problem of sharing the outcomes of a parametric source with an untrusted party while ensuring the privacy of the parameters. They develop tools for privacy preservation in this setup and investigate the properties of the proposed mechanisms by particularly focusing on utility-privacy trade-off.

In Chap. 5, the authors consider the privacy of a partially observable stochastic system in the presence of an eavesdropper with the knowledge of the system's model. They propose a notion of privacy using the intruder's belief about the system. The privacy preservation problem can then be cast as a reachability problem with respect to the unsafe subset of beliefs defined by the privacy requirement. They then use control-theoretic tools to ensure privacy for the system.

In Chap. 6, the authors consider maximizing privacy of datasets using additive noise generated by synchronized chaotic oscillators. First, the authors compute the distribution of noise by minimizing the mutual information between private entries of the dataset and the obfuscated responses. They then propose an algorithm to generate pseudo-random realizations of this distribution using trajectories of a chaotic oscillator. At the other end of the channel, a second chaotic oscillator can be used to generate realizations from the same distribution for decoding the responses.

In Chap. 7, the authors consider privacy of transportation data. They first use mobility data to illustrate the difficulty of publishing location data with privacy guarantees. They then use differential privacy to publish aggregate statistics of the data. They present a dynamic estimator for traffic along a highway segment in real time from single-loop detector and floating car data while providing privacy guarantees for the vehicle trajectories. They use a nonlinear Kalman filter to mitigate the effect of privacy noise in their estimator.

In Chap. 8, the authors consider multi-agent systems in which the agents wish to maintain privacy of their states. The agents use differential privacy for maintaining their privacy. The authors determine the performance degradation due to the noise and show that depending on the desired level of privacy, the system performance can be optimized by changing the level of cooperation among the agents.

In Chap. 9, the authors explore privacy-preserving computation for networked controllers using homomorphic encryption and secret sharing. They particularly discuss these techniques in the context of implementation of a model predictive controller. This is done by computing stabilizing control actions for the system by solving an optimization problem on encrypted data.

In Chap. 10, the authors use a post-quantum homomorphic cryptosystem based on learning-with-errors for secure and private control over networks. This cryptosystem cannot handle unlimited recursive operation of homomorphic arithmetics on ciphertexts. However, the authors overcome this limitation for dynamic feedback controllers and guarantee stability of the closed-loop system if the controller's state-space representation only consists of integer numbers.

In Chap. 11, the author also investigates encrypted model predictive control. In addition to presenting a survey of the methods available for encrypted model predictive control, the author presents a novel encrypted control using the alternating direction method of multipliers. Finally, the author theoretically and experimentally compares the methods for encrypted model predictive control.

In Chap. 12, the author uses homomorphic public-key ElGamal encryption system to encrypt a linear controller. The use of homomorphic encryption ensures privacy of controller parameters, references, measurements, control commands, and parameters of plant models in the internal model principle while maintaining the original function of the controller.

Melbourne, Australia
August 2019

Farhad Farokhi

Acknowledgements

There are a few people without whom this book would not have been finished in time or even published.

First, I would like to thank all the authors for their timely submissions, revisions, and assistance. This book is made possible with their efforts.

I would also like to thank Christoph Baumann, N. S. Pandian, Sriram Srinivas, and S. Priyadharshini from Springer Nature for their assistance in getting the book accepted and ready for publication.

Numerous people volunteered to review the chapters anonymously. Their efforts have resulted in improving the contents of this book significantly. I would also like to thank them.

My part-time research fellow position at the University of Melbourne is supported by the Deputy Vice-Chancellor Research (DVCR). I am grateful to their contribution that has allowed me to pursue my research in the privacy and cyber-security, and to create collaboration and connections with international researchers in this field.

Melbourne, Australia   Farhad Farokhi
August 2019

Contents

Part II Encryption-Based Privacy

Contributors

Andreea B. Alexandru Department of Electrical Engineering, University of Pennsylvania, Philadelphia, PA, USA

Germán Bassi KTH Royal Institute of Technology, Stockholm, Sweden

Pier Luigi Dragotti Department of Electrical and Electronic Engineering, Imperial College London, London, UK

Ecenaz Erdemir Department of Electrical and Electronic Engineering, Imperial College London, London, UK

Farhad Farokhi The University of Melbourne and CSIRO's Data61, Melbourne, Australia;
Department of Electrical and Electronic Engineering, University of Melbourne, Melbourne, VIC, Australia;
The Commonwealth Scientific and Industrial Research Organisation (CSIRO), Data61, Canberra, Australia

Deniz Gündüz Department of Electrical and Electronic Engineering, Imperial College London, London, UK

Vijay Gupta Department of Electrical Engineering, University of Notre Dame, Notre Dame, IN, USA

Kyoohyung Han Department of Mathematical Sciences, Seoul National University, Seoul, South Korea

Karl H. Johansson KTH Royal Institute of Technology, Stockholm, Sweden

Vaibhav Katewa Department of Mechanical Engineering, University of California Riverside, Riverside, CA, USA

Junsoo Kim ASRI, Department of Electrical and Computer Engineering, Seoul National University, Seoul, South Korea

Kiminao Kogiso The University of Electro-Communications, Chofu, Tokyo, Japan

Jerome Le Ny Department of Electrical Engineering, Polytechnique Montreal, Montreal, QC, Canada

Zuxing Li KTH Royal Institute of Technology, Stockholm, Sweden

Hai Lin Department of Electrical Engineering, University of Notre Dame, Notre Dame, IN, USA

Carlos Murguia Department of Electrical and Electronic Engineering, University of Melbourne, Melbourne, VIC, Australia

Ehsan Nekouei City University of Hong Kong, Kowloon Tong, Hong Kong

Dragan Nešić Department of Electrical and Electronic Engineering, University of Melbourne, Melbourne, VIC, Australia

Tobias J. Oechtering KTH Royal Institute of Technology, Stockholm, Sweden

George J. Pappas Department of Electrical Engineering, University of Pennsylvania, Philadelphia, PA, USA

Fabio Pasqualetti Department of Mechanical Engineering, University of California Riverside, Riverside, CA, USA

Mathilde Pelletier Department of Civil Engineering, Polytechnique Montreal, Montreal, QC, Canada

Henrik Sandberg KTH Royal Institute of Technology, Stockholm, Sweden

Nicolas Saunier Department of Civil Engineering, Polytechnique Montreal, Montreal, QC, Canada

Moritz Schulze Darup Automatic Control Group, Department of Electrical Engineering and Information Technology, Universität Paderborn, Paderborn, Germany

Iman Shames Department of Electrical and Electronic Engineering, University of Melbourne, Melbourne, VIC, Australia

Hyungbo Shim ASRI, Department of Electrical and Computer Engineering, Seoul National University, Seoul, South Korea

Mikael Skoglund KTH Royal Institute of Technology, Stockholm, Sweden

Ufuk Topcu Department of Aerospace Engineering and Engineering Mechanics, Oden Institute for Computational Engineering and Sciences, University of Texas at Austin, Austin, TX, USA

Bo Wu Oden Institute for Computational Engineering and Sciences, University of Texas at Austin, Austin, TX, USA

Yang You KTH Royal Institute of Technology, Stockholm, Sweden

Part I
Statistical Data Privacy

Chapter 1
Fisher Information Privacy with Application to Smart Meter Privacy Using HVAC Units

Farhad Farokhi and Henrik Sandberg

Abstract In this chapter, we use Heating, Ventilation, and Air Conditioning (HVAC) units to preserve the privacy of households with smart meters in addition to regulating indoor temperature. We model the effect of the HVAC unit as an additive noise in the household consumption. The Cramér-Rao bound is used to relate the inverse of the trace of the Fisher information matrix to the quality of an adversary's estimation error of the household private consumption from the aggregate consumption of the household with the HVAC unit. This establishes the Fisher information as the measure of privacy leakage. We compute the optimal privacy-preserving policy for controlling the HVAC unit through minimizing a weighted sum of the Fisher information and the cost operating the HVAC unit. The optimization problem also contains the constraints on the temperatures of the house.

1.1 Introduction

Smart meters are commonly used in modern electricity networks for metering consumption and sensing variables-of-interest for feedback control. The data gathered by the smart meters might also be used by adversaries to infringe on the privacy of households [14, 19, 26]. Therefore, appropriate privacy-preserving mechanisms must be developed and implemented to ensure the privacy of households with smart meters.

A common tool in privacy analysis is differential privacy [3, 4, 13, 15, 17]. Differential privacy requires that outputs, such as smart meter readings, do not noticeably change if an individual's data, such as its habits and appliance usage, changes. Privacy-preserving mechanisms that meet the standards of differential privacy often rely on additive noises with slow-decaying distributions, such as the Laplace noise.

F. Farokhi (✉)
The University of Melbourne and CSIRO's Data61, Melbourne, Australia
e-mail: farhad.farokhi@unimelb.edu.au; farhad.farokhi@data61.csiro.au

H. Sandberg
KTH Royal Institute of Technology, Stockholm, Sweden
e-mail: hsan@kth.se

© Springer Nature Singapore Pte Ltd. 2020

F. Farokhi (ed.), *Privacy in Dynamical Systems*,
https://doi.org/10.1007/978-981-15-0493-8_1

Previously, differentially-private mechanisms have been previously used in private smart metering [1, 21].

Although potentially powerful when dealing with consumption aggregates across multiple households, implementing differentially-private mechanisms at household level remains impractical due to multiple reasons. First, in addition sensing and control purposes by the grid operator, smart meters are used for metering and billing purposes and thus adding noise might not be desirable from the perspective of the retailer and the household. Second, an additive Laplace noise might result in an inferior utility [2, 10, 20, 21] and increase the computational complexity of optimal filtering [7]. Differential privacy is also susceptible to adversarial attacks [12]. Finally, and perhaps possibly most importantly, because of the lack of a systematic way for setting the so-called privacy parameter or budget, differential privacy has shown to be sometimes ineffective in practice [11, 23].

Fueled by these criticisms, information-theoretic privacy has gathered momentum as a rival approach for measuring private information leakage and designing privacy-preserving policies [18]. Information-theoretic privacy dates back to wiretap channels [24] and their extensions [25]. Information-theoretic measures of privacy leakage rely on mutual information (or relative entropy) [18], Fisher information [8, 9], maximin information [6] for measuring private information leakage and cast the privacy problem as a generalized rate-distortion problem. The use of entropy as a measure of privacy forces the private dataset to be statistically distributed with known distributions. This motivates the extension of information-theoretic privacy to use the Fisher information and maximin information as measures of private information leakage. This is because these information measures does not impose statistical assumptions on the private dataset.

In this paper, we use Heating, Ventilation, and Air Conditioning (HVAC) units to preserve the privacy of households. The effect of the HVAC unit can be modeled as an additive noise in the household consumption but there are several important distinctions with the case of additive measurement noise. Firstly, the HVAC units cannot be run arbitrarily as their main goal is to keep the temperature of the house within a comfortable zone and thus noises with unbounded support, such as the Gaussian and Laplace noises, cannot be implemented. Secondly, the measurements of the voltage and the current provided by the smart meter in this case are exact and thus estimation complexities and metering issues are entirely avoided. We consider an adversary that aims to estimate the household's consumption, as closely as possible, from the aggregate consumption, i.e., the combined private consumption of the household and the consumption of the HVAC unit. We use the Cramér-Rao bound [22, p. 169] to relate the variance of the estimation error of unbiased estimators of the household consumption from the aggregate consumption by the adversary to the inverse of the trace of the Fisher information matrix. This enables the use of the Fisher information as a measure of privacy. We can then find the optimal policies for controlling the HVAC unit through minimizing the Fisher information. We also consider the cost operating the HVAC unit and add that to the optimization problem in order to capture the trade-off between privacy and utility/comfort. We do this while enforcing the constraints on the indoor temperature of the house.

The design of privacy-preserving policies based on the Fisher information and its interpretation through the Cramér-Rao bound are have been previously studied in [8, 9]. However, using this metric for designing a privacy-preserving control unit for HVAC units, in this chapter, is entirely novel.

The rest of the chapter is organized as follows. In Sect. 1.2, we present the problem formulation by discussing the merits of using the Fisher information as a measure of privacy and modeling HVAC units. We present optimal privacy-preserving policies in Sect. 1.3. Finally, we present a numerical example in Sect. 1.4 and conclude the paper in Sect. 1.5.

1.2 Problem Formulation

Let $x[k] \in \mathbb{R}^n$, for all $k \in \mathfrak{T} := \{0, \ldots, T\}$ with T denoting the privacy-preservation horizon, be an arbitrary deterministic sequence which is desired to be kept private. At every time instant k, the following report is made:

$$y[k] = x[k] + w[k], \tag{1.1}$$

where $w[k]$ is an additive noise for privacy preservation. We assume that the additive noise over the horizon $\mathfrak{T}$ must satisfy $Cw[0:T] \leq d$ with probability one, where $w[0:T] = [w[0]^\top \cdots w[T]^\top]^\top$. We assume that the probability measure of the additive noise $w[0:T]$ is absolutely continuous with respect to the Lebesgue measure μ. Therefore, there exists probability density function γ such that

$$\mathbb{P}\{w[0:T] \in \mathfrak{W}\} = \int_{\mathfrak{W}} \gamma(w)\mu(dw),$$

where $\mathfrak{W} \subseteq \mathbb{R}^{nT}$ is any Borel-measurable set. Further, we may define $\Omega := \{w[0:T] \mid Cw[0:T] \leq d\}$. In this chapter, we restrict ourselves to additive noise $w[k]$ that is independent of the data $x[k]$. One could imagine that a causal stochastic map from $x[k]$ to $w[k]$ may yield an even higher level of privacy in some cases but that is out of our scope. We restrict our attention for selecting the design variable γ to the set of density functions that fulfill the following standing assumption.

Assumption 1 (i) $\mathbb{P}\{w[0:T] \in \Omega\} = 1$, (ii) $\gamma(w[0:T]) > 0$ for all $w[0:T] \in$ int Ω, and (iii) $\gamma(w[0:T]) = 0$ for all $w[0:T] \in \partial\Omega$.

The first part of the assumption is required for the feasibility of the additive noise. This will be discussed thoroughly in the case of HVAC systems at the end of this section. The second and the third parts of the assumption ensure the validity of the Cramér-Rao bound for relating the estimation error an adversary to the Fisher information. In what follows, Γ denotes the set of all density functions γ that meet Assumption 1.

In order to keep $x[k]$, $\forall k \in \mathfrak{T}$, private, we opt to make the estimation error $\sum_{k=0}^{T} \mathbb{E}\{\|x[k] - \hat{x}[k]\|_2^2\}$ as large as possible, where $\hat{x}[0 : T]$ is an estimate of $x[0 : T]$ based on the transmitted measurements $y[0 : T]$. This is however a function of the adversary's estimation policy, which might not be available at the time of computing the privacy-preserving additive noise. Therefore, we use the Cramér-Rao bound [22, p. 169] to relate the estimation error to a fundamental property of the additive noise $w[0 : T]$, known as the Fisher information. Under Assumption 1, for any unbiased estimate of $x[0 : k]$ given measurements $y[0 : T]$ denoted by $\hat{x}[0 : T]$, we have

$$\frac{1}{T+1} \sum_{k=0}^{T} \mathbb{E}\{\|x[k] - \hat{x}[k]\|_2^2\} = \frac{1}{T}\mathbb{E}\{\|x[0 : k] - \hat{x}[0 : T]\|_2^2\}$$

$$\geq \frac{n}{\text{trace}(\mathfrak{J}(\gamma))},$$

where

$$\mathfrak{J}(\gamma) := \int_{\Omega} \left\| \frac{1}{\gamma(w[0 : T])} \nabla \gamma(w[0 : T]) \right\|_2^2 \gamma(w[0 : T])\mu(dw[0 : T]))$$

$$= \int_{\Omega} \frac{1}{\gamma(w[0 : T])} \sum_{k=0}^{T} \sum_{i=1}^{n} \left(\frac{\partial}{\partial w_i[k]} \gamma(w[0 : T]) \right)^2 \mu(dw[0 : T])). \quad (1.2)$$

We consider unbiased estimators; however, biased estimators can be treated similarly to [9]. We can also define a measure of cost for implementing the additive noise $w[0 : T]$ by the privacy-preserving agent for balancing against privacy as

$$\mathfrak{Q}(\gamma) := \frac{1}{T+1} \int_{\Omega} f(w[0 : T])\gamma(w[0 : T])\mu(dw[0 : T]),$$

where $f : \Omega \to \mathbb{R}$ is a function determining the cost of the realization $w[0 : T]$. To extract the optimal privacy-preserving policy, we can solve the optimization problem

$$\min_{\gamma \in \Gamma} \quad \mathfrak{J}(\gamma) + \varrho\mathfrak{Q}(\gamma), \quad (1.3)$$

where $\varrho > 0$ determines the balance between privacy and quality. Note, here, ϱ can be seen as a Lagrange multiplier for the following optimization problem:

$$\min_{\gamma \in \Gamma} \quad \mathfrak{J}(\gamma), \quad (1.4a)$$

$$\text{s.t.} \quad \mathfrak{Q}(\gamma) \leq \rho. \quad (1.4b)$$

In fact, it can be shown that, for any $\varrho > 0$, there exists $\rho > 0$ such that the solutions of Problems (1.3) and (1.4) are equal due to the convexity of the optimization problem and linearity of the constraints [16].

1.2.1 Motivating Example

Assume that the private energy consumption of a household at time instant $k \in \mathfrak{T}$ by $x[k]$. The house has a heating, ventilation, and air conditioning (HVAC) unit that the occupants can use for providing a comfortable leaving place and masking their private energy consumption profile and thus preserving their privacy. The total energy consumption of the house is given by

$$y[k] = x[k] + w[k],$$

where $w[k]$ denotes electricity consumption of the unit. Privacy can be achieved by ensuring that the total energy consumption of the house (i.e., the private profile plus the HVAC's consumption) $y[k]$ cannot be used to construct an accurate estimate of the household's private consumption $x[k]$. Assume that

$$z[k + 1] = Az[k] + Bw[k], \quad z[0] = z_0,$$

where $z[k] \in \mathbb{R}^m$ denotes the states of the HVAC unit and the temperature in the house. An example of a simple model for the HVAC unit is

$$\begin{aligned} \tau[k + 1] &= z[k] - a(\tau[k] - T_a) + bw[k] \\ &= (1 - a)\tau[k] + aT_a + bw[k], \end{aligned}$$

where $\tau[k]$ is the average temperature of the house and T_a is the ambient temperature. Defining $z[k] = \tau[k] - T_a$ results in

$$z[k + 1] = Az[k] + Bw[k], \tag{1.5}$$

where $A = (1 - a)$. The primary use of an air conditioning system is to provide a comfortable environment for the occupants. This can be guaranteed by ensuring that

$$\bar{C}z[k] \leq \bar{d}, \quad \forall k \in \mathfrak{T}.$$

An example of constraints on the temperature is that $T_{\min} \leq \tau[k] \leq T_{\max}$, which can be captured by

$$\begin{bmatrix} 1 \\ -1 \end{bmatrix} z[k] \leq \begin{bmatrix} T_{\max} - T_a \\ T_a - T_{\min} \end{bmatrix}. \tag{1.6}$$

Note that

$$\begin{bmatrix} z[1] \\ z[2] \\ \vdots \\ z[T+1] \end{bmatrix} = \begin{bmatrix} A \\ A^2 \\ \vdots \\ A^{T+1} \end{bmatrix} z_0 + \begin{bmatrix} B & 0 & \cdots & 0 \\ AB & B & \cdots & 0 \\ \vdots & \vdots & \ddots & \vdots \\ A^T B & A^{T-1} B & \cdots & B \end{bmatrix} \begin{bmatrix} w[0] \\ w[1] \\ \vdots \\ w[T] \end{bmatrix}.$$

Therefore, the constraints on the state of the HVAC unit can be rewritten as

$$\underbrace{\begin{bmatrix} \bar{C}B & 0 & \cdots & 0 \\ \bar{C}AB & \bar{C}B & \cdots & 0 \\ \vdots & \vdots & \ddots & \vdots \\ \bar{C}A^T B & \bar{C}A^{T-1} B & \cdots & \bar{C}B \end{bmatrix}}_{=:C} \begin{bmatrix} w[0] \\ w[1] \\ \vdots \\ w[T] \end{bmatrix} \leq \underbrace{\begin{bmatrix} \bar{d} \\ \bar{d} \\ \vdots \\ \bar{d} \end{bmatrix} - \begin{bmatrix} \bar{C}A \\ \bar{C}A^2 \\ \vdots \\ \bar{C}A^{T+1} \end{bmatrix} z_0}_{=:d}. \qquad (1.7)$$

The problem of finding the optimal privacy-preserving control input for an HVAC unit can then be cast in the form of (1.3).

1.3 Privacy-Preserving Policies

We slightly restrict the set of policies over which we search for the solution of (1.3) to be able to use tools from calculus of variations. To do so, we define the complete normed space $\mathcal{C}^1(\Omega)$ as the set of continuously differentiable functions with the norm $\sup_{w \in \Omega} |\gamma(w)| + \sup_{w \in \Omega} \|\nabla \gamma(w)\|_\infty$. For this set, we can prove the following result.

Theorem 1.1 *Let $\beta : \mathbb{R}^{nT} \to \mathbb{R}$ and $\lambda \in \mathbb{R}$ be solutions of the following partial differential equation:*

$$\nabla^2 \beta(w[0:T]) + \frac{1}{4}\left(\lambda - \frac{\varrho}{T+1} f(w[0:T])\right)\beta(w[0:T]) = 0,$$
$$\forall w[0:T] \in \Omega, \qquad (1.8a)$$
$$\beta(w[0:T]) = 0, \qquad \forall w[0:T] \in \partial\Omega, \qquad (1.8b)$$
$$\beta(w[0:T]) \neq 0, \qquad \forall w[0:T] \in \text{int } \Omega, \qquad (1.8c)$$
$$\int_\Omega \beta(w[0:T])^2 \mu(dw[0:T]) = 1. \qquad (1.8d)$$

Then, $\gamma(w[0:T]) = \beta(w[0:T])^2$ is a solution of (1.3) over $\Gamma \cap \mathcal{C}^1(\Omega)$. If there are multiple solutions over $\Gamma \cap \mathcal{C}^1(\Omega)$, all exhibit the same cost in the sense of (1.3).

Proof First, we can see that the cost function and the constraint sets are convex [9]. Now, following the result of [16], the Lagrangian can be constructed as

$$\mathcal{L} := \int_\Omega \frac{1}{\gamma(w[0:T])} \sum_{k=0}^{T} \sum_{i=1}^{n} \left(\frac{\partial}{\partial w_i[k]} \gamma(dw[0:T]) \right)^2 \mu(dw[0:T]))$$

$$+ \frac{\varrho}{T+1} \int_\Omega f(w[0:T]) \gamma(w[0:T]) \mu(dw[0:T])$$

$$- \lambda \left(\int_\Omega \gamma(w[0:T]) \mu(dw[0:T]) - 1 \right)$$

where $\lambda \in \mathbb{R}$ is the Lagrange multiplier corresponding to the equality constraint that

$$\int_\Omega \gamma(w[0:T]) \mu(dw[0:T]) = 1.$$

Using Theorem 5.3 in [5, p. 440], it can be seen that the extrema must satisfy

$$-\lambda + \frac{\varrho}{T+1} f(w[0:T]) - \frac{1}{\gamma(w[0:T])^2} \sum_{k=0}^{T} \sum_{i=1}^{n} \left(\frac{\partial \gamma(w[0:T])}{\partial w_i[k]} \right)^2$$

$$- 2 \sum_{k=0}^{T} \sum_{i=1}^{n} \frac{\partial}{\partial w_i[k]} \left[\frac{1}{\gamma(w[0:T])} \frac{\partial \gamma(w[0:T])}{\partial w_i[k]} \right] = 0. \tag{1.9}$$

Using the change of variable $\gamma(w[0:T]) = \beta(w[0:T])^2$, we can rewrite the optimality condition in (1.9) as

$$-\lambda + \frac{\varrho}{T+1} f(w[0:T]) - \frac{4}{\beta(w[0:T])^2} \sum_{k=0}^{T} \sum_{i=1}^{n} \left(\frac{\partial \beta(w[0:T])}{\partial w_i[k]} \right)^2$$

$$- 2 \sum_{k=0}^{T} \sum_{i=1}^{n} \frac{\partial}{\partial w_i[k]} \left[\frac{2}{\beta(w[0:T])} \frac{\partial \beta(w[0:T])}{\partial w_i[k]} \right] = 0,$$

which is equivalent to

$$-\lambda + \frac{\varrho}{T+1} f(w[0:T]) - \frac{4}{\beta(w[0:T])} \sum_{k=0}^{T} \sum_{i=1}^{n} \frac{\partial^2 \beta(w[0:T])}{\partial w_i[k]^2} = 0.$$

This concludes the proof. $\qquad\square$

The partial differential equations in Theorem 1.1 can be solved to find the optimal privacy-preserving policy. However, simultaneously solving the partial differential equation with boundary conditions in (1.8a)–(1.8c) and the non-linear equation in (1.8d) is complex task in general, except for specific costs and constraints studied in [8, 9]. In what follows, a more tractable algorithm for finding the optimal privacy-preserving policy is proposed for the case where $f(w[0:T]) = 0$. Note that the inclusion of the cost function f is not the only way for trading off privacy and performance since the constraint $Cw[0:T] \leq d$ can also be used to ensure the additive noise is "small".

Theorem 1.2 *Assume* $f(w[0:T]) = 0$. *Let* $A_{\tilde{\xi}} \geq 0$, $\phi_{\tilde{\xi}} \in (-\pi, +\pi]$, *and* $\lambda \geq 0$ *be solutions of the following nonlinear complex equations:*

$$\sum_{\tilde{\xi} \in \aleph} A_{\tilde{\xi}} \cos\left(2\pi \sum_{k=0}^{T} \sum_{i=1}^{n} \tilde{\xi}_i[k] w_i[k] + \phi_{\tilde{\xi}}\right) = 0, \quad w[0:T] \in \partial\Omega, \quad (1.10a)$$

$$\sum_{\tilde{\xi} \in \aleph} A_{\tilde{\xi}} \cos\left(2\pi \sum_{k=0}^{T} \sum_{i=1}^{n} \tilde{\xi}_i[k] w_i[k] + \phi_{\tilde{\xi}}\right) \neq 0, \quad w[0:T] \in \text{int } \Omega, \quad (1.10b)$$

$$\sum_{\tilde{\xi} \in \aleph} A_{\tilde{\xi}}^2 = 1, \quad (1.10c)$$

with

$$\aleph := \left\{ \tilde{\xi} \in \mathbb{R}^{nT} \,\middle|\, \sum_{k=0}^{T} \sum_{i=1}^{n} \tilde{\xi}_j[\ell]^2 = \frac{\lambda}{16\pi^2} \right\}. \quad (1.10d)$$

Then,

$$\gamma(w[0:T]) = \left(\sum_{\tilde{\xi} \in \aleph} A_{\tilde{\xi}} \cos\left(2\pi \sum_{k=0}^{T} \sum_{i=1}^{n} \tilde{\xi}_i[k] w_i[k] + \phi_{\tilde{\xi}}\right)\right)^2$$

is a solution of (1.3) *over* $\Gamma \cap \mathfrak{C}^1(\Omega)$.

Proof Let the Fourier transform of β be given by

$$\hat{\beta}(\xi) := \int_{\Omega} \exp(-2\pi\iota\xi^{\top} w[0:T])\beta(w[0:T])\mu(dw[0:T]), \quad \forall \xi \in \mathbb{R}^{nT}.$$

We can decompose ξ as $(\xi_i[k])_{i,k}$ according to the structure of $w[0:T]$. Hence, we can rewrite the Fourier transform as

$$\hat{\beta}(\xi) = \int_{\Omega} \exp\left(-2\pi\iota \sum_{k=0}^{T} \sum_{i=1}^{n} \xi_i[k] w_i[k]\right) \beta(w[0:T])\mu(dw[0:T]).$$

Similarly, the inverse Fourier transform is given by

$$\beta(w[0:T]) := \int_{\Omega} \exp\left(2\pi\iota \sum_{k=0}^{T}\sum_{i=1}^{n} \xi_i[k]w_i[k]\right) \hat{\beta}(\xi)\mu(\mathrm{d}w[0:T]).$$

Therefore,

$$\frac{\partial^2 \beta(w[0:T])}{\partial w_j[\ell]^2} = \int_{\Omega} \exp\left(2\pi\iota \sum_{k=0}^{T}\sum_{i=1}^{n} \xi_i[k]w_i[k]\right)$$
$$\times (-4\pi^2 \xi_j[\ell]^2 \hat{\beta}(\xi))\mu(\mathrm{d}w[0:T]).$$

This implies that the Fourier transform of $\partial^2\beta(w[0:T])/\partial w_j[\ell]^2$ is given by $-4\pi^2 \xi_j[\ell]^2 \hat{\beta}(\xi)$. Taking Fourier transform from the partial differential equation in Theorem 1.1 results in

$$\left(\frac{\lambda}{4} - 4\pi^2 \sum_{k=0}^{T}\sum_{i=1}^{n} \xi_j[\ell]^2\right) \hat{\beta}(\xi) = 0.$$

This implies that

$$\hat{\beta}(\xi) = \sum_{\tilde{\xi}:\sum_{k=0}^{T}\sum_{i=1}^{n} \tilde{\xi}_j[\ell]^2 = \lambda/(16\pi^2)} c_{\tilde{\xi}}\delta(\xi - \tilde{\xi}).$$

Taking inverse Fourier transform from this solution results in

$$\beta(w[0:T]) = \sum_{\tilde{\xi}:\sum_{k=0}^{T}\sum_{i=1}^{n} \tilde{\xi}_j[\ell]^2 = \lambda/(16\pi^2)} c_{\tilde{\xi}} \exp\left(2\pi\iota \sum_{k=0}^{T}\sum_{i=1}^{n} \tilde{\xi}_i[k]w_i[k]\right).$$

Furthermore, the Parseval's Theorem shows that

$$\int_{\Omega} \beta(w[0:T])^2 \mu(\mathrm{d}w[0:T]) = \int_{\mathbb{R}^{nT}} |\hat{\beta}(\xi)|^2 \mu(\mathrm{d}\xi)$$
$$= \sum_{\tilde{\xi}:\sum_{k=0}^{T}\sum_{i=1}^{n} \tilde{\xi}_j[\ell]^2 = \lambda/(16\pi^2)} |c_{\tilde{\xi}}|^2.$$

Noting that $\beta(w[0:T]) = \beta(w[0:T])^*$ (because $\beta(w[0:T])$ is real), we get $c_{\tilde{\xi}}^* = c_{-\tilde{\xi}}$, which implies that $\angle c_{\tilde{\xi}} = -\angle c_{-\tilde{\xi}}$. Also,

$$\beta(w[0:T]) = \sum_{\tilde{\xi}:\sum_{k=0}^{T}\sum_{i=1}^{n}\tilde{\xi}_j[\ell]^2=\lambda/(16\pi^2)} |c_{\tilde{\xi}}|\exp(\iota\angle c_{\tilde{\xi}})\exp\left(2\pi\iota\sum_{k=0}^{T}\sum_{i=1}^{n}\tilde{\xi}_i[k]w_i[k]\right)$$

$$= \sum_{\tilde{\xi}:\sum_{k=0}^{T}\sum_{i=1}^{n}\tilde{\xi}_j[\ell]^2=\lambda/(16\pi^2)} |c_{\tilde{\xi}}|\left(\cos\left(2\pi\sum_{k=0}^{T}\sum_{i=1}^{n}\tilde{\xi}_i[k]w_i[k]+\angle c_{\tilde{\xi}}\right)\right.$$

$$\left. + \iota\sin\left(2\pi\sum_{k=0}^{T}\sum_{i=1}^{n}\tilde{\xi}_i[k]w_i[k]+\angle c_{\tilde{\xi}}\right)\right)$$

$$= \sum_{\tilde{\xi}:\sum_{k=0}^{T}\sum_{i=1}^{n}\tilde{\xi}_j[\ell]^2=\lambda/(16\pi^2)} |c_{\tilde{\xi}}|\cos\left(2\pi\sum_{k=0}^{T}\sum_{i=1}^{n}\tilde{\xi}_i[k]w_i[k]+\angle c_{\tilde{\xi}}\right),$$

where the last equality follows from that $\angle c_{\tilde{\xi}} = -\angle c_{-\tilde{\xi}}$. $\square$

We can develop a numerical method for approximately solving (1.3) by discretizing the frequency space and enforcing the conditions in Theorem 1.2. The results of Theorem 1.2 can be seen as the extension of the results of [8] on optimality of square cosine additive noise for privacy preservation to more general support sets for the additive noise.

Finally, we find the optimal privacy-preserving HVAC control in the next corollary.

Corollary 1.1 *Assume $f = 0$. For the HVAC unit in (1.5) with the temperature constraint in (1.6), the optimal privacy-preserving policy is given by*

$$w[k] = \begin{cases} \dfrac{1}{b}v[k], & k = 0 \\ -\dfrac{1-a}{b}v[k-1] + \dfrac{1}{b}v[k], & k > 0, \end{cases} \quad (1.11)$$

where $v[k]$, for all $k \in \mathfrak{T}$, are i.i.d. distributed according to the probability density function:

$$p(v[k]) = \frac{2}{T_{\max} - T_{\min}}\cos^2\left(\frac{\pi}{T_{\max} - T_{\min}}\left[v[k] - \frac{T_{\max} + T_{\min}}{2} + c_k\right]\right), \quad (1.12)$$

with $c_k = T_a + (1-a)^{k+1}(\tau[0] - T_a)$.

Proof Define the change of variable

$$v[0:T] := \begin{bmatrix} v[0] \\ v[1] \\ \vdots \\ v[T] \end{bmatrix} = \underbrace{\begin{bmatrix} B & 0 & \cdots & 0 \\ AB & B & \cdots & 0 \\ \vdots & \vdots & \ddots & \vdots \\ A^T B & A^{T-1}B & \cdots & B \end{bmatrix}}_{=:\Psi} \begin{bmatrix} w[0] \\ w[1] \\ \vdots \\ w[T] \end{bmatrix}$$

Clearly, the constraints in (1.7) can be rewritten as

$$(I \otimes C)v[0:k] \leq d,$$

where $\otimes$ denotes the Kronecker product. The constraints are separable. Let $\beta(w[0:T]) = \beta'(\Psi^{-1}w[0:T])$. In what follows, we show that

$$\beta'(v[0:T]) = \beta_0'(v[0]) \ldots \beta_T'(v[T])$$

is a solution of (1.3). With this change of variable, we get

$$
\nabla^2 \beta(w[0:T]) \Big|_{w[0:T]=\Psi^{-1}v[0:T]} = \mathrm{trace}(\Psi^{-\top} D^2 \beta'(\Psi^{-1}v[0:k])\Psi^{-1})
$$
$$
= \mathrm{trace}(D^2 \beta'(\Psi^{-1}v[0:k])(\Psi^{\top}\Psi)^{-1})
$$
$$
= \sum_{k=0}^{T} \frac{d^2 \beta_k'(v[k])}{dv[k]^2} \left(\prod_{\ell \neq k} \beta_\ell'(v[\ell]) \right) [(\Psi^{\top}\Psi)^{-1}]_{kk}.
$$

Therefore, the optimality condition in Theorem 1.1 can be decomposed into

$$\frac{d^2 \beta_k'(v[k])}{dv[k]^2} + \frac{1}{4}\lambda_k \beta_k'(v[k]) = 0,$$

where λ_k, $\forall k$, are constants such that $\sum_{k=0}^{T} \lambda_k[(\Psi^{\top}\Psi)^{-1}]_{kk} = \lambda$. The rest follows from the same line of reasoning as in [8]. □

Note that separation of the variables in the proof of Corollary 1.1 is only possible if there are constraint on either $z[k]$ or $w[k]$. This method cannot be perused if there are constraints on both $z[k]$ and $w[k]$. In this case, the solution must be approximated by implementing a discretized version of Theorem 1.2. Let $f(w) = w^{\top}w$. For the privacy-preserving policy in Corollary 1.1, we have

$$
\mathfrak{Q}(\gamma) = \frac{1}{T+1} \int w[0:T]^{\top}w[0:T]\gamma(w[0:k])\mu(dw[0:T])
$$
$$
= \frac{1}{T+1} \int v[0:T]^{\top}\Psi^{-\top}\Psi^{-1}v[0:T]\gamma(v[0:k])\mu(dw[0:T])
$$
$$
= \frac{1}{T+1} \left(\frac{1}{b^2} \left[\left(\frac{1}{12} - \frac{1}{2\pi^2} \right)(T_{\max} - T_{\min})^2 + \left(\frac{T_{\max} + T_{\min}}{2} - c_T \right)^2 \right] \right.
$$
$$
+ \frac{(1-a)^2 + 1}{b^2} \sum_{k=0}^{T-1} \left[\left(\frac{1}{12} - \frac{1}{2\pi^2} \right)(T_{\max} - T_{\min})^2 \right.
$$
$$
\left. \left. + \left(\frac{T_{\max} + T_{\min}}{2} - c_k \right)^2 \right] \right).
$$

Therefore,

$$\lim_{T \to \infty} \mathfrak{Q}(\gamma) = \frac{(1-a)^2 + 1}{b^2} \left[\left(\frac{1}{12} - \frac{1}{2\pi^2} \right) (T_{max} - T_{min})^2 \right.$$
$$\left. + \left(\frac{T_{max} + T_{min}}{2} - T_a \right)^2 \right]. \qquad (1.13)$$

If privacy was not an issue, any sequence of control actions for the HVAC unit $w[0 : T]$ could be selected so long as the temperature in the house is kept between T_{min} and T_{max}. An example of such control actions is to ensure that the temperature is constantly kept at $(T_{min} + T_{max})/2$. In this case, we get

$$\lim_{T \to \infty} \mathfrak{Q}(\gamma) = \frac{(1-a)^2 + 1}{b^2} \left(\frac{T_{max} + T_{min}}{2} - T_a \right)^2. \qquad (1.14)$$

The ratio of costs in (1.13) and (1.14) captures the price of privacy (PoP), which is given by

$$\text{PoP} := 1 + \left(\frac{1}{3} - \frac{2}{\pi^2} \right) \left(\frac{T_{max} - T_{min}}{T_{max} + T_{min} - 2T_a} \right)^2$$
$$\approx 1 + 0.1307 \left(\frac{T_{max} - T_{min}}{T_{max} + T_{min} - 2T_a} \right)^2.$$

1.4 Numerical Example

Consider the motivating example in Sect. 1.2.1 with horizon $T = 100$. We set $a = 0.1$ and $b = 0.5$. Further, we assume that constraints such as (1.6) must be enforced with $T_{min} = 20$ and $T_{max} = 24$, capturing the comfort zone of the occupants. We also assume that the starting temperature is given by $\tau[0] = 20$, which is within the comfort zone. Figure 1.1 shows the temperature $\tau[k]$ versus time k for the HVAC system with the optimal privacy-preserving controller in Corollary 1.1 with the dashed lines capturing the constraints on the temperature. Noting that the optimal privacy-preserving controller in Corollary 1.1 is stochastic, it makes sense to also observe the statistics of the temperature. Figure 1.2 illustrates the statistics of the temperature $\tau[k]$ versus time k extracted from one hundred-thousand runs of the HVAC system with the optimal privacy-preserving controller in Corollary 1.1. The boxes, i.e., the vertical lines at each iterations, illustrate the range of 25–75% percentiles of the temperature and the thin vertical gray lines illustrate the maximum and minimum temperatures in all the runs. The black line show the median temperature. The dashed lines illustrate the boundary of the comfort zone. Evidently, the temperature always stays within the comfort zone and hovers around 22°. For this example, PoP = 1.0082

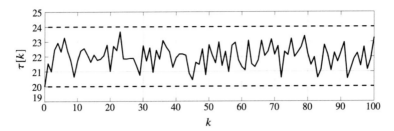

Fig. 1.1 The temperature $\tau[k]$ versus time k for the HVAC system with the optimal privacy-preserving controller in Corollary 1.1. The dashed lines show the constraints on the temperature

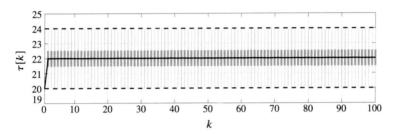

Fig. 1.2 Statistics of the temperature $\tau[k]$ versus time k extracted from hundred-thousands runs of the HVAC system with the optimal privacy-preserving controller in Corollary 1.1. The boxes, i.e., the vertical lines at each iterations, illustrate the range of 25–75% percentiles of the temperature and the thin gray lines illustrate the maximum and minimum temperatures in all the runs. The black line show the median temperature. The dashed lines show the constraints on the temperature

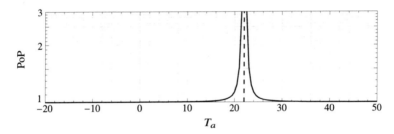

Fig. 1.3 Price of privacy versus the ambient temperature

(i.e., 0.82% increase in consumption due to privacy), which is a negligible price to pay in average for privacy.

Figure 1.3 shows the price of privacy PoP versus the ambient temperature T_a. Note that as the ambient temperature moves towards 22°, the price of privacy goes to infinity. This is because at 22°, there is no need to use the HVAC unit for staying comfortable and all the energy consumption is essentially used for preserving privacy.

1.5 Conclusions

We used HVAC units to preserve the privacy of households with smart meters by modelling the effect of the HVAC unit as an additive noise in the household consumption. We computed the optimal privacy-preserving policy for controlling the HVAC unit by minimizing a weighted sum of the measure of privacy, based on the Fisher information, and the cost operating the HVAC unit.

References

1. Ács G, Castelluccia C (2011) I have a DREAM! (DiffeRentially privatE smArt Metering). In: Filler T, Pevný T, Craver S, Ker A (eds) Information Hiding: 13th International Conference, IH 2011, Prague, Czech Republic, May 18–20, 2011, Revised Selected Papers. Springer, Berlin, pp 118–132
2. Bambauer J, Muralidhar K, Sarathy R (2013) Fool's gold: an illustrated critique of differential privacy. Vanderbilt J Entertain Technol Law 16:701
3. Dwork C (2008) Differential privacy: a survey of results. In: Agrawal M, Du D, Duan Z, Li A (eds) 5th Proceedings of international conference theory and applications of models of computation, TAMC 2008, Xi'an, China, 25–29 April 2008, pp. 1–19. Springer, Berlin
4. Dwork C (2011) Differential privacy. In: van Tilborg HCA, Jajodia S (eds) Encyclopedia of Cryptography and Security. Springer, US, Boston, MA
5. Edwards CH (1973) Advanced Calculus of Several Variables. Academic Press
6. Farokhi F (2019) Development and analysis of deterministic privacy-preserving policies using non-stochastic information theory. IEEE Trans Inf Forensics Secur 14(10):2567–2576
7. Farokhi F, Milosevic J, Sandberg H (2016) Optimal state estimation with measurements corrupted by Laplace noise. In: Proceedings of the 55th IEEE conference on decision and control
8. Farokhi F, Sandberg H (2018) Fisher information as a measure of privacy: preserving privacy of households with smart meters using batteries. IEEE Trans Smart Grid 9(5):4726–4734
9. Farokhi F, Sandberg H (2019) Ensuring privacy with constrained additive noise by minimizing Fisher information. Automatica 99:275–288
10. Garfinkel SL, Abowd JM, Powazek S (2018) Issues encountered deploying differential privacy. In: Proceedings of the 2018 workshop on privacy in the electronic society, pp 133–137
11. Greenberg A (2017) How one of apple's key privacy safeguards falls short. https://www.wired.com/story/apple-differential-privacy-shortcomings/
12. Haeberlen A, Pierce BC, Narayan A (2011) Differential privacy under fire. In: USENIX security symposium
13. Han S, Topcu U, Pappas GJ (2014) Differentially private convex optimization with piecewise affine objectives. In: Proceedings of the 53rd IEEE conference on decision and control, pp 2160–2166
14. Hart GW (1989) Residential energy monitoring and computerized surveillance via utility power flows. IEEE Technol Soc Mag 8(2):12–16
15. Huang Z, Wang Y, Mitra S, Dullerud GE (2014) On the cost of differential privacy in distributed control systems. In: Proceedings of the 3rd international conference on high confidence networked systems, pp 105–114
16. Jeyakumar V, Wolkowicz H (1990) Zero duality gaps in infinite-dimensional programming. J Optim Theory Appl 67(1):87–108
17. Le Ny J, Pappas GJ (2014) Differentially private filtering. IEEE Trans Autom Control 59(2):341–354
18. Liang Y, Poor HV, Shamai S (2009) Information theoretic security. Found Trends® Commun Inf Theory 5(4–5), 355–580

19. McDaniel P, McLaughlin S (2009) Security and privacy challenges in the smart grid. IEEE Secur Priv 7(3):75–77
20. Muralidhar K, Sarathy R (2010) Does differential privacy protect terry gross' privacy? In: Domingo-Ferrer J, Magkos E (eds) Privacy in statistical databases. Springer, Berlin, pp 200–209
21. Sandberg H, Dán G, Thobaben R (2015) Differentially private state estimation in distribution networks with smart meters. In: Proceedings of the 54th IEEE conference on decision and control, pp 4492–4498
22. Shao J (2003) Mathematical Statistics. Springer Texts in Statistics. Springer, New York
23. Tang J, Korolova A, Bai X, Wang X, Wang X (2017) Privacy loss in Apple's implementation of differential privacy on macos 10.12. arXiv preprint arXiv:1709.02753
24. Wyner AD (1975) The wire-tap channel. Bell Syst Tech J 54(8):1355–1387
25. Yamamoto H (1983) A source coding problem for sources with additional outputs to keep secret from the receiver or wiretappers. IEEE Trans Inf Theory 29(6):918–923
26. Zoha A, Gluhak A, Imran MA, Rajasegarar S (2012) Non-intrusive load monitoring approaches for disaggregated energy sensing: a survey. Sensors 12(12):16838–16866

Chapter 2
Smart Meter Privacy

Ecenaz Erdemir, Deniz Gündüz and Pier Luigi Dragotti

Abstract The new generation electricity supply network, called the smart grid (SG), provides consumers with an active management and control of the power. By utilizing digital communications and sensing technologies which make the grid *smart*, SGs yield more efficient electricity transmission, reduced peak demand, improved security and increased integration of renewable energy systems compared to the traditional grid. Smart meters (SMs) are one of the core enablers of SG systems; they measure and record the high resolution electricity consumption information of a household almost in a real time basis, and report it to the utility provider (UP) at regular time intervals. SM measurements can be used for time-of-use pricing, trading user-generated energy, and mitigating load variations. However, real-time SM readings can also reveal sensitive information about the consumer's activities which the user may not want to share with the UP, resulting in serious privacy concerns. SM privacy enabling techniques proposed in the literature can be categorized as SM data manipulation and demand shaping. While the SM data is modified before being reported to the UP in the former method, the latter requires direct manipulation of the real energy consumption by exploiting physical resources, such as a renewable energy source (RES) or a rechargeable battery (RB). In this chapter, a data manipulation privacy-enabling technique and three different demand shaping privacy-enabling techniques are presented, considering SM with a RES and an RB, SM with only an RB and SM with only a RES. Information theoretic measures are used to quantify SM privacy. Optimal energy management strategies and bounds which are obtained using control theory, specifically Markov decision processes (MDPs), and rate distortion theory are analyzed.

E. Erdemir (✉) · D. Gündüz · P. L. Dragotti
Department of Electrical and Electronic Engineering,
Imperial College London, London, UK
e-mail: e.erdemir17@imperial.ac.uk

D. Gündüz
e-mail: d.gunduz@imperial.ac.uk

P. L. Dragotti
e-mail: p.dragotti@imperial.ac.uk

2.1 Introduction

An electrical grid is a network that distributes electricity to consumers. The foundations of the current electrical grid were laid out in the late 19th century as a centralized unidirectional transmission and distribution system. However, the current grid has reached its capacity and is not fit to manage the growing energy demand [11].

Developing technology has led to an increasing number of electronic appliances, electrical vehicles and integration of renewable energy sources. In order to handle load imbalance, inefficient usage of energy, and blackouts with domino effect a new energy grid is currently being introduced. Smart grid (SG) is an energy grid which controls energy generation, distribution, transmission and consumption using advanced communication and sensing technologies. SGs are developed to increase the efficiency of energy infrastructure, reliability against attacks, flexibility with bidirectional energy flows and the load balancing against variations [14]. For instance, thanks to SG's ability to support customer energy generation, farms that produce electricity using methane generators, consumers with solar panels or wind turbines can sell excess generated energy back to the utility provider (UP).

One of the main enablers of SGs are the *smart meters* (SMs), computerized replacement of the traditional analog electrical meters attached to the exterior of households [27]. Unlike traditional electrical meters, which measure only the total consumption, SMs can monitor fine grained electricity usage of a household and report it to the UP. This provides efficient use of energy resources since the SM owners can track and control their consumption almost real-time. SM data can also be used for time-of-usage pricing, which can reduce peak electricity demands by controlling customer behavior. Moreover, SMs also facilitate detecting energy theft, trading user-generated energy to increase grid efficiency, and mitigating effects of load variations [14].

2.1.1 Privacy and Security Concerns

SM measurements contain detailed information related to the real-time state of the customers. The UP or a third party can deduce power signatures of specific home appliances by using non-intrusive load monitoring (NILM) techniques [19]. NILM systems identify appliances by using a series of changes in their power draw. For instance, appliances such as kitchen ovens, tumble dryers and dishwashers go through a number of states, where heaters and fans are turned on and off in various combinations. Such appliances are modelled as finite state machines. On the other hand, when on, a light bulb draws power continuously.

In Fig. 2.1, an example of the 24 h period of SM measurements for a household is illustrated. Specific appliances with distinguishable power signatures are highlighted with different colors. As in Fig. 2.1, the high resolution consumption data can reveal

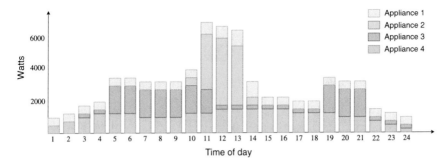

Fig. 2.1 Electricity consumption profile of a household for 24 h period

details about private activities of the user. This real-time data might enable a malicious eavesdropper to learn user's presence at home, illnesses, disabilities and even political views due to the TV channel the user is watching [26]. SM privacy becomes even more critical when we consider businesses, since their power consumption might reveal the state of their business to competitors. The controversy about SM roll-out plans due to privacy concerns have attracted public and political attention across the world. In 2009, a court in Netherlands decided that mandatory installation of SMs would be a violation to the customer's right to privacy, and would be in breach of the European Convention of Human Rights [8]. In 2018, in the case of Naperville Smart Meter Awareness v. City of Naperville, a court in the United States has agreed that the Fourth Amendment protects user's energy consumption data collected by SMs. That is, user's expectation for SMs data privacy is reasonable and the government's access to this private information constitutes a search [1]. SM privacy concerns can be a major roadblock on the path of achieving worldwide SM usage.

2.2 Smart Meter Privacy Techniques

Various SM privacy enabling techniques have been proposed in the literature [2, 3, 6, 12, 13, 15, 16, 18, 23, 24, 35, 38], which can be categorized as into two groups (see Fig. 2.2): those based on SM data manipulation and those based on demand shaping. While the techniques in the former group focus on modifying SM measurements [12, 23], there in the latter group directly manipulate user's energy consumption exploiting physical resources, such as a rechargeable battery (RB) [2, 3, 18, 24, 38] or a renewable energy source (RES) [6, 13, 15, 16, 35]. Representative works for each group are briefly explained in the following sections.

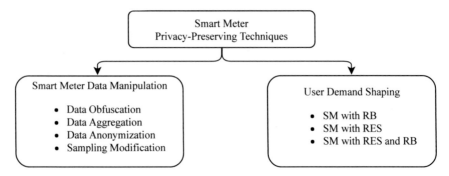

Fig. 2.2 SM privacy enabling techniques

2.2.1 Data Manipulation

Data manipulation techniques modify SM measurements before sending them to the UP. There are many different approaches to SM data manipulation in the literature, such as data obfuscation, data aggregation, data anonymization and down-sampling.

SM data obfuscation can be performed by corrupting the SM measurements with additive noise. A cooperative state estimation technique is proposed in [23] to preserve privacy by obfuscating the power consumption measurement. As the amount of noise added increases, information leaked to the UP decreases. However, such a modification makes SM measurements less relevant to the UP for prediction and control purposes, which contradicts the purpose of installing SMs. In [32], a general theoretical framework is proposed for both utility and privacy requirements of data release mechanisms using information theoretic tools. In this context, SM measurements are perturbed before being reported to the UP. The goal is to minimize the information leakage rate between the perturbed data and the private data of user's choice while keeping the distortion between the real and perturbed meter measurements below a certain level. For a stationary Gaussian Markov model of the electricity load, the optimal utility-privacy trade-off is characterized in [30] using the framework proposed in [32].

Data aggregation, on the other hand, proposes sending aggregated SM readings, instead of individual readings, to the UP. In [12], data aggregation is used in combination with homomorphic encryption and secret sharing techniques. The UP has access only to encrypted SM readings and the total consumption. Moreover, the users send their random shares to the UP after encrypting with each others public keys and aggregating. Hence, the UP does not have access to individual consumption information. However, encryption methods increase the computational complexity substantially [33].

Data anonymization approach in [9], instead, considers utilizing pseudonyms rather than the real identities of consumers such that information gleaned from the SM cannot be easily associated with an identified person.

In [37], two data manipulation techniques are combined, namely down-sampling and noise addition. The SM data is first down-sampled by summing up n consecutive samples, then noise is added to the down-sampled data. Similarly to [23], perturbation of SM readings can cause undesired data loss.

2.2.2 Demand Shaping

Manipulating SM readings reduces the relevance of the reported values for grid management and load prediction, limiting the benefits of SMs. Moreover, the grid operator can place sensors outside a household or a business, and obtain the real consumption data, since they own and control the infrastructure. Therefore, data manipulation cannot provide strict privacy against UPs. Demand shaping tackles these issues by manipulating the real energy consumption. Unlike in data manipulation, UP receives accurate measurements of the energy taken from the grid. However, these measurements do not belong to the actual energy consumption of the household; and therefore, only provide very limited information about user behavior. Instead, the energy demand of the appliances are supplied either by alternative energy sources, such as renewable energy sources, or from rechargeable energy storage devices. Hence, the instantaneous energy demand of an appliance is supplied only partially by the power grid, if at all, and the rest can be provided by the RB or RES. This effectively filters the real energy consumption time series, and creates a new time series for the energy received from the grid, which has only limited correlation with the original time series, and consequently reduces the information leakage to the UP. Note that, in the extreme cases of unlimited RES or RB capacity, the two time series can be made completely independent, leading to zero information leakage, i.e., perfect privacy [31]. The objective of the SM privacy demand shaping problem is to determine the optimal energy management strategy between the grid, RB, RES and the appliances, under given physical limitations, such as RB capacity, RES generation rate, peak power constraints, etc. which provide the maximum privacy.

In [24], information theoretic privacy in an SM system with an RB is formulated as a Markov decision process (MDP). Markovian energy demand is considered, and the minimum leakage is obtained numerically through dynamic programming (DP), while a single-letter expression is obtained for an independent and identically distributed (i.i.d) demand. This approach is extended to the scenario with a RES in [13], which considers both cases where the energy generation process is private and known to the UP. When the energy generation process is known by the UP, it is numerically shown in [10] that the infinite-horizon MDP performance can be achieved by a low complexity algorithm under the assumption of a special energy generation process.

Privacy-cost trade-off is examined in an SM system with an RB in [38]. Due to Markovian demand and price processes, the problem is formulated as a partially observable Markov decision process (POMDP) with belief-dependent rewards. Bellman equation for stationary strategies is provided. However, due to the non-linear and belief-dependent reward, the Bellman equation corresponds to a continuous state, continuous action, continuous reward MDP. Obtaining optimal policies using continuous action Bellman equation is computationally complex. Therefore, the authors provided upper and lower bounds, and presented numerical results using classical rate-distortion theory in [38].

Information theoretic SM privacy with RB and RES is studied in [15] with average and peak power constraints on RES. While closed-form expressions are obtained for the scenarios with zero and infinite capacity RB, low complexity energy management policies are proposed for finite capacity. For a zero-capacity RB, rate-distortion theory is used to obtain a single letter expression under the assumption of independent and identically distributed (i.i.d.) demand process. The SM privacy problem in the existence of an alternative energy source, e.g. RES, is also studied in [16] exploiting rate-distortion theory, and numerical results are obtained by Blahut Arimoto algorithm. This approach is extended to multiple-user scenario in [17].

2.3 Information Theoretic Smart Meter Privacy

In this section, SM privacy problem is examined from information theory perspective. Information theoretic privacy measures and Markov decision processes are introduced in Sect. 2.3.1 as preliminary information. SM privacy enabling techniques with data manipulation and demand shaping are presented in detail in Sects. 2.3.2 and 2.3.3, respectively.

2.3.1 Preliminaries

2.3.1.1 Privacy Measures

In information theory, entropy is a measure of the uncertainty of a random variable (r.v.). Let X be a discrete r.v. with probability mass function $p(x) = Pr\{X = x\}$ over alphabet $\mathcal{X}$. The entropy of X is denoted by

$$H(X) = -\sum_{x \in \mathcal{X}} p(x) \log P(x). \tag{2.1}$$

The mutual information (MI) between two r.v.'s X and Y is the relative entropy between the joint probability mass function, $p(x, y)$, and the product of their marginal probability mass functions, $p(x)p(y)$, and is given by,

$$I(X;Y) = \sum_{x \in \mathcal{X}, y \in \mathcal{Y}} p(x,y) \log \frac{p(x,y)}{p(x)p(y)} \tag{2.2}$$

$$= E_{p(x,y)} \log \frac{p(X,Y)}{p(X)p(Y)}. \tag{2.3}$$

Mutual information can also be written as the reduction in the uncertainty of X due to the knowledge of Y, i.e., $I(X;Y) = H(X) - H(X|Y)$, where $H(X|Y)$ is the conditional entropy.

In information theoretic SM privacy framework, the user load and RES statistics are assumed to be stationary and known by both the energy management unit and the attacker. Then, the mutual information $I(X;Y)$ is used as a privacy measure, representing the information leakage about X through Y. Even though a malicious third party can obtain the statistics of the random processes in the SM system causally in an infinite time horizon, he cannot infer the realizations of the private information due to the privacy measure based on uncertainty. Since information theoretic metrics are independent of the attacker behavior, they are preferably used as privacy measures.

As a result of the memory effect introduced by the RB and RES in the demand shaping scenario, information leakage should be considered over a time horizon of reasonable length. Masking the energy usage over a short period of time by supplying all the demand from the RB and/or the RES might cause full leakage in the future. Thus, information theoretic SM privacy is typically measured as average information leakage over a time horizon T, i.e., $\frac{1}{T}I(X^T;Y^T)$. However, the analysis becomes computationally complex as the horizon increases as the mutual information involves an increasing number of r.v.'s. In the literature, single-letter expressions for the information leakage in SM privacy problems have been obtained considering i.i.d. or Markov demand and/or renewable energy generation. Single letter expressions guarantee that, no matter how long the problem horizon is, the minimal leakage can be written as a function of the joint distribution of the involved r.v.'s single realization.

2.3.1.2 MDP Formulation

Consider a sequential decision making problem under uncertainty. At each time instance, an external decision maker (agent, controller, etc.) observes the state of the system and takes an action accordingly. As a result of the action taken at a particular state, a reward is received by the decision maker. The goal of such a problem is to find the decision rules that specify the best actions to take at each system state, such that the maximum total reward is accrued by the decision maker under the system constraints [29].

Markov property is based on the idea that the future is independent of the past given the present. Since considering the effect of entire time horizon in a decision problem is computationally complex, these problems are modeled for Markovian state space. Hence, MDPs are discrete time stochastic control processes which are used to model sequential decision problems with uncertainty. MDPs take into account both the short-term outcomes of current decisions and the possible future gain. An MDP is formally defined as a 4-tuple $< \mathcal{S}, \mathcal{A}, \mathcal{T}, \mathcal{R} >$, which represent the *state space* $\mathcal{S}$, *action space* $\mathcal{A}$, *transition probabilities* reflecting the system dynamics $\mathcal{T}$, and *reward* (or, inversely, *cost*) $\mathcal{R}$ of taking a certain action at a certain state [21]. The state is Markov if

$$P(S_{t+1}|S_t) = P(S_{t+1}|S_t, S_{t-1}, \ldots, S_1), \tag{2.4}$$

where $S_t \in \mathcal{S}$. For deterministic policies, transition probabilities are the mappings from each state-action pair to the next state,

$$\mathcal{T} : \mathcal{S} \times \mathcal{A} \to \mathcal{S}, \tag{2.5}$$

whereas for stochastic policies, each state-action pair is mapped to a probability distribution over the next states,

$$\mathcal{T} : \mathcal{S} \times \mathcal{A} \to P(\mathcal{S}). \tag{2.6}$$

The goal of an MDP is to obtain a set of decision rules, so called *policy*, that performs optimally with respect to a certain performance criterion. A policy q is a function specifying the action that the decision maker takes in a particular state, i.e., $q : \mathcal{S} \to \mathcal{A}$. The objective functions of MDPs map infinite or finite sequences of rewards (or costs) to a single real number. MDPs can have objectives, such as *discounted*, *expected-total* or *average* costs (rewards) to minimize (maximize) over a specified duration, i.e., *finite-horizon* setting, or an indefinite time, i.e., *infinite-horizon* setting [29]. To solve MDPs optimally, Bellman optimality equations are used. Value function of the decision problem in state s is denoted by $V_q(s)$ which represents the expected reward/cost obtained following the policy q in state s. Bellman optimality equation for a Markov reward process is denoted by

$$V_q(s) = \max_{q(s) \in \mathcal{A}} \left\{ r(s, q(s)) + \sum_{s' \in \mathcal{S}} T(s, q(s), s') V_q(s') \right\}, \tag{2.7}$$

where the maximization is over all the possible actions induced by the policy q for each state s. The optimal value can be achieved by maximizing/minimizing the right hand side of (2.7) using dynamic programming, which is an optimization method used to avoid redundant calculations in recursive problems with additive objective function [5].

A POMDP is a generalization of MDP when the decision maker might not have complete information about the system state. Instead, she can maintain a belief which is a conditional probability distribution over the possible states given the past observations from the environment. POMDPs can be modeled as belief MDPs by inducing a continuous belief state. In the literature, there are various approaches to solve belief MDPs using finite-state MDP solution methods, e.g. value iteration, policy iteration and gradient-based methods. These are based on the discretization of continuous belief states to obtain a finite state MDP [34].

2.3.2 SM Privacy with Data Perturbation

Privacy aware data release mechanism was studied in [32] for the first time, in which a theoretical framework of privacy-utility trade-off for data manipulation is proposed. This framework was later applied to SMs in [30]. In SM systems, load measurements are complex valued including real and reactive components. The empirical load measurements are shown to be approximately Gaussian in [22]; hence the continuous valued discrete-time SM data can be modelled as a sequence $\{Y_t\}$ of r.v.'s $Y_t \in \mathcal{Y}$, $t = \{\ldots, -1, 0, 1, \ldots\}$, generated by a stationary continuous Gaussian source with memory.

Total energy consumption data of a household contains individual consumption information of each home appliance which can be inferred from Y_t. Let $\{X_t\}$ be a private information of the data collector's choice, such as the energy consumption of a particular home appliance. This private information $X_t \in \mathcal{X}$ is correlated with and can be inferred from Y_t. Since continuous SM data Y_t cannot be transmitted losslessly over finite capacity links, its sampled sequence is compressed before transmission. In addition to the distortion by compression, discrete samples of Y_t can even be perturbed to guarantee extra privacy on the sensitive data X_t. Formally, the encoding function on SM side is a mapping from the meter reading sequence $Y^n = (Y_1, Y_2, \ldots, Y_n)$, where $Y_t \in \mathbb{R}$, to an index $Z_n \in \mathcal{Z}_n = \{1, 2, \ldots, Z_{max}\}$ given by

$$F_{enc} : \mathcal{Y}^n \to \mathcal{Z}_n, \tag{2.8}$$

where each index is a quantized sequence. The decoder at the UP side computes a distorted output sequence $\hat{Y}^n = (\hat{Y}_1, \hat{Y}_2, \ldots, \hat{Y}_n)$, $\hat{Y}_t \in \mathbb{R}$, using the decoding function,

$$F_{dec} : \mathcal{Z} \to \hat{Y}^n. \tag{2.9}$$

To obtain a certain level of privacy in the SM problem, the encoding function F_{enc} is chosen such that the private information X_t cannot be inferred from the distorted output $\hat{Y}_t$. However, the distortion level must be kept limited such that the UP can still achieve utility from the distorted SM readings. The utility is measured by the mean-square error (MSE) distortion function,

$$D_n = \frac{1}{n} \sum_{t=1}^{n} \mathbb{E}\left[(Y_t - \hat{Y}_t)^2 \right], \tag{2.10}$$

where the expectation is over the joint distribution $p(y^n, \hat{y}^n) = P(y^n) p_t(\hat{y}^n | y^n)$. Privacy leakage is measured by the mutual information rate between the SM measurements received by the UP $\{\hat{Y}_t\}$ and the private information sequence $\{X_t\}$,

$$L_n = \frac{1}{n} I(X^n; \hat{Y}^n). \tag{2.11}$$

For a coding scheme given by (2.8) and (2.9) which satisfies (2.10) and (2.11), the SM utility-privacy trade-off region is a set of all (D, L) pairs, where D and L are the limit values of D_n and L_n as $n \to \infty$, respectively. However, this utility-privacy trade-off region does not bound the number of encoded sequences. The rate-distortion-leakage (RDL) trade-off region is the set of all (R, D, L) triplets for which there exists a sequence of coding schemes with (2.8), (2.9), each with a bounded number of encoded sequences

$$Z_{\max} \leq 2^{n(R_n+\epsilon)}, \tag{2.12}$$

where $\epsilon > 0$, $R_n = (\log Z_{max})/n$ and $D_n \leq D + \epsilon$ while we have $R = \lim_{n \to \infty} R_n$. Under the constraints (2.10) and (2.11), SM utility-privacy trade-off can be quantified by the RDL trade-off region, where the rate-distortion and minimal leakage functions are denoted by

$$R(D, L) = \lim_{n \to \infty} \inf_{p(y^n, x^n) p(\hat{y}^n | y^n)} \frac{1}{n} I(Y^n; \hat{Y}^n), \tag{2.13}$$

$$\lambda(D) = \lim_{n \to \infty} \inf_{p(y^n, x^n) p(\hat{y}^n | y^n)} \frac{1}{n} I(X^n; \hat{Y}^n). \tag{2.14}$$

The set of all distributions minimizing (2.13) and (2.14) follows the Markov relationship $X^n \to Y^n \to \hat{Y}^n$. Rate-distortion function for Gaussian sources is well known [7] and can be obtained from the covariance matrix which is obtained by transforming the correlated source sequence into its eigen-space where the MSE function and the mutual information leakage are invariant. For example, the optimal encoding strategy for independent Gaussian r.v.'s can be obtained using the reverse water-filling algorithm [30].

2.3.3 SM Privacy with Demand Shaping

In this section, SM privacy problem is examined in the presence of a renewable energy source and a rechargeable battery, which allow the user to physically manipulate its consumption [13, 15, 35]. A discrete time model of the SM system is illustrated in Fig. 2.3, in which the energy demand of the user and energy requested from the grid at time slot t are denoted by $X_t \in \mathcal{X}$ and $Y_t \in \mathcal{Y}$, respectively, where $(|\mathcal{X}|, |\mathcal{Y}| < \infty)$.

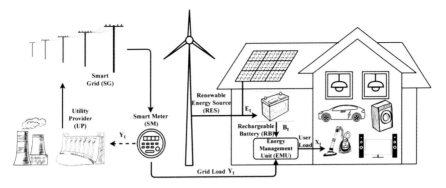

Fig. 2.3 Illustration of the SM system model with RB and RES

The RB state of charge at the beginning of time slot t is denoted by $B_t \in \mathcal{B} :=$ $\{0, \ldots, B_m\}$, in which the initial state B_1 is distributed with probability p_{B_1}. The battery charging and discharging process is assumed ideal without any losses (see [3] for a model with energy losses). $E_t \in \mathcal{E} := \{0, \ldots, E_m\}$ units of energy are generated by the RES at the beginning of each time slot t, and these can be used by the appliances only through the RB. The E_t process is assumed to be independent of X_t. Unless it is stated in the following sections, the realizations of E_t are not known by the UP.

The appliances' energy demand is always satisfied by assuming $E_t + B_t + Y_t \geq X_t$, $\forall t$. In addition, intentional energy waste to provide privacy, or selling energy to the grid are not allowed.

2.3.3.1 SM Privacy with a RES

First, we consider the special case where the RB capacity of the SM system illustrated in Fig. 2.3 is zero, i.e., $B_m = 0$. Here the energy from the grid or RES cannot be stored in an RB to provide additional privacy. Since the UP cannot access the amount of energy generated by the RES at a particular time instant, users can achieve a certain level of privacy depending on the amount of energy they can receive from the RES. Assume that the RES is limited in terms of the average and peak power it can provide. Therefore, the objective of the SM privacy problem with RES is to obtain the optimal policy providing the best privacy under the average and peak power constraints of the RES. Information leakage rate to be minimized can be written as the mutual information rate between the user demand and the grid energy, i.e.,

$$I_T = \frac{1}{T} I(X^T; Y^T). \tag{2.15}$$

The maximum power which can be received from the RES in a time slot t is denoted by $\hat{P}$, and must satisfy $0 \leq X_t - Y_t \leq \hat{P}$. Moreover, the average power, $\bar{P}_T$, that the RES can provide over a finite horizon T is defined by,

$$\bar{P}_T = \mathbb{E}\left[\frac{1}{T}\sum_{t=1}^{T}(X_t - Y_t)\right], \tag{2.16}$$

where the expectation is taken over the joint probability distribution of the user demand and the grid power. Under these constraints, the asymptotic performance limit of the n-letter problem becomes an infinite dimensional optimization problem. On the other hand, using single-letter r.v.'s allows achieving the optimal solution solving a finite-dimensional optimization problem. A single letter expression for the minimum information leakage rate can be obtained under the assumption of i.i.d. demand X_t, and it can be characterized by the *privacy-power function* defined as,

$$\mathcal{I}(\bar{P}, \hat{P}) = \inf_{P_{Y|X} \in \mathcal{F}} I(X; Y), \tag{2.17}$$

where $\mathcal{F} := \{P_{Y|X} : y \in \mathcal{Y}, \mathbb{E}[(X - Y)] \leq \bar{P}, 0 \leq X - Y \leq \hat{P}\}$. Here, the energy constraints are not affected by the past, since there is no battery, and thus no memory in the system. The optimal energy management policy minimizing (2.17) is stochastic and memoryless, and depends only on the current demand.

We note that the objective function (2.17) is similar to *rate-distortion function* $R(D)$ in information theory, which describes the minimum required compression rate R, in bits per sample, for an i.i.d. source sequence X^T with distribution p_X such that the receiver can reconstruct the source sequence achieving a particular expected distortion level D [7]. Average distortion between sequences X^T and $\hat{X}^T$ is denoted by $D = \frac{1}{T}\sum_{t=1}^{T} d(x_t, \hat{x}_t)$, where $d(x, \hat{x})$ and $\hat{X}$ represent the distortion measure used, and the reconstruction alphabet, respectively. The *information rate-distortion function* $R^{(I)}(D)$ for a source X with distortion measure $d(x, \hat{x})$ is defined by Shannon as [7],

$$R^{(I)}(D) = \min_{p(x|\hat{x}) \in \tilde{\mathcal{F}}} I(X; \hat{X}), \tag{2.18}$$

where $\tilde{\mathcal{F}} := \{P(x|\hat{x}) : \sum_{(x,\hat{x})} P(x)P(\hat{x}|x)d(x, \hat{x}) \leq D\}$. The analogy between the privacy-power function (2.17) and the rate-distortion function (2.18) can be made assuming the distortion measure:

$$d(x, y) = \begin{cases} x - y, & \text{if } 0 \leq x - y \leq \hat{P}, \\ \infty, & \text{otherwise.} \end{cases} \tag{2.19}$$

Hence, this enables us to use tools from rate-distortion theory to examine SM privacy problems with RES [15–17]. However, there are two major differences between the rate-distortion and SM privacy problems, namely (i) grid energy Y^T is the direct output of the "encoder", which is represented by the EMU in the SM problem, rather than the reconstruction of the decoder, and (ii) unlike the lossy encoder, EMU determines the output load Y_t instantaneously after receiving the demand. Since the mutual information is a convex function of the distribution $P_{Y|X}$, the privacy-power function can be written as a constrained convex optimization problem and solved numerically using the Blahut Arimoto algorithm [7].

2.3.3.2 SM Privacy with an RB

Here, we consider another special case where the SM system illustrated in Fig. 2.3 has an RB, and no RES, i.e., $E_t = 0$ for all t. The energy demand of the user is supplied by the grid energy through the RB, and charging of the RB can only be performed by the energy grid. X_t is assumed to be first-order time-homogeneous Markov chain with the transition probability q_X and initial state distribution p_{X_1}. This scenario is studied in [24, 38], where both recast the problem as an MDP.

The battery state of charge is updated by,

$$B_{t+1} = B_t + Y_t - X_t, \tag{2.20}$$

where Y_t is chosen such that $B_{t+1} \leq B_m$.

The amount of energy requested from the grid is determined by a randomized battery charging policy $q = \{q_t\}_{t=1}^{\infty}$, where q_t is a conditional probability distribution $q_t(Y_t|X^t, B^t, Y^{t-1})$ which randomly decides on the amount of energy received from the grid at time t given the histories of demand $X^t := \{X_1, \ldots, X_t\}$, battery charge B^t and grid energy Y^{t-1}, i.e.,

$$q_t : \mathcal{X}^t \times \mathcal{B}^t \times \mathcal{Y}^{t-1} \rightarrow \mathcal{Y}. \tag{2.21}$$

The goal of the SM privacy problem is to find an energy management policy, $\{q_t^*\}_{t=1}^{\infty}$, which provides the best privacy.

Privacy of an energy management policy over a time period T can be measured by the *information leakage rate*, which is defined as the average mutual information between the demand side load (X^T, B_1), and SM readings Y^T:

$$L_q(T) := \frac{1}{T} I(X^T, B_1; Y^T), \tag{2.22}$$

where B_1 is the initial RB state of charge. In [24], it is proved that there is no loss of optimality in considering policies of the form $q_t(Y_t|X_t, B_t, Y^{t-1})$; that is, it is sufficient to consider only the current demand and battery state. Hence, (2.22) can be rewritten in an additive form

$$L_q(T) = \frac{1}{T} \sum_{t=1}^{T} I(X_t, B_t; Y_t|Y^{t-1}). \tag{2.23}$$

Markovity of optimal actions and the additive objective function of information leakage rate enable this problem to be cast as a stochastic control problem, which can be formulated as an MDP.

SM privacy problem in the existence of RB can be cast as an average cost, infinite-horizon MDP with state $S_t = \{X_t, B_t\} \in \mathcal{S}$. However, the leakage at time t depends on Y^{t-1}, which leads to a growing state space in time. Therefore, the problem is formulated as a belief MDP and belief state $\beta_t(s_t)$ is defined as the causal posterior probability distribution over the state space of (X_t, B_t) given Y^{t-1}:

$$\beta_t(s_t) = P^q(S_t = s_t | Y^{t-1} = y^{t-1}). \tag{2.24}$$

The control actions chosen by randomized policies are the conditional probabilities of energy received from the grid given the state and belief, denoted by $a_t(y_t|s_t) = P^q(Y_t = y_t | S_t = s_t, \beta_t)$, where $a_t \in \mathcal{A}$ [24]. As a result of the action taken at time t, belief is updated for the next time interval as follows:

$$\beta(s_{t+1}) = p(s_{t+1}|y^t) \tag{2.25a}$$

$$= \frac{\sum_{s_t} p(s_{t+1}, s_t, y_t | y^{t-1})}{p(y_t | y^{t-1})} \tag{2.25b}$$

$$= \frac{\sum_{s_t} p(s_t | y^{t-1}) p(y_t | s_t, y^{t-1}) p(s_{t+1} | y_t, s_t)}{\sum_{s_t, s_{t+1}} p(s_t | y^{t-1}) p(y_t | s_t, y^{t-1}) p(s_{t+1} | y_t, s_t)} \tag{2.25c}$$

$$= \frac{\sum_{s_t} \beta(s_t) a_t(y_t | s_t) q_X(x_{t+1} | x_t)}{\sum_{s_t, s_{t+1}} \beta(s_t) a_t(y_t | s_t) q_X(x_{t+1} | x_t)} \times \frac{1_{b_{t+1}} \{b_t + y_t - x_t\}}{1_{b_{t+1}} \{b_t + y_t - x_t\}}. \tag{2.25d}$$

where (2.25c) follows from the Bayes rule and the Markov chain $Y^{t-1} \to (S_t, Y_t) \to S_{t+1}$; and (2.25d) from the definitions of β and a_t. Given Y^{t-1}, per-step leakage of taking action $a_t(y_t|s_t)$ due to policy q is,

$$l_t(s_t, a_t, y^t; q) := \log \frac{a_t(y_t|s_t)}{P^q(y_t|y^{t-1})}. \tag{2.26}$$

Taking the expectation of the per-step leakage over a finite-horizon T, $\frac{1}{T}\mathbb{E}_q[\sum_{t=1}^{T} l_t(s_t, a_t, y^t)]$, results in an objective function equivalent to the original formulation in (2.23). Given belief and action probabilities, average information leakage at time t is formulated as,

$$\mathbb{E}_q[l_t(s_t, a_t, y^t)] = I(S_t; Y_t | Y^{t-1} = y^{t-1})$$

$$= \sum_{s_t \in \mathcal{S}, y_t \in \mathcal{Y}} \beta_t(s_t) a_t(y_t|s_t) \log \frac{a_t(y_t|s_t)}{\sum_{\hat{s}_t \in \mathcal{S}} \beta_t(\hat{s}_t) a_t(y_t|\hat{s}_t)}.$$

$$= I(S_t; Y_t | \beta_t, a_t). \tag{2.27}$$

SM privacy problem which is cast as an average cost belief-MDP can be solved by DP. While an exact DP solution cannot be achieved due to the continuous belief state, approximate numerical solutions can be obtained by using belief quantization methods [34]. To formulate the corresponding Bellman equation, which is a necessary condition for the optimality of DP [4], Bellman operator T is written as,

$$[T_a v](\beta) = l(s, q(\beta), \beta) + \sum_{s \in \mathcal{S}, y \in \mathcal{Y}} \beta(s) a(y|s) v(\phi(\beta, y, a)), \qquad (2.28)$$

where v is the value function and the updated belief state is represented by $\beta_{t+1} = \phi(\beta_t, y_t, a_t)$. Implementation of DP for the finite-horizon and infinite-horizon settings are given follows:

Finite horizon DP:

- For $v_{n+1}(\beta) = 0$ and $t \in \{n, \ldots, 1\}$, value functions, v_t, are recursively defined [29]:

$$v_t(\beta) = \min_{a \in A} [T_a v_{t+1}](\beta). \qquad (2.29)$$

  Optimal leakage rate is given by $v_1(\beta_1)/n$, where $\beta_1(s) = p_{X_1} p_{B_1}$.
- The optimal policy minimizing the right hand side of (2.29) is denoted by $\mathbf{q}^* = (q_1^*, \ldots, q_n^*)$:

$$q_t^*(y_t|s_t, \beta) = a_t(y_t|s_t). \qquad (2.30)$$

Infinite horizon DP:

- For λ constant [29], the value function v is time-homogeneous and defined iteratively:

$$\lambda + v(\beta) = \min_{a \in A} [T_a v](\beta). \qquad (2.31)$$

  Optimal leakage rate is given by λ.
- Time-homogeneous optimal policy, $\mathbf{q}^* = (q^*, q^*, \ldots)$,

$$q^*(y_t|s_t, \beta) = a(y_t|s_t). \qquad (2.32)$$

Single Letter Expression for I.I.D. Demand.

When the Markov assumption for the demand is relaxed and X_t is assumed to be i.i.d. with probability distribution p_X, it is possible to achieve the optimal policy by solving a cost function in a single letter form. Consider an auxiliary state variable

$W_t = B_t - X_t$, where $w \in \{b - x : b \in \mathcal{B}, x \in \mathcal{X}\}$. Then, the single letter minimum information leakage rate is given by [24],

$$J^* = \min_{\theta \in \mathcal{P}_B} I(B - X; X) = \min_{\theta \in \mathcal{P}_B} \{H(B - X) - H(B)\}, \qquad (2.33)$$

where r.v.'s X and B are independent; θ is the probability distribution over B given the past observations and actions, i.e., $\theta := p(b_t|y^{t-1}, a^{t-1})$; and actions a_t are the conditional probabilities of grid load given the current demand, battery charge and the belief. Contrary to the Markovian demand case, here belief states are on W_t. Since the objective function (2.33) is convex over θ, the optimal policy can be obtained by Blahut-Arimoto algorithm [7]. The resulting grid load is i.i.d., and the optimal charging policy is memoryless and time-invariant.

Privacy-cost Trade-Off

In practice, in addition to privacy, energy cost is an important concern. Indeed, home energy storage devices are mainly installed to reduce energy consumption by storing energy during off-peak price periods [20, 28]. It is possible to maximize privacy by constantly purchasing high amount of energy from the grid and wasting the extra energy. However, this is against the purpose of SM from both the user and the UP point of view.

The same as minimizing the mutual information to maximize the achievable privacy, the conditional entropy of the Markovian demand process given the observations of UP can also be used as a privacy measure to maximize. In [38], the authors take both privacy and cost into account and recast the problem as an MDP. The privacy is formulated as,

$$\mathcal{P}(q) := \frac{1}{T} H(X^T|Y^T, P^T), \qquad (2.34)$$

where $P^T = (P_1, \ldots, P_T)$ is the price of the energy purchased from the grid for $t = \{1, \ldots, T\}$. Unlike privacy, energy cost has an additive formulation and can be easily incorporated into the MDP formulation. Following policy q, the average cost savings per time slot are defined by,

$$\mathcal{C}(q) := \frac{1}{T} \sum_{t=1}^{T} c(X_t, B_{t+1}, Y_t, P_t), \qquad (2.35)$$

where $c(X_t, B_{t+1}, Y_t, P_t) = (X_t - Y_t)P_t$, $\forall B_{t+1} \in \mathcal{B}$. The objective of the SM privacy-cost trade-off problem with RB is considered as the weighted sum of privacy, $\mathcal{P}$, and average cost savings per time slot, $\mathcal{C}$. That is, the weighted reward function to be maximized is given by $\mathcal{R}(q, \lambda) = \lambda \mathcal{P}(q) + (1 - \lambda)\mathcal{C}(q)$, where $\lambda \in [0, 1]$ denotes user's choice regarding the balance between privacy and cost. If $\lambda = 0$, only the cost savings are maximized, whereas if $\lambda = 1$, only the privacy is maximized. The problem in [38] is reformulated as a belief MDP. A Bellman equation which corresponds to a continuous state, continuous action, continuous reward MDP is written for stationary policies. However, due to the high computational complexity, only the privacy of cost-optimal, deterministic and greedy policies are studied in [38]. Optimal privacy-cost trade-off bounds are also obtained using rate distortion theory.

2.3.3.3 SM Privacy with a RES and an RB

In this section, we consider a more general case in which the SM system is equipped with a finite capacity RB and a RES with non-zero energy generation (see Fig. 2.3) [10, 13]. Unless otherwise is stated, both X_t and E_t are assumed to be first-order time-homogeneous Markov chains with transition probabilities q_X and q_E, and initial state distributions p_{X_1} and p_{E_1} for their initial states X_1 and E_1, respectively. While the RB provides demand shifting, the RES supplies alternative energy to mask the energy consumption of the appliances. However, the memory introduced by the RB and the additional randomness due to the energy generation process of the RES, the SM privacy problem becomes more complicated than the previous cases with only RB or RES.

Here, the battery state of charge is updated by,

$$B_{t+1} = \min(E_t + B_t - X_t, B_m) + Y_t, \qquad \forall t, \tag{2.36}$$

where Y_t is chosen such that $B_{t+1} \leq B_m$. When the realizations of the energy generation process E_t are not known by the UP, information leakage rate of the SM system with RB and RES is defined by

$$L_q(T) := \frac{1}{T} I(X^T, B_1, E^T; Y^T). \tag{2.37}$$

Randomized battery charging policies in the existence of an RB and a RES are defined such that $q_t : \mathcal{X}^t \times \mathcal{E}^t \times \mathcal{B}^t \times \mathcal{Y}^{t-1} \to \mathcal{Y}$. Similarly to Sect. 2.3.3.2, there is no loss of optimality in considering battery charging policies of the form $q_t(Y_t | X_t, B_t, E_t, Y^{t-1})$. Therefore, (2.37) can be rewritten in an additive form

$$L_q(T) = \frac{1}{T} \sum_{t=1}^{T} I(X_t, B_t, E_t; Y_t | Y^{t-1}). \tag{2.38}$$

Employing Markovian actions and additive objective function, SM privacy problem with RB and RES can be cast as an average cost MDP with states $S_t = \{X_t, B_t, E_t\} \in \mathcal{S}$. As before, the history dependence of the information leakage due to RB causes a growing state space in time. Hence, the problem is formulated as a belief MDP and belief state $\beta_t(s_t)$ is defined as the causal posterior probability distribution over the state space of (X_t, B_t, E_t) given Y^{t-1}. As a result of the action $a_t(y_t | s_t) \in \mathcal{A}$ taken at time t, belief is updated for the next time interval as follows:

$$\beta(s_{t+1}) =$$

$$\frac{\sum_{s_t} \beta(s_t)a_t(y_t|s_t)q_E(e_{t+1}|e_t)q_X(x_{t+1}|x_t)1_{b_{t+1}}\{\min(e_t + b_t - x_t, B_m) + y_t\}}{\sum_{s_t,s_{t+1}} \beta(s_t)a_t(y_t|s_t)q_E(e_{t+1}|e_t)q_X(x_{t+1}|x_t)1_{b_{t+1}}\{\min(e_t + b_t - x_t, B_m) + y_t\}}.$$

$$(2.39)$$

The derivation of the intermediate steps can be performed following (2.25) with the corresponding modifications. Given $Y^{t-1} = y^{t-1}$, the average information leakage in (2.38) can be written in terms of belief and actions by averaging the per-step leakage in (2.26) over the belief and action probabilities, when $S_t = \{X_t, B_t, E_t\}$. With the integration of renewable energy generation, the resulting objective (2.27) can be minimized by following the DP steps (2.29)–(2.32).

Renewable Energy Known by the UP

Here, we consider a special case of the SM privacy problem with an RB and a RES, in which the UP knows the realizations of E_t. In this scenario, energy management policies of the form $q_t(Y_t|X_t, B_t, E^t, Y^{t-1})$ are taken into account, and the information leakage rate induced by policy **q** is denoted by,

$$L_q(T) := \frac{1}{T}I(X^T, B_1; Y^T|E^T) = \frac{1}{T}\sum_{t=1}^{T} I(X_t, B_t; E_t, Y_t|Y^{t-1}, E^{t-1}). \quad (2.40)$$

Similarly to the E_t unknown case, the problem can be reformulated as a belief MDP. The belief state is defined as the conditional probability on the system state $S_t :=$ (X_t, B_t), given the observation history (Y^{t-1}, E^{t-1}), i.e., $\beta(s_t) := p(s_t|y^{t-1}, e^{t-1})$. As a result of the action $a_t(y_t|s_t, e_t) = P^q(Y_t = y_t|S_t=s_t, E_t = e_t, \beta_t)$, belief is updated as follows,

$$\beta(s_{t+1}) =$$

$$\frac{\sum_{s_t} \beta(s_t)a_t(y_t|s_t, e_t)q_E(e_t|e_{t-1})q_X(x_{t+1}|x_t)1_{b_{t+1}}\{\min(e_t + b_t - x_t, B_m) + y_t\}}{\sum_{s_t,s_{t+1}} \beta(s_t)a_t(y_t|s_t, e_t)q_E(e_t|e_{t-1})q_X(x_{t+1}|x_t)1_{b_{t+1}}\{\min(e_t + b_t - x_t, B_m) + y_t\}},$$

$$(2.41)$$

where the intermediate steps can be derived from the Bayes rule, Markovity of E_t, and the Markov chain $(Y^{t-1}, E^{t-1}) \rightarrow (S_t, Y_t, E_t) \rightarrow S_{t+1}$. Unlike the E_t unknown scenario, energy generation process is not included in belief since the UP has the exact information about E_t realizations. Given (Y^{t-1}, E^{t-1}), per-step information leakage of taking action $a_t(y_t|s_t, e_t)$ incurred by policy **q** is,

$$l_t(s_t, e^t, a_t, y^t; q) := \log \frac{a_t(y_t|s_t, e_t)q_E(e_t|e_{t-1})}{P^q(y_t, e_t|y^{t-1}, e^{t-1})}. \quad (2.42)$$

Taking average leakage over a finite-horizon T, $\frac{1}{T}\mathbb{E}_q[\sum_{t=1}^{T} l_t(s_t, e^t, a_t, y^t)]$, is equal to the original formulation in (2.40). Given belief and action probabilities, average information leakage at time t is denoted by:

$$\mathbb{E}_q[l_t(s_t, e^t, a_t, y^t)] = I(S_t; E_t, Y_t | Y^{t-1} = y^{t-1}, E^{t-1} = e^{t-1})$$

$$= \sum_{\substack{s_t \in \mathcal{S} \\ e_t \in \mathcal{E} \, y_t \in \mathcal{Y}}} \beta_t(s_t) a_t(y_t | s_t, e_t) q_E(e_t | e_{t-1}) \log \frac{a_t(y_t | s_t, e_t) q_E(e_t | e_{t-1})}{\sum_{\hat{s}_t \in \mathcal{S}} \beta_t(\hat{s}_t) a_t(y_t | \hat{s}_t, \hat{e}_t) q_E(e_t | e_{t-1})}$$

$$= I(S_t; E_t, Y_t | \beta_t, q_E, a_t). \tag{2.43}$$

The problem is recast as a belief MDP, and the Bellman equation to be used in DP is modified with the integration of observed energy generation process,

$$[T_a v](\beta) = l(s, q(\beta), \beta, q_E) + \sum_{\substack{s \in \mathcal{S} \\ e \in \mathcal{E} \, y \in \mathcal{Y}}} \beta(s) a(y | s, e) q_E(e | \hat{e}) v(\phi(\beta, y, a, e)),$$

$$\tag{2.44}$$

where $\hat{e}$ is the energy generated in the previous step and the updated belief state is represented by $\beta_{t+1} = \phi(\beta_t, y_t, a_t, e_t)$. Finite-horizon and infinite-horizon MDP steps can be followed from (2.29) to (2.32).

Special Renewable Energy Generation Process

Here, we propose low complexity policies and numerical solutions for SM privacy-cost trade-off in the existence of both RES and RB. We relax Markov assumption of energy generation process and introduce a special i.i.d. E_t process that fully recharges the battery at random time instances, i.e, $E_t \in \{0, B_m\}$. The realizations of the renewable energy generation process E_t is assumed to be known by the UP. Due to the special energy arrival process, the problem is an episodic MDP, which resets to an initial state of full RB at every renewable energy instant. Between two consecutive energy arrivals, energy transitions occur only between the grid, the battery and the home appliances. An example for the RB state of charge for $B_m = 5$ under the special energy generation process assumption is given in Fig. 2.4. Red bars express the fully charged battery state at time instances $t = 0, 5, 8, 12$, when the renewable energy is generated. Between two consecutive energy arrivals, the RB state of charge is represented by grey bars. Hence, for each time period between two RES charging instants, the system can be modeled as an SM with only an RB and no RES. Accordingly, a finite-horizon privacy-cost trade-off problem is formulated for an SM system with an initially full RB, which is used to propose a low-complexity policy as well as a lower bound for the original problem. Between two RES charging instants, battery update is performed according to (2.20) and the finite-horizon average information leakage is formulated as in (2.23). Energy cost has an additive formulation and can be incorporated into the MDP formulation. Price process of the energy purchased from the grid at time t is defined as P_t. Following policy q, the average energy cost per time slot is defined by,

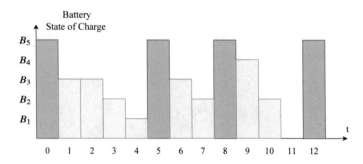

Fig. 2.4 Illustration of the RB state of charge under the special energy generation process assumption in [10]

$$C_q(T) := \frac{1}{T} \sum_{t=1}^{T} Y_t P_t. \tag{2.45}$$

Due to the growing space of observations of the UP, belief states are defined and the problem is recast as a belief-MDP. The weighted objective function is given by $U_q(\lambda, T) = \lambda L_q(T) + (1 - \lambda)C_q(T)$, where $\lambda \in [0, 1]$ denotes user's choice regarding the privacy-cost balance, is represented in terms of belief states and actions, and minimized over the action space. The optimal policy for each episode is obtained by applying finite-horizon DP via the Bellman operator given in (2.28).

According to the low complexity threshold policy (TP) in [10], after each RB recharge instance, the optimal policy obtained for a fixed finite-horizon n is employed. The optimal policy for horizon n is followed until either the battery is recharged again, in which case the algorithm restarts with the same policy, or the time horizon n is reached. If the RB is not recharged at time $(n + 1)$, it is assumed that all the energy demand is directly supplied by the grid, resulting in full information leakage. The intuition behind this scheme follows from the law of large numbers, which suggests that, with high probability, the RB will be charged after $n = \frac{1}{P_E}$ time slots, where P_E is the energy generation probability at any t. We consider policies with a fixed time horizon of $n = \frac{1}{P_E}$, as well as those with an optimized time horizon. The numerical results show that the performance of TP with optimized but fixed time horizon closely follows that of the infinite-horizon MDP solution given in Sect. 2.3.3.3.

2.4 Concluding Remarks

SMs are end user interfaces that monitor the energy consumption of users. SMs provide accurate, high frequency consumption data to the UPs, and they are being widely deployed around the world. The adoption of SM s has created a multi-billion dollar business. However, private information about user's personal lives can be inferred from detailed SM readings by the UP, which has led to significant consumer outrage, creating a serious roadblock in front of the widespread deployment of SMs. Therefore, enabling privacy-aware SM technology has an undeniable importance both for consumers and for other stakeholders in this multi-billion dollar industry.

In this chapter, SM privacy-preserving techniques have been discussed. They are classified into two: data manipulation methods which modify SM measurements and, demand shaping methods which manipulate the energy received from the grid physically. The second group of methods, which use physical resources, such as RB and RES, have been examined in detail as they provide privacy without compromising the role of SM in providing timely and accurate energy consumption information. Unlike SM data manipulation, demand shaping methods report accurate and real consumption measurements to the UP, which maintains the benefits of the SG concept. We have mainly focused on information theoretic privacy measures, in particular the mutual information between the real energy consumption and the energy received from the grid, which is also what the SMs report to the UP. Other measures have also been considered in the literature, see for example [25, 36]. Rate-distortion theory and MDPs have been used as mathematical tools to study the fundamental information theoretic privacy measures.

Although there are a vast number of solutions which have been proposed in the literature, SM privacy problem still has many challenges to be addressed. Among the various privacy metrics defined, there is still lack of a privacy measure which is generic, device-independent and well suited to various privacy-preserving methods. Information theoretic privacy metrics provide solutions independent of the attacker behavior, such as the particular detection technology employed by the attacker; however, they depend on an underlying statistical model governing the various processes involved. The assumed statistical models may not be valid in practice, or more involved models might be needed, under which clean optimal solutions may not be possible, requiring computationally limited sub-optimal solution that can provide reasonable privacy guarantees. Moreover, the cost of privacy-preserving techniques and installation of RB or RES is still considerably high compared to cost savings due to SM usage. However, this cost may reduce as renewable energy becomes more widespread making RES and RBs more commonly available to households.

References

1. Naperville smart meter awareness v. city of Naperville 16:3766 (7th cir. 2018)
2. Arrieta M, Esnaola I (2017) Smart meter privacy via the trapdoor channel. In: 2017 IEEE international conference on smart grid communications (SmartGridComm), pp 277–282
3. Avula RR, Oechtering TJ, Månsson D (2018) Privacy-preserving smart meter control strategy including energy storage losses. ArXiv e-prints
4. Bellman R (1952) On the theory of dynamic programming. Proc Natl Acad Sci:716–71
5. Bertsekas DP (2007) Dynamic programming and optimal control, vol II, 3rd edn. Athena Scientific
6. Chin J, Tinoco De Rubira T, Hug G (2017) Privacy-protecting energy management unit through model-distribution predictive control. IEEE Trans Smart Grid 8(6):3084–3093

7. Cover TM, Thomas JA (2006) Elements of information theory (Wiley series in telecommunications and signal processing). Wiley-Interscience, New York, NY, USA
8. Cuijpers C, Koops BJ (2013) Smart metering and privacy in Europe: lessons from the Dutch Case. Springer Netherlands, Dordrecht, pp 269–293
9. Efthymiou C, Kalogridis G (2010) Smart grid privacy via anonymization of smart metering data. In: 2010 first IEEE international conference on smart grid communications, pp 238–243
10. Erdemir E, Dragotti PL, Gündüz D (2019) Privacy-cost trade-off in a smart meter system with a renewable energy source and a rechargeable battery. In: IEEE international conference on acoustics, speech, and signal processing (ICASSP)
11. Farhangi H (2010) The path of the smart grid. IEEE Power Energy Mag 8(1):18–28
12. Garcia FD, Jacobs B (2011) Privacy-friendly energy-metering via homomorphic encryption. In: Cuellar J, Lopez J, Barthe G, Pretschner A (eds) Security and Trust Management. STM 2010. Lecture Notes in Computer Science, vol 6710. Springer, Berlin, Heidelberg
13. Giaconi G, Gündüz D (2016) Smart meter privacy with renewable energy and a finite capacity battery. In: 2016 IEEE 17th international workshop on signal processing advances in wireless communications (SPAWC), pp 1–5
14. Giaconi G, Gündüz D, Poor HV (2018) Privacy-aware smart metering: progress and challenges. IEEE Signal Process Mag 35(6):59–78
15. Giaconi G, Gündüz D, Poor HV (2018) Smart meter privacy with renewable energy and an energy storage device. IEEE Trans Inf Forensics Secur 13(1):129–142
16. Gomez-Vilardebo J, Gündüz D (2015) Smart meter privacy for multiple users in the presence of an alternative energy source. IEEE Trans Inf Forensics Secur 10(1):132–141
17. Gomez-Vilardebo J, Gündüz D (2013) Smart meter privacy for multiple users in the presence of an alternative energy source. In: 2013 IEEE global conference on signal and information processing, pp 859–862
18. Han S, Topcu U, Pappas GJ (2016) Event-based information-theoretic privacy: a case study of smart meters. In: American control conference (ACC), pp 2074–2079
19. Hart GW (Dec, 1992) Nonintrusive appliance load monitoring. Proc IEEE:1870–1891
20. Hubert T, Grijalva S (2012) Modeling for residential electricity optimization in dynamic pricing environments. IEEE Trans Smart Grid 3(4):2224–2231
21. Kallenberg L (2011) Lecture notes in markov decision processes
22. Kim H, Marwah M, Arlitt M, Lyon G, Han J (2011) Unsupervised disaggregation of low frequency power measurements. In: Proceedings of the 2011 SIAM international conference on data mining, vol 11, pp 747–758
23. Kim Y, Ngai ECH, Srivastava MB (2011) Cooperative state estimation for preserving privacy of user behaviors in smart grid. In: 2011 IEEE international conference on smart grid communications (SmartGridComm), pp 178–183
24. Li S, Khisti A, Mahajan A (2018) Information-theoretic privacy for smart metering systems with a rechargeable battery. IEEE Trans Inf Theory 64(5):3679–3695
25. Li Z, Oechtering TJ, Gündüz D (2019) Privacy against a hypothesis testing adversary. IEEE Trans Inf Forensics Secur 14(6):1567–1581
26. Makonin S (2012) Approaches to non-intrusive load monitoring (NILM) in the home. Technical report
27. McDaniel P, McLaughlin S (2009) Security and privacy challenges in the smart grid. IEEE Secur Priv 7(3):75–77
28. Mishra A, Irwin D, Shenoy P, Kurose J, Zhu T (2012) Smartcharge: cutting the electricity bill in smart homes with energy storage. In: Proceedings of the 3rd international conference on future energy systems: where energy, computing and communication meet, e-energy 2012, pp 29:1–29:10. ACM, New York, NY, USA
29. Puterman ML (1994) Markov decision processes: discrete stochastic dynamic programming, 1st edn. Wiley, New York, NY, USA
30. Rajagopalan SR, Sankar L, Mohajer S, Poor HV (2011) Smart meter privacy: a utility-privacy framework. In: 2011 IEEE international conference on smart grid communications (SmartGridComm), pp 190–195

31. Rassouli B, Gunduz D (2018) On perfect privacy. In: IEEE international symposium on information theory (ISIT)
32. Rebollo-Monedero D, Forne J, Domingo-Ferrer J (2010) From t-closeness-like privacy to postrandomization via information theory. IEEE Trans Knowl Data Eng 22(11):1623–1636
33. Sakuma J, Kobayashi S, Wright RN (2008) Privacy-preserving reinforcement learning. In: Proceedings of the 25th international conference on machine learning, ICML 2008, pp 864–871. ACM, New York, NY, USA
34. Saldi N, Linder T, Yuksel S (2018) Approximations for partially observed markov decision processes. In: Finite approximations in discrete-time stochastic control, chap. 5. Birkhäuser, Cham, pp 99–124
35. Tan O, Gunduz D, Poor HV (2013) Increasing smart meter privacy through energy harvesting and storage devices. IEEE J Sel Areas Commun 31(7):1331–1341
36. Tan O, Gómez-Vilardebó J, Gündüz D (2017) Privacy-cost trade-offs in demand-side management with storage. IEEE Trans Inf Forensics Secur 12(6):1458–1469
37. Yang H, Cheng L, Chuah MC (2016) Evaluation of utility-privacy trade-offs of data manipulation techniques for smart metering. In: 2016 IEEE conference on communications and network security (CNS), pp 396–400
38. Yao J, Venkitasubramaniam P (2013) On the privacy-cost tradeoff of an in-home power storage mechanism. In: Annual Allerton conference on communication, control, and computing (Allerton), pp 115–122

Chapter 3
Privacy Against Adversarial Hypothesis Testing: Theory and Application to Smart Meter Privacy Problem

Zuxing Li, Yang You and Tobias J. Oechtering

Abstract Hypothesis testing is the fundamental theory behind decision-making and therefore plays a critical role in information systems. A prominent example is machine learning, which is currently developed and applied to a wide range of applications. However, besides the utilities, hypothesis testing can also be implemented for an illegitimate purpose to infer on people's privacy. Thus, the development of hypothesis testing techniques further increases the privacy leakage risks. Accordingly, the research on privacy-by-design techniques that enhance the privacy against adversarial hypothesis testing receives more and more attention recently. In this chapter, the problem of privacy against adversarial hypothesis testing is formulated in the presence of a distortion source. Information-theoretic fundamental bounds on the optimal privacy performance and corresponding privacy-enhancing technologies are first discussed under the assumption of independent and identically distributed adversarial observations. The discussion is then extended to considering a privacy problem model with memory. In the end, applications of the theoretic results and privacy-enhancing technologies to the smart meter privacy problem are illustrated.

3.1 Introduction

Nowadays, a variant of forms of decision-making are implemented in information systems to realize smart applications. For instance, in a sensor network, local decisions of sensors are transmitted to a fusion node to make the final sensing decision; in a smart grid, the energy provider decides the energy generation and distribution based on smart meters' feedbacks; and in a game with a human champion, Alpha Go makes a decision based on the locations of pieces. The development of

Z. Li (✉) · Y. You · T. J. Oechtering
KTH Royal Institute of Technology, SE-10044 Stockholm, Stockholm, Sweden
e-mail: zuxing@kth.se

Y. You
e-mail: youy@kth.se

T. J. Oechtering
e-mail: oech@kth.se

© Springer Nature Singapore Pte Ltd. 2020
F. Farokhi (ed.), *Privacy in Dynamical Systems*,
https://doi.org/10.1007/978-981-15-0493-8_3

information technologies enables more precise decision-makings but also leads to higher privacy leakage risks. A prominent example is smart meter privacy problem, where high time-resolution feedbacks of smart meters can be used to infer on privacy-sensitive energy consumption behaviors of consumers by an adversary using a powerful machine learning algorithm.

Hypothesis testing is the fundamental theory behind decision-making schemes and has been well-established [1, 2]. Depending on their optimization formulations, different hypothesis testing problems have been introduced, among which Bayesian and Neyman–Pearson hypothesis testing problems are commonly discussed in the literature. It has been shown that the optimal hypothesis testing strategies for Bayesian and Neyman–Pearson hypothesis testing problems depend on the likelihoods [1]. The optimal hypothesis testing performances have been characterized in the asymptotic regime from the large deviations perspective [3, 4].

Hypothesis testing has been studied in the privacy problems [5–8], where the objectives are to enhance the hypothesis testing performance and meanwhile to degrade the privacy leakage risk usually measured by an information-theoretic metric. On the contrary, the privacy problem can also be formulated to enhance an utility performance and meanwhile to degrade the privacy leakage risk measured by a hypothesis testing metric [9–16]. In this chapter, the privacy problem is modeled as an adversarial hypothesis testing and the privacy is enhanced by optimally utilizing a limited-capability distortion source.

Nowadays, smart meter privacy receives more and more attention due to fast deployments of smart grid and increasing concerns of privacy. In the literature, there are two main classes of privacy-enhancing techniques: smart meter reading manipulation and energy supply shaping [17], which share the same idea of reducing the correlation between smart meter readings and privacy. Smart meter reading manipulation has been realized through obfuscation [18], aggregation [19], anonymization [20]. However, these techniques lead to distorted smart meter readings and further result in degradation of utility performances, e.g., energy consumption prediction and grid failure detection. On the contrary, the energy supply shaping techniques report true smart meter readings and enhance privacy by making use of alternative energy sources, e.g., renewable energy sources [16, 21–23] and energy storage devices [11, 14, 23–26], to manipulate the smart meter readings from the real energy consumption. Regarding the privacy-enhancing objective, most literature focuses on minimizing the mutual information between the energy consumption data and the smart meter readings [21–24, 26] or realizing differential privacy [27–29]. In [11, 16], the framework of privacy against adversarial hypothesis testing has been used in the smart meter privacy context.

The works of privacy against adversarial hypothesis testing in [11, 16] are summarized in this chapter: Hypothesis testing elements are briefly recapitulated first; the fundamental bounds on the privacy performance against adversarial hypothesis testing subject to an average distortion constraint are characterized; the design of a privacy-enhancing management policy subject to the causal operation of a limited-capability distortion source is studied; and finally these results are applied to the smart meter privacy problem.

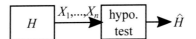

Fig. 3.1 Hypothesis testing problem, where H is a binary hypothesis; the hypothesis testing decision $\hat{H}$ is made based on the data sequence $(X_1, \ldots, X_n)$

In the following, unless otherwise specified, a random variable is denoted by a capital letter, e.g., X; its realization is denoted by the lower-case letter, e.g., x; and its alphabet is denoted by the calligraphic letter, e.g., $\mathcal{X}$. Let X_t^k, x_t^k, and $\mathcal{X}_t^k$ denote a random sequence $(X_t, \ldots, X_k)$, its realization $(x_t, \ldots, x_k)$, and its alphabet $\mathcal{X}_t \times \cdots \times \mathcal{X}_k$. For simplification, X^k, x^k, and $\mathcal{X}^k$ are used when $t = 1$. Expectation is denoted by $E[\cdot]$. Kullback–Leibler divergence is denoted by $D(\cdot||\cdot)$. Chernoff information is denoted by $C(\cdot, \cdot)$. Alphabet cardinality is denoted by $|\cdot|$.

3.2 Hypothesis Testing Elements

Consider a binary hypothesis testing shown in Fig. 3.1, where the hypothesis H can be h_0 or h_1, e.g., watching TV or not, with prior distribution denoted as p_H; the random observation X_i is defined on a finite support set $\mathcal{X}$ and is i.i.d. generated following $p_{X|h_0}$ under hypothesis h_0 or following $p_{X|h_1}$ under hypothesis h_1; on receiving an observation sequence x^n, a randomized hypothesis testing strategy $\phi_{(n)} : \mathcal{X}^n \rightarrow \{h_0, h_1\}$ is implemented to make a random decision $\hat{H} = \phi_{(n)}(x^n)$ to infer the true hypothesis. There are two commonly-used hypothesis testing problems formulated for the hypothesis testing strategy optimization, which are briefly recapitulated in the following. Note that the results derived in the considered binary hypothesis testing can be extended to more general hypothesis testings.

3.2.1 Bayesian Hypothesis Testing

For the Bayesian hypothesis testing problem, the hypothesis testing strategy is optimized to minimize the Bayesian risk:

$$\phi_{(n)}^* = \arg \min_{\phi_{(n)} \in \Phi_{(n)}} r\left(\phi_{(n)}\right), \tag{3.1}$$

where $\Phi_{(n)}$ denotes the set of all hypothesis testing strategies; the Bayesian risk is defined as

$$r\left(\phi_{(n)}\right) = E\left[c\left(H, \phi_{(n)}(X^n)\right)\right]; \tag{3.2}$$

and $c : \{h_0, h_1\} \times \{h_0, h_1\} \rightarrow \mathbb{R}_{\geq 0}$ denotes a deterministic cost function.

When the cost function satisfies $c(h_1, h_0) > c(h_1, h_1)$ and $c(h_0, h_1) > c(h_0, h_0)$, an optimal hypothesis testing strategy for (3.1) is the following deterministic likelihood-ratio test (LRT) [2]:

$$\frac{p_{X^n|h_0}(x^n)}{p_{X^n|h_1}(x^n)} \underset{\phi_{(n)}^*(x^n)=h_0}{\overset{\phi_{(n)}^*(x^n)=h_1}{\lessgtr}} \frac{p_H(h_1)(c(h_1, h_0) - c(h_1, h_1))}{p_H(h_0)(c(h_0, h_1) - c(h_0, h_0))}. \tag{3.3}$$

The minimum Bayesian risk is

$$r\left(\phi_{(n)}^*\right) = \sum_{x^n \in \mathcal{X}^n} \min_{\hat{h} \in \{h_0, h_1\}} \sum_{h \in \{h_0, h_1\}} c(h, \hat{h}) p_{X^n|h}(x^n) p_H(h). \tag{3.4}$$

When a particular cost function is used with $c(h_0, h_0) = c(h_1, h_1) = 0$ and $c(h_1, h_0) = c(h_0, h_1) = 1$, the Bayesian risk reduces to the hypothesis testing error probability $e(\phi_{(n)})$. As the observation sequence length n increases, the minimum error probability $e\left(\phi_{(n)}^*\right)$ monotonically decreases. In this case, an alternative measure of the optimal hypothesis testing performance is the error exponent $\frac{1}{n} \log \frac{1}{e(\phi_{(n)}^*)}$. In the asymptotic regime as $n \to \infty$, the asymptotic error exponent is characterized by the Chernoff information [3] as

$$\lim_{n \to \infty} \frac{1}{n} \log \frac{1}{e(\phi_{(n)}^*)} = C(p_{X|h_0}, p_{X|h_1})$$

$$= \max_{0 \leq \tau \leq 1} -\log\left(\sum_{x \in \mathcal{X}} \left(p_{X|h_0}(x)\right)^\tau \left(p_{X|h_1}(x)\right)^{1-\tau}\right). \tag{3.5}$$

3.2.2 Neyman–Pearson Hypothesis Testing

Note that the hypothesis prior distribution and the decision cost function in the Bayesian hypothesis testing problem are not always available. If the priors and costs are not available, Neyman–Pearson hypothesis testing problem is formulated to optimally tradeoff the type I error $e_I(\phi_{(n)})$ and type II error $e_{II}(\phi_{(n)})$ as

$$\phi_{(n)}^* = \arg \min_{\phi_{(n)} \in \Phi_{(n)}} e_{II}(\phi_{(n)}), \text{ s.t., } e_I(\phi_{(n)}) \leq \lambda, \tag{3.6}$$

where $e_I(\phi_{(n)}) = \Pr(\phi_{(n)}(X^n) = h_1 | H = h_0)$ and $e_{II}(\phi_{(n)}) = \Pr(\phi_{(n)}(X^n) = h_0 | H = h_1)$.

The Neyman–Pearson problem is a convex optimization and can be solved using the method of Lagrange multiplier. Different from the Bayesian problem, the optimal hypothesis testing strategy of (3.6) is generally a randomized strategy of two deterministic LRTs, which achieves the type I error upper bound, i.e., $e_I(\phi_{(n)}^*) = \lambda$.

Likewise, the minimum type II error $e_{II}(\phi^*_{(n)})$ monotonically decreases as the observation sequence length n increases. When the deterministic hypothesis testing strategies are considered and in the asymptotic regime as $n \to \infty$, the asymptotic type II error exponent, subject to any type I error upper bound $\lambda \in (0, 1)$, is characterized by the Kullback–Leibler divergence [4] as

$$\lim_{n\to\infty} \frac{1}{n} \log \frac{1}{e_{II}\left(\phi^*_{(n)}\right)} = D(p_{X|h_0}||p_{X|h_1})$$

$$= \sum_{x \in \mathcal{X}} p_{X|h_0}(x) \log \frac{p_{X|h_0}(x)}{p_{X|h_1}(x)}. \qquad (3.7)$$

3.2.3 Adversarial Hypothesis Testing

Suppose that the considered hypothesis is a privacy-sensitive information, e.g., the energy consumption behavior. When an optimal hypothesis testing strategy is implemented by an informed and powerful adversary, the corresponding optimal hypothesis testing performance then measures the privacy leakage risk in the worst case. Although such a worst case does not necessarily happen in practice, the privacy performance of a privacy-enhancing technique considered in the worst case can be seen as a privacy guarantee or performance bound, which is of high practical relevance.

Different privacy-enhancing techniques against an adversarial hypothesis testing have been proposed and studied [9–16]. In this chapter, we consider the method used in [11, 16], which enhances the privacy by optimally exploiting a distortion source to degrade the adversarial hypothesis testing performance, while satisfies a certain utility constraint.

3.3 Privacy Against Adversarial Hypothesis Testing

The considered privacy problem is shown in Fig. 3.2, which extends the adversarial hypothesis testing (AHT) problem in Fig. 3.1 and includes a management unit (MU) to manipulate the adversarial observation sequence Y^n from the data sequence X^n by exploiting a distortion source (DS). The same settings are applied on the binary privacy-sensitive hypothesis H and the i.i.d. data sequence X^n as used in Sect. 3.2. The adversarial observation sequence Y^n is determined by the implemented management policy $\gamma_{(n)} : \mathcal{X}^n \times \mathcal{H} \to \mathcal{Y}^n$, where $\mathcal{Y}$ is the finite alphabet of the adversarial observation. In practice, an instantaneous information processing is needed in every time slot. Let $\gamma_i : \mathcal{X}^i \times \mathcal{Y}^{i-1} \times \mathcal{H} \to \mathcal{Y}$ denote an instantaneous management policy in the i-th slot, i.e., the adversarial observation Y_i is determined based on the data sequence x^i, the previous adversarial observation sequence y^{i-1}, and the

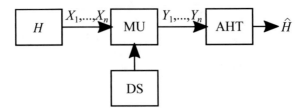

Fig. 3.2 Privacy problem, where H is a binary hypothesis; the management unit (MU) manipulates the data sequence $(X_1, \ldots, X_n)$ to an adversarial observation sequence $(Y_1, \ldots, Y_n)$ by exploiting the distortion source (DS); the adversarial hypothesis testing (AHT) decision $\hat{H}$ is made based on the observation sequence

true hypothesis h. Then implementing an n-slot policy $\gamma_{(n)}$ means implementing the instantaneous policies $\gamma_1, \ldots, \gamma_n$ successively. Furthermore, the management policy is subject to certain distortion constraints, which are imposed either to guarantee a utility requirement from the data, or due to the availability of limited DS to manipulate the data sequence. Here, the following average distortion constraint is imposed on an n-slot policy as

$$\mathrm{E}\left[\left.\frac{1}{n}\sum_{i=1}^{n}d(X_i, Y_i)\right| h_j\right] \leq s, \ j = 0, 1, \tag{3.8}$$

where $d : \mathcal{X} \times \mathcal{Y} \to \mathbb{R}_{\geq 0}$ is an additive distortion measure. An n-slot policy that satisfies (3.8) is denoted by $\gamma_{(n,s)}$, which is in the policy set $\Gamma_{(n,s)}$.

An informed adversary is assumed to have access to the observations y^n, to know p_H, $p_{X|h_0}$, $p_{X|h_1}$, $\gamma_{(n,s)}$, the resulting adversarial observation statistics $p_{Y^n|h_0}$, $p_{Y^n|h_1}$, and to make an optimal AHT on the binary privacy-sensitive hypothesis. In the following, the optimal privacy performances are studied under an adversarial Neyman–Pearson hypothesis testing and an adversarial Bayesian hypothesis testing, respectively.

3.3.1 Privacy Performances

3.3.1.1 Privacy Performance Under Adversarial Neyman–Pearson Hypothesis Testing

Under an adversarial Neyman–Pearson hypothesis testing, the privacy leakage risk is measured by the minimum type II error subject to a type I error upper bound λ. When an n-slot management policy $\gamma_{(n,s)}$ is used, the observation statistics $p_{Y^n|h_0}$, $p_{Y^n|h_1}$ are known by the adversary. Consider the deterministic hypothesis testing strategies and let $e_{\mathrm{II}}\left(\gamma_{(n,s)}, \lambda\right)$ denote the minimum type II error as

$$e_{\text{II}}\left(\gamma_{(n,s)}, \lambda\right) = \min_{\mathcal{A}_{(n)} \subseteq \mathcal{Y}^n} p_{Y^n|h_1}\left(\mathcal{A}_{(n)}\right), \tag{3.9a}$$

$$\text{s.t.} \quad p_{Y^n|h_0}\left(\mathcal{A}_{(n)}^c\right) \leq \lambda, \tag{3.9b}$$

where $\mathcal{A}_{(n)}$ is a decision set of h_0 and the complement $\mathcal{A}_{(n)}^c$ is a decision set of h_1. An optimal privacy-enhancing policy is designed to maximize the minimum type II error of the adversary as

$$e_{\text{II}}(n, s, \lambda) = \max_{\gamma_{(n,s)} \in \Gamma_{(n,s)}} e_{\text{II}}\left(\gamma_{(n,s)}, \lambda\right), \tag{3.10}$$

which measures the optimal privacy performance in the worst privacy leakage scenario.

By solving the maximin optimization problem, the optimal privacy performance and the corresponding optimal management policy are obtained. Unfortunately, a closed-form solution of the optimization can be extremely difficult when there are a large number of divisions of decision sets. Instead, an elegant evaluation of the asymptotic privacy performance is studied in the following from a large deviations perspective.

The asymptotic privacy performance under adversarial Neyman–Pearson hypothesis testing is characterized by the following Kullback–Leibler divergence rate:

$$\theta(s) = \inf_{k \in \mathbb{Z}_{\geq 1}} \inf_{\gamma_{(k,s)} \in \Gamma_{(k,s)}} \frac{1}{k} D(p_{Y^k|h_0} \| p_{Y^k|h_1})$$

$$= \lim_{k \to \infty} \inf_{\gamma_{(k,s)} \in \Gamma_{(k,s)}} \frac{1}{k} D(p_{Y^k|h_0} \| p_{Y^k|h_1}). \tag{3.11}$$

Theorem 3.1 *Given* $s > 0$,

$$\limsup_{n \to \infty} \frac{1}{n} \log \frac{1}{e_{\text{II}}(n, s, \lambda)} \leq \theta(s), \quad \forall \lambda \in (0, 1), \tag{3.12}$$

$$\lim_{\lambda \to 1} \liminf_{n \to \infty} \frac{1}{n} \log \frac{1}{e_{\text{II}}(n, s, \lambda)} \geq \theta(s). \tag{3.13}$$

The proof ideas of Theorem 3.1 are summarized as follows: The equality (3.11) is derived by showing $\left(\inf_{\gamma_{(k,s)} \in \Gamma_{(k,s)}} \frac{1}{k} D(p_{Y^k|h_0} \| p_{Y^k|h_1})\right)_{k \in \mathbb{Z}_{\geq 1}}$ is subadditive and applying the Fekete's Lemma [30]; the upper bound in (3.12) is derived from the maximization in the definition of $e_{\text{II}}(n, s, \lambda)$ and the Stein's Lemma [4] shown in (3.7); the lower bound in (3.13) is derived from the minimization in the definition of $e_{\text{II}}(n, s, \lambda)$. The complete proof can be found in [16].

From Theorem 3.1, there are some interesting observations: $\theta(s)$ is an infimum Kullback–Leibler divergence rate, where the Kullback–Leibler divergence rate can be intuitively seen as the asymptotic adversarial hypothesis testing performance and the infimum over management policies of all time-slot lengths accounts for the optimal privacy enhancement; $\theta(s)$ is generally not achievable since the infimum over time-slot lengths is taken at the limit; the limit of optimal privacy performance exits when $\lambda \to 1$, i.e.,

$$\lim_{\lambda \to 1} \lim_{n \to \infty} \frac{1}{n} \log \frac{1}{e_{\mathrm{II}}(n, s, \lambda)} = \theta(s); \tag{3.14}$$

$\theta(s)$ accounts for the worst asymptotic privacy performance, i.e., $\lambda \to 1$ is the worst privacy leakage case under an adversarial Neyman–Pearson hypothesis testing.

3.3.1.2 Privacy Performance Under Adversarial Bayesian Hypothesis Testing

Under an adversarial Bayesian hypothesis testing, the privacy leakage risk is measured by the minimum error probability, which is a particular Bayesian risk. When an n-slot management policy $\gamma_{(n,s)}$ is used, the observation statistics $p_{Y^n|h_0}$, $p_{Y^n|h_1}$ are assumed to be known by the informed adversary. As shown in Sect. 3.2, the optimal adversarial hypothesis testing strategy is a deterministic LRT. Let $e\left(\gamma_{(n,s)}\right)$ denote the minimum error probability as

$$e\left(\gamma_{(n,s)}\right) = \min_{\mathcal{A}_{(n)} \subseteq \mathcal{Y}^n} p_{Y^n|h_1}\left(\mathcal{A}_{(n)}\right) p_H(h_1) + p_{Y^n|h_0}\left(\mathcal{A}^c_{(n)}\right) p_H(h_0)$$

$$= \sum_{y^n \in \mathcal{Y}^n} \min_{h \in \{h_0, h_1\}} p_{Y^n|h}(y^n) p_H(h). \tag{3.15}$$

An optimal privacy-enhancing policy is designed to maximize the minimum error probability of the adversary as

$$e(n, s) = \max_{\gamma_{(n,s)} \in \Gamma_{(n,s)}} e\left(\gamma_{(n,s)}\right), \tag{3.16}$$

which measures the optimal privacy performance under an adversarial Bayesian hypothesis testing.

Similarly, the following Chernoff information rate characterizes the asymptotic privacy performance under an adversarial Bayesian hypothesis testing:

$$\mu(s) = \inf_{k \in \mathbb{Z}_{\geq 1}} \inf_{\gamma_{(k,s)} \in \Gamma_{(k,s)}} \frac{1}{k} C(p_{Y^k|h_0}, p_{Y^k|h_1})$$

$$= \lim_{k \to \infty} \inf_{\gamma_{(k,s)} \in \Gamma_{(k,s)}} \frac{1}{k} C(p_{Y^k|h_0}, p_{Y^k|h_1}). \tag{3.17}$$

Theorem 3.2 *Given $s > 0$,*

$$\lim_{n \to \infty} \frac{1}{n} \log \frac{1}{e(n, s)} = \mu(s). \tag{3.18}$$

The proof ideas of Theorem 3.2 are similar to those used in the proof of Theorem 3.1. To prove the equality in (3.18), the Chernoff information rate $\mu(s)$ is first shown to be an upper bound and a lower bound on the asymptotic privacy performance. The upper bound follows from the maximization in the definition of $e(n, s)$ and the Chernoff Theorem [3] shown in (3.5). For more proof details, please refer to [16].

The Chernoff information rate $\mu(s)$ measures the limit of optimal privacy performance under an adversarial Bayesian hypothesis testing and has an intuitive interpretation as: The Chernoff information rate measures the asymptotic adversarial hypothesis testing performance; and the infimum over management policies means the optimization of privacy-enhancing management policy.

3.3.2 Memoryless Policy

Generally, the asymptotic privacy performances and the optimal privacy-enhancing management policies are difficult to evaluate and implement since the infimum over time-slot lengths is taken at the limit. On the other hand, a simple privacy-enhancing policy is needed in practice. Therefore, memoryless policies are considered here.

An n-slot memoryless policy is denoted as $\pi_{(n)} : \mathcal{X}^n \times \mathcal{H} \rightarrow \mathcal{Y}^n$, which consists of memoryless instantaneous policies $\pi_1, \dots, \pi_n$. In the i-th slot, a memoryless instantaneous policy $\pi_i : \mathcal{X} \times \mathcal{H} \rightarrow \mathcal{Y}$ randomly decides the adversarial observation Y_i based on the current data x_i and the true hypothesis h. Compared with an instantaneous policy γ_i, π_i does not need a memory to store the previous data sequence and previous adversarial observation sequence. Furthermore, the design, implementation, and evaluation of the optimal memoryless policy will be shown to be much simpler. An n-slot memoryless policy satisfying the average distortion constraint (3.8) is denoted as $\pi_{(n,s)}$, which is in the policy set $\Pi_{(n,s)}$.

3.3.2.1 Memoryless Management Against Adversarial Neyman–Pearson Hypothesis Testing

Under an adversarial Neyman–Pearson hypothesis testing, an optimal privacy-enhancing memoryless policy is designed to maximize the minimum type II error of the adversary as

$$e_{\mathrm{II}}^{\mathrm{ML}}(n, s, \lambda) = \max_{\pi_{(n,s)} \in \Pi_{(n,s)}} e_{\mathrm{II}}\left(\pi_{(n,s)}, \lambda\right). \tag{3.19}$$

The asymptotic results of the optimal privacy performance in Theorem 3.1 can be directly applied to the optimal memoryless management policy.

Corollary 3.1 *Given s > 0,*

$$\lim_{\lambda \to 1} \lim_{n \to \infty} \frac{1}{n} \log \frac{1}{e_{\mathrm{II}}^{\mathrm{ML}}(n, s, \lambda)} = \inf_{k \in \mathbb{Z}_{\geq 1}} \inf_{\pi_{(k,s)} \in \Pi_{(k,s)}} \frac{1}{k} \mathrm{D}(p_{Y^k|h_0} \| p_{Y^k|h_1}). \qquad (3.20)$$

Although the infimum Kullback–Leibler divergence rate can be taken at the limit as $k \to \infty$, it can also be shown that the infimum Kullback–Leibler divergence rate over memoryless policies reduces to a single-letter minimum Kullback–Leibler divergence over single-slot memoryless policies. Then, the evaluation of asymptotic privacy performance and the design of optimal memoryless policy become feasible.

Note that a single-slot memoryless policy can be alternatively represented by the corresponding conditional probability distribution $p_{Y|X,H}$, which has to satisfy the average distortion constraint such that $\mathrm{E}\left[d(X, Y)|h_j\right] \leq s$ for $j = 0, 1$. Let $\mathcal{P}(s)$ denote the set of conditional probability distributions of single-slot memoryless policies:

$$\mathcal{P}(s) = \left\{ p_{Y|X,H} : \mathrm{E}\left[d(X, Y)|h_j\right] \leq s, \ j = 0, 1 \right\}. \qquad (3.21)$$

Then, the following theorem shows the asymptotic privacy performance can be characterized by a single-letter Kullback–Leibler divergence.

Theorem 3.3 *Given s > 0,*

$$\lim_{\lambda \to 1} \lim_{n \to \infty} \frac{1}{n} \log \frac{1}{e_{\mathrm{II}}^{\mathrm{ML}}(n, s, \lambda)} = \min_{p_{Y|X,H} \in \mathcal{P}(s)} \mathrm{D}(p_{Y|h_0} \| p_{Y|h_1}). \qquad (3.22)$$

The proof of Theorem 3.3 is equivalent to show

$$\inf_{k \in \mathbb{Z}_{\geq 1}} \inf_{\pi_{(k,s)} \in \Pi_{(k,s)}} \frac{1}{k} \mathrm{D}(p_{Y^k|h_0} \| p_{Y^k|h_1}) = \min_{p_{Y|X,H} \in \mathcal{P}(s)} \mathrm{D}(p_{Y|h_0} \| p_{Y|h_1}).$$

The complete proof is shown in [16]. Here, the proof ideas are summarized as follows:

- Show that $\min_{p_{Y|X,H} \in \mathcal{P}(s_0, s_1)} \mathrm{D}(p_{Y|h_0} \| p_{Y|h_1})$, which allows different average distortion constraints with $\mathcal{P}(s_0, s_1) = \left\{ p_{Y|X,H} : \mathrm{E}\left[d(X, Y)|h_j\right] \leq s_j, \ j = 0, 1 \right\}$, is non-increasing, continuous, and jointly convex in $s_0 > 0, s_1 > 0$;
- Use these properties to show that $\inf_{k \in \mathbb{Z}_{\geq 1}} \inf_{\pi_{(k,s)} \in \Pi_{(k,s)}} \frac{1}{k} \mathrm{D}(p_{Y^k|h_0} \| p_{Y^k|h_1})$ is lower bounded by $\min_{p_{Y|X,H} \in \mathcal{P}(s)} \mathrm{D}(p_{Y|h_0} \| p_{Y|h_1})$;
- Use the definitions to show that $\inf_{k \in \mathbb{Z}_{\geq 1}} \inf_{\pi_{(k,s)} \in \Pi_{(k,s)}} \frac{1}{k} \mathrm{D}(p_{Y^k|h_0} \| p_{Y^k|h_1})$ is upper bounded by $\min_{p_{Y|X,H} \in \mathcal{P}(s)} \mathrm{D}(p_{Y|h_0} \| p_{Y|h_1})$.

From Theorem 3.3, there are some interesting observations: The asymptotic privacy performance can be evaluated by solving the convex optimization (3.22); the asymptotic privacy performance can be achieved by implementing the single-slot memoryless policy corresponding to the optimizer of the convex optimization (3.22) in all time slots, i.e., the optimal memoryless policy is an i.i.d. policy.

3.3.2.2 Memoryless Management Against Adversarial Bayesian Hypothesis Testing

Under an adversarial Bayesian hypothesis testing, an optimal privacy-enhancing memoryless policy is designed to maximize the minimum error probability of the adversary as

$$e^{\mathrm{ML}}(n, s) = \max_{\pi_{(n,s)} \in \Pi_{(n,s)}} e\left(\pi_{(n,s)}\right). \tag{3.23}$$

The asymptotic results of the optimal privacy performance in Theorem 3.2 can be directly applied to the optimal memoryless management policy.

Corollary 3.2 *Given $s > 0$,*

$$\lim_{n \to \infty} \frac{1}{n} \log \frac{1}{e^{\mathrm{ML}}(n, s)} = \inf_{k \in \mathbb{Z}_{\geq 1}} \inf_{\pi_{(k,s)} \in \Pi_{(k,s)}} \frac{1}{k} C(p_{Y^k|h_0}, p_{Y^k|h_1}). \tag{3.24}$$

Similarly, the following theorem shows that the infimum Chernoff information rate over memoryless policies reduces to a single-letter minimum Chernoff information over single-slot memoryless policies.

Theorem 3.4 *Given $s > 0$,*

$$\lim_{n \to \infty} \frac{1}{n} \log \frac{1}{e^{\mathrm{ML}}(n, s)} = \min_{p_{Y|X,H} \in \mathcal{P}(s)} C(p_{Y|h_0}, p_{Y|h_1}). \tag{3.25}$$

To prove Theorem 3.4, similar proof ideas as Theorem 3.3 are used to show the following equation:

$$\inf_{k \in \mathbb{Z}_{\geq 1}} \inf_{\pi_{(k,s)} \in \Pi_{(k,s)}} \frac{1}{k} C(p_{Y^k|h_0}, p_{Y^k|h_1}) = \min_{p_{Y|X,H} \in \mathcal{P}(s)} C(p_{Y|h_0}, p_{Y|h_1}),$$

where the Chernoff information can be represented as a maximization as

$$C(p_{Y|h_0}, p_{Y|h_1}) = \max_{0 \leq \tau \leq 1} C_\tau(p_{Y|h_0}, p_{Y|h_1})$$

with

$$C_\tau(p_{Y|h_0}, p_{Y|h_1}) = -\log \left(\sum_{y \in \mathcal{Y}} (p_{Y|h_0}(y))^\tau (p_{Y|h_1}(y))^{1-\tau} \right).$$

More specifically,

- Show that $\min_{p_{Y|X,H} \in \mathcal{P}(s_0, s_1)} C_\tau(p_{Y|h_0}, p_{Y|h_1})$, with $0 \leq \tau \leq 1$, is non-increasing, continuous, and jointly convex in $s_0 > 0$, $s_1 > 0$;

- Use these properties to show that $\inf_{k \in \mathbb{Z}_{\geq 1}} \inf_{\pi_{(k,s)} \in \Pi_{(k,s)}} \frac{1}{k} C(p_{Y^k|h_0}, p_{Y^k|h_1})$ is lower bounded by $\max_{0 \leq \tau \leq 1} \min_{p_{Y|X,H} \in \mathcal{P}(s)} C_\tau(p_{Y|h_0}, p_{Y|h_1})$;
- Use the definitions to show that $\inf_{k \in \mathbb{Z}_{\geq 1}} \inf_{\pi_{(k,s)} \in \Pi_{(k,s)}} \frac{1}{k} C(p_{Y^k|h_0}, p_{Y^k|h_1})$ is upper bounded by $\min_{p_{Y|X,H} \in \mathcal{P}(s)} C(p_{Y|h_0}, p_{Y|h_1})$;
- Finally show that the upper and lower bounds are equal by von Neumann's Minimax Theorem [31].

From Theorem 3.4, the asymptotic privacy performance can be evaluated and the optimal i.i.d. memoryless policy can be designed by solving the convex optimization (3.25).

3.3.3 Privacy Against AHT in the Presence of Memory

The considered privacy problem makes assumptions of a time-invariant binary hypothesis, an i.i.d. data sequence, and a powerful distortion source subject to a simple average distortion constraint. These assumptions lead to elegant theoretic results but are often not satisfied in practice. In this section, a more general privacy problem model is considered with a time-variant multi-hypothesis, memory in the data sequence, and causal processing of the distortion source.

The modified privacy problem is shown in Fig. 3.3: The time-variant multi-hypothesis H_i is defined on the alphabet $\mathcal{H} = \{h_0, \ldots, h_{m-1}\}$ and is randomly evolved from the previous hypothesis h_{i-1} following a time-invariant transition $p_{H_i|H_{i-1}}$; the original data X_i defined on a finite alphabet $\mathcal{X}$ is randomly generated depending on the current hypothesis h_i and the previous data x_{i-1} as $p_{X_i|h_i,x_{i-1}}$; the MU randomly manipulates the original data x_i to the adversarial observation Y_i, which is defined on a finite alphabet $\mathcal{Y}$, by exploiting the DS of state s_i, which is defined on a finite alphabet $\mathcal{S}$, following an instantaneous policy $\omega_i : \mathcal{X} \times \mathcal{S} \to \mathcal{Y}$ in the i-th slot, i.e., the adversarial observation Y_i is determined based on the current data x_i and the current distortion state s_i; the next distortion state is determined by the current distortion state, current data, and current adversarial observation as $s_{i+1} = g(s_i, x_i, y_i)$. Furthermore, the management policy is subject to the limited

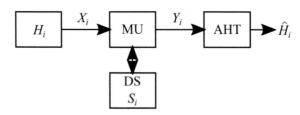

Fig. 3.3 Privacy problem, where H_i is a time-variant privacy-sensitive multi-hypothesis; the management unit manipulates the original data X_i to an adversarial observation Y_i by exploiting the distortion source of state S_i; the adversarial hypothesis testing decision $\hat{H}_i$ is made based on the observation Y_i

capacity of the DS and the causal processing: Given any $s_i \in \mathcal{S}$ and $x_i \in \mathcal{X}$, after the composition processing of policy ω_i and distortion state update g,

$$\Pr\left(g \circ \omega_i(s_i, x_i) \notin \mathcal{S}\right) = 0. \tag{3.26}$$

An adversarial Bayesian hypothesis testing is assumed here: In the i-th time slot, the informed adversary knows the needed statistics $p_{Y_i|H_i}$ and uses the optimal hypothesis testing strategy, which minimizes the error probability of inferring the current hypothesis H_i based on the current observation y_i. These privacy problem settings are summarized in the following equation:

$$p_{\hat{H}^n, H^n, S^n, X^n, Y^n} = p_{H_1, S_1, X_1} \prod_{i=1}^{n} p_{\hat{H}_i|Y_i} p_{Y_i|S_i, X_i}$$

$$\times \prod_{j=2}^{n} p_{S_j|S_{j-1}, X_{j-1}} p_{X_j|H_j, X_{j-1}} p_{H_j|H_{j-1}}, \tag{3.27}$$

where $p_{\hat{H}_i|Y_i}$ represents the adversarial hypothesis testing strategy; $p_{Y_i|S_i, X_i}$ and $p_{S_j|S_{j-1}, X_{j-1}}$ represent management policies in the i-th and $(j-1)$-th time slots, respectively.

3.3.3.1 Privacy Leakage and Privacy Performance

In the i-th slot, the adversarial Bayesian hypothesis testing strategy is a deterministic likelihood test. Given initial statistics $b_1 = p_{H_1, S_1, X_1}$ and an i-slot management policy $\omega_{(i)}$, from (3.4), the minimum error probability of adversary in the i-th slot is

$$e_i\left(\omega_{(i)}, b_1\right) = e_i\left(\omega_i, b_i\right)$$

$$= \sum_{y_i \in \mathcal{Y}} \min_{\hat{h}_i \in \mathcal{H}} \sum_{h_i \in \mathcal{H}, \, h_i \neq \hat{h}_i} p_{Y_i|h_i}(y_i) p_{H_i}(h_i)$$

$$= \sum_{y_i \in \mathcal{Y}} \min_{\hat{h}_i \in \mathcal{H}} \sum_{h_i \in \mathcal{H}, \, h_i \neq \hat{h}_i} \sum_{s_i \in \mathcal{S}, x_i \in \mathcal{X}} p_{Y_i|s_i, x_i}(y_i) p_{H_i, S_i, X_i}(h_i, s_i, x_i). \tag{3.28}$$

Note that the statistics $b_i = p_{H_i, S_i, X_i}$ depend on the initial statistics $b_1 = p_{H_1, S_1, X_1}$ and the previous policies $\omega_{(i-1)}$, which will be discussed in detail later.

The instantaneous privacy leakage risk in the i-th slot is measured by $e_i\left(\omega_{(i)}, b_1\right)$. Given b_1, over n time slots, the privacy performance of management policy $\omega_{(n)}$ is measured by

$$V\left(\omega_{(n)}, \beta, b_1\right) = \sum_{i=1}^{n} \beta^{i-1} e_i\left(\omega_{(i)}, b_1\right), \tag{3.29}$$

which is an accumulated discounted privacy leakage risk with a discount coefficient $0 \leq \beta < 1$. This measure of privacy performance means that the privacy-enhancing concern degrades as time goes on. For instance, the adversary also has an exposure risk, which increases as the time goes on.

The optimal n-slot privacy-enhancing management policy is designed to maximize the accumulated discounted privacy leakage risk and the optimal privacy performance is

$$V(n, \beta, b_1) = \max_{\omega_{(n)} \in \Omega_{(n)}} V\left(\omega_{(n)}, \beta, b_1\right). \tag{3.30}$$

3.3.3.2 Privacy-Enhancing Management Policy

The optimal n-slot policy and the corresponding privacy performance can be obtained and evaluated by solving the optimization problem (3.30). However, it is a difficult task because of the memory in the system. As shown in (3.28), the instantaneous privacy leakage risk is determined not only by the current management policy but also the previous management policies. Therefore, the design of an instantaneous policy needs to take into account its impact on the future. Instead of solving (3.30) directly, the privacy-enhancing management policy is designed from a perspective of Markov decision process (MDP) in the following.

There are two important observations from the problem settings:

- In (3.28), the instantaneous privacy leakage risk $e_i(\omega_i, b_i)$ is determined by the instantaneous policy $\omega_i = p_{Y_i | S_i, X_i}$ and the statistics $b_i = p_{H_i, S_i, X_i}$, which are determined by the previous policies $\omega_{(i-1)}$ and the initial statistics b_1 as implied by the second observation;
- The other observation is

$$\begin{aligned} & p_{H_{i+1}, S_{i+1}, X_{i+1} | h_i, s_i, x_i}(h_{i+1}, s_{i+1}, x_{i+1}) \\ & = p_{S_{i+1} | s_i, x_i}(s_{i+1}) p_{X_{i+1} | h_{i+1}, x_i}(x_{i+1}) p_{H_{i+1} | h_i}(h_{i+1}), \end{aligned} \tag{3.31}$$

$$\begin{aligned} & p_{H_{i+1}, S_{i+1}, X_{i+1}}(h_{i+1}, s_{i+1}, x_{i+1}) \\ & = \sum_{h_i \in \mathcal{H}, s_i \in \mathcal{S}, x_i \in \mathcal{X}} p_{H_{i+1}, S_{i+1}, X_{i+1} | h_i, s_i, x_i}(h_{i+1}, s_{i+1}, x_{i+1}) p_{H_i, S_i, X_i}(h_i, s_i, x_i), \end{aligned} \tag{3.32}$$

which means that the next statistics $b_{i+1} = p_{H_{i+1}, S_{i+1}, X_{i+1}}$ are determined by the current instantaneous policy $\omega_i = p_{S_{i+1} | S_i, X_i}$ and the current statistics $b_i = p_{H_i, S_i, X_i}$.

Based on these observations, the design of privacy-enhancing management policy can be cast to an MDP problem, whose elements are specified as follows.

- Belief state: $b_i = p_{H_i, S_i, X_i}$
- Action: $\omega_i = p_{Y_i | S_i, X_i}$
- Reward: $e_i(\omega_i, b_i)$
- Belief state update following (3.32): $b_{i+1}(\omega_i, b_i)$

- Action selection policy: $\delta_i : \mathcal{B} \rightarrow \Omega$, which maps the belief state b_i to an action $\omega_i = \delta_i(b_i)$

Remark 3.1 When the optimal action selection policy δ_i^* is obtained, an optimal privacy-enhancing policy is designed as follows: On observing a belief state b_i, the optimal management policy ω_i^* is selected based on δ_i^*; the next belief state b_{i+1} can be calculated/observed based on the optimal management policy ω_i^* and the belief state b_i.

The remaining problem is the optimization of action selection policies. Here, the problem is simplified by considering the asymptotic regime as $n \rightarrow \infty$, which leads to time-invariant optimal action selection policies [32]. Let $V_a(\beta, b_1)$ denote $V(n, \beta, b_1)$ as $n \rightarrow \infty$. An optimal action selection policy satisfies the Bellman equation [33]: For all $i \in \mathbb{Z}_{\geq 1}$ and $b_i \in \mathcal{B}$,

$$V_a(\beta, b_i) = \max_{\omega_i \in \Omega} e_i(\omega_i, b_i) + \beta \cdot V_a(\beta, b_{i+1}(\omega_i, b_i)), \tag{3.33}$$

$$\omega_i^* = \delta_i^*(b_i) = \arg\max_{\omega_i \in \Omega} e_i(\omega_i, b_i) + \beta \cdot V_a(\beta, b_{i+1}(\omega_i, b_i)). \tag{3.34}$$

Value iteration and policy iteration algorithms [32] can be used to solve the Bellman equation to obtain the time-invariant optimal action selection policies. The optimization of management policies can be done following Remark 3.1.

3.4 Application to Smart Meter Privacy

In a smart meter privacy problem, the high time-resolution smart meter readings, which consist of privacy-sensitive information, e.g., energy consumption behaviors, are intercepted for an illegitimate purpose by an informed adversary, e.g., a manager of the energy provider (EP). A popular privacy-enhancing method is to exploit a renewable energy source (RES) or an energy storage (ES) at the premise of a consumer to manipulate the smart meters readings of energy supply from the EP from the real energy consumption.

Such a problem can be cast to the discussed privacy against adversarial hypothesis testing: The privacy-sensitive information can be represented by a hypothesis; the energy consumption can be seen as original data, which is generated depending on the hypothesis; the RES and ES can be used as different distortion sources; the smart meter readings of energy supply from the EP can be seen as adversarial observation data; and the adversary is assumed to infer the privacy-sensitive hypothesis based on the observation. Then, the obtained results of privacy performance and privacy-enhancing management policy can be directly applied to the smart meter privacy problem. In the smart meter privacy context, privacy enhancements by exploiting an RES or an ES are illustrated in the following.

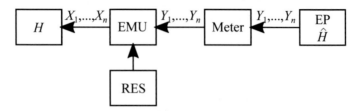

Fig. 3.4 Smart meter privacy problem, where the binary hypothesis H represents the privacy; the energy management unit (EMU) follows a memoryless energy management policy and manipulates the energy consumption sequence $(X_1, \ldots, X_n)$ to an energy supply sequence $(Y_1, \ldots, Y_n)$ from the energy provider (EP) by exploiting the renewable energy source (RES); the EP is assumed to an informed adversary and to make an adversarial Neyman–Pearson hypothesis testing decision $\hat{H}$ based on the energy supply sequence. Note that the arrows represent the energy flows

3.4.1 Privacy Enhancement Using RES

Consider the problem model shown in Fig. 3.4. The binary hypothesis H represents the privacy, e.g., the consumer is at home or not. The i.i.d. non-negative energy consumption data X_i in the i-th time slot is from a finite alphabet $\mathcal{X} \subset \mathbb{R}_{\geq 0}$ and is generated following $p_{X|h_0}$ or $p_{X|h_1}$. Assume that the EP is an informed adversary, which makes an adversarial Neyman–Pearson hypothesis testing based on the non-negative smart meter readings, or equivalently the energy supply sequence from the EP, Y^n. The finite energy supply alphabet $\mathcal{Y}$ satisfies $\mathcal{X} \subseteq \mathcal{Y} \subset \mathbb{R}_{\geq 0}$. In every time slot, the energy management unit (EMU) follows a memoryless instantaneous energy management policy $\pi_i : \mathcal{X} \times \mathcal{H} \rightarrow \mathcal{Y}$ to determine Y_i based on x_i and h. It is assumed that the remaining energy consumption, $x_i - y_i$, is satisfied by the RES, which can be solar panels. The following instantaneous constraint is imposed on the policy:

$$p_{Y_i|X_i}(y_i|x_i) = 0, \text{ if } y_i > x_i, \tag{3.35}$$

i.e., the RES cannot be charged with energy from the grid. The RES is assumed to have a positive average energy generation rate s and is equipped with a sufficiently large ES. Accordingly, the following average energy constraint is imposed on an n-slot energy management policy $\pi_{(n,s)}$ for the long-term availability of renewable energy supply:

$$\mathrm{E}\left[\frac{1}{n}\sum_{i=1}^{n}(X_i - Y_i)\middle| h_j\right] \leq s, \; j = 0, 1. \tag{3.36}$$

Note that an n-slot memoryless energy management policy satisfying (3.35) and (3.36) satisfies the general constraint (3.8).

The privacy-enhancing energy management is to manipulate the energy consumption sequence X^n into an energy supply sequence Y^n by exploiting the RES in an online manner to prevent the EP from correctly inferring the privacy of consumer. Since the memoryless energy management policy against an adversarial Neyman–

Pearson hypothesis testing is considered here, Theorem 3.3 can be directly applied to obtain the asymptotic privacy performance and the optimal memoryless energy management policy.

Under the constraint on the alphabets of energy consumption and supply $\mathcal{X} \subseteq \mathcal{Y} \subset \mathbb{R}_{\geq 0}$, the cardinality of energy consumption alphabet $|\mathcal{X}|$ can be small due to the limited number of operation modes of house appliances while the cardinality of energy supply alphabet $|\mathcal{Y}|$ can be arbitrarily large. From the definition of privacy performance $e_{\mathrm{II}}^{\mathrm{ML}}(n, s, \lambda)$ and the data processing inequality of Kullback–Leibler divergence [34], it can be shown that the energy supply alphabet can be constrained to the energy consumption alphabet without loss of optimality [16].

Here, a numerical experiment using energy data from the REDD dataset [35] is presented. Consider a dishwasher, which can be type A (h_0) or type B (h_1). From the energy data, four operation modes of a dishwasher can be identified. The corresponding statistics obtained through training under the assumption of i.i.d. energy consumption are listed in Table 3.1. In Fig. 3.5, it shows the asymptotic privacy performance improvement with the increasing average energy generation rate. Given three average renewable energy generation rates, $s = 0, 3000, 6000$ [W], the optimal memoryless energy management policies are implemented and the manipulated energy supply sequences are shown in Fig. 3.6. Although Fig. 3.5 shows very good privacy performance to be achieved given $s = 500$ [W], it is easy to identify h_0 from h_1 based on the manipulated energy supply sequences given $s = 3000$ [W] in Fig. 3.6. These conflicting observations are because of non-i.i.d. energy consumption in practice. When $s = 6000$ [W], it becomes very difficult for the adversary to identify the type of dishwasher from the energy supply data.

Table 3.1 The operation modes under each hypothesis

x [W]	0	200	500	1200
$p_{X|h_0}(x)$	0.2528	0.3676	0	0.3796
$p_{X|h_1}(x)$	0.1599	0.0579	0.2318	0.5504

Fig. 3.5 Asymptotic privacy performances against adversarial Neyman–Pearson hypothesis testing when the optimal memoryless energy management policies are used with different average renewable energy generation rates

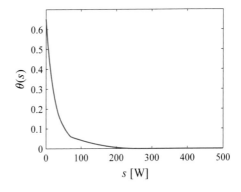

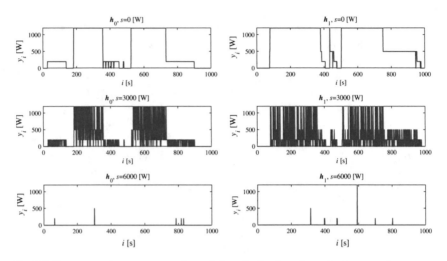

Fig. 3.6 Energy supply sequences under each hypothesis when the optimal memoryless energy management policies are used with different average renewable energy generation rates

3.4.2 Privacy Enhancement Using ES

The other smart meter privacy problem is shown in Fig. 3.7. The time-variant multi-hypothesis H_i is defined on $\mathcal{H} = \{h_0, \ldots, h_{m-1}\}$, represents the privacy in the i-th time slot, and is generated depending on the previous hypothesis h_{i-1} following $p_{H_i|H_{i-1}}$. The energy consumption X_i, energy storage state S_i, and energy supply Y_i are defined on finite non-negative integer alphabets[1]: $\mathcal{X} = \{0, \ldots, m_x\}$, $\mathcal{S} = \{0, \ldots, m_s\}$, $\mathcal{Y} = \{0, \ldots, m_y\}$, with $m_x + m_s = m_y$. The energy consumption data X_i in the i-th time slot is generated depending on the current hypothesis h_i and the previous energy consumption data x_{i-1} and following $p_{X_i|h_i,x_{i-1}}$. In the i-th time slot, the EMU follows an instantaneous energy management policy $\omega_i : \mathcal{X} \times \mathcal{S} \to \mathcal{Y}$ to determine Y_i based on x_i and s_i. The next energy storage state is determined by the current energy storage state, current energy consumption, and current energy supply as $s_{i+1} = s_i + y_i - x_i$, i.e., when $y_i - x_i > 0$, the ES is charged with the extra energy from the EP; when $y_i - x_i < 0$, the ES is discharged to satisfy partial energy consumption. Because of the finite non-negative energy storage alphabet $\mathcal{S}$, the energy management policy ω_i has to satisfy

$$p_{Y_i|s_i,x_i}(y_i) = 0, \text{ if } y_i < \max\{0, x_i - s_i\} \text{ or } y_i > m_s + x_i - s_i. \quad (3.37)$$

Assume that the EP is an informed adversary, which makes an adversarial Bayesian hypothesis testing to minimize the error probability of inferring the current hypothesis H_i based on the current energy supply data y_i.

---

[1]This setting assumes an energy measure precision.

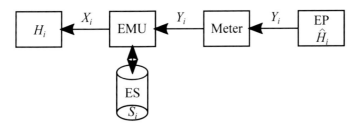

Fig. 3.7 Smart meter privacy problem, where the time-variant multi-hypothesis H_i represents the privacy in the i-th slot; the EMU follows an energy management policy and manipulates the energy consumption data X_i to an energy supply data Y_i from the EP by exploiting the energy storage (ES) of state S_i; the EP is assumed to an informed adversary and to make an adversarial Bayesian hypothesis testing decision $\hat{H}_i$ based on the energy supply data Y_i. Note that the arrows represent the energy flows

Table 3.2 Settings of parameters

Parameter	Value	Parameter	Value
m	2	$p_{X_i|H_i,X_{i-1}}(0|h_0,0)$	0.7
m_x	1	$p_{X_i|H_i,X_{i-1}}(0|h_0,1)$	0.3
m_s	1	$p_{X_i|H_i,X_{i-1}}(0|h_1,0)$	0.7
m_y	2	$p_{X_i|H_i,X_{i-1}}(0|h_1,1)$	0.3
β	0.5	$b(1)$	$p_{H_1,S_1,X_1}(h_0,1,0)=0.5$
			$p_{H_1,S_1,X_1}(h_1,1,1)=0.5$
$p_{H_i|H_{i-1}}(h_0|h_0)$	0.9	$b(2)$	$p_{H_1,S_1,X_1}(h_0,0,0)=0.5$
			$p_{H_1,S_1,X_1}(h_1,1,1)=0.5$
$p_{H_i|H_{i-1}}(h_0|h_1)$	0.2	$b(3)$	$p_{H_1,S_1,X_1}(h_0,1,1)=0.5$
			$p_{H_1,S_1,X_1}(h_1,0,1)=0.5$

As Sect. 3.3.3, the optimal privacy performance is measured by $V_a(\beta, b_1)$, where β denotes the discount coefficient and b_1 denotes the initial belief state p_{H_1,S_1,X_1}. The privacy-enhancing energy management policy can be optimized by firstly solving the Bellman equation (3.34) to obtain the optimal action selection policy, and secondly selecting the optimal instantaneous energy management policies as Remark 3.1.

A simple experiment is presented here. The settings of parameters are listed in Table 3.2, where three initial belief states are considered and denoted as $b(1)$, $b(2)$, and $b(3)$.

A benchmark energy management policy is the instantaneous optimal policy, which is designed to maximize the instantaneous minimum error probability of adversary without considering the impact on the future. In Fig. 3.8, the optimal privacy performance $V_a(\beta, b_1)$ is compared with the privacy performance using instantaneous optimal policy $J_i(\beta, b_1)$, which is the achieved accumulated discounted privacy leakage risk until the i-th time slot. The optimal privacy performance $V_a(\beta, b_1)$ should not be worse than $J_i(\beta, b_1)$ achieved by instantaneous optimal policy, since the opti-

Fig. 3.8 Comparison of
privacy performances of
using optimal energy
management policy and
using instantaneous optimal
energy management policy

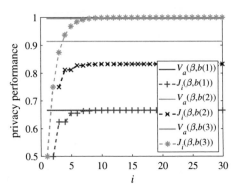

mal energy management policy is designed for long-term privacy enhancement while
the instantaneous optimal policy is designed for myopic privacy enhancement. From
the numerical results, although it is not always true, the optimal privacy performance
can be approached using instantaneous optimal policy for some initial belief states,
e.g., $b_1 = b(1)$ or $b_1 = b(3)$.

3.5 Conclusion

Privacy against adversarial hypothesis testing by exploiting a distortion source has
been studied in this chapter. Under the assumption of i.i.d. original data and average distortion constraint, the asymptotic privacy performance is characterized by a
Kullback–Leibler divergence rate or a Chernoff information rate. The optimal memoryless privacy-enhancing policy is an i.i.d. policy, which makes the policy design
and privacy performance evaluation easier. The other studied privacy model takes
into account constraints on the original data and distortion source in practice. The
privacy-enhancing policy design is then reformulated as a Markov decision process
problem. The ideas and results are applied to smart meter privacy problem. Furthermore, privacy against adversarial hypothesis testing can also be applied to other
privacy problems, e.g., multimedia forensics, adversarial machine learning, etc.

References

1. Van Trees HL (2001) Detection, estimation, and modulation theory, Part I. Wiley
2. Varshney PK (1996) Distributed detection and data fusion. Springer, New York
3. Chernoff H (1952) A measure of asymptotic efficiency for tests of a hypothesis based on the
   sum of observations. Ann Math Stat 23:493–507

4. Cover TM, Thomas JA (2006) Elements of information theory. Wiley
5. Mhanna M, Piantanida P (2015) On secure distributed hypothesis testing. In: Proceeding of IEEE ISIT 2015, pp 1605–1609
6. Barni M, Tondi B (2016) Source distinguishability under distortion-limited attack: an optimal transport perspective. IEEE Trans Inf Forensics Secur 11:2145–2159
7. Sreekumar S, Gündüz D, Cohen A (2018) Distributed hypothesis testing under privacy constraints. Proc IEEE ITW 2018:470–474
8. Liao J, Sankar L, Tan VYF, Calmon FP (2018) Hypothesis testing under mutual information privacy constraints in the high privacy regime. IEEE Trans Inf Forensics Secur 13:1058–1071
9. Calmon FP, Fawaz N (2012) Privacy against statistical inference. Proc Allerton 2012:1401–1408
10. Li Z, Oechtering TJ (2015) Privacy-aware distributed Bayesian detection. IEEE J Sel Top Signal Process 9:1345–1357
11. Li Z, Oechtering TJ, Skoglund M (2016) Privacy-preserving energy flow control in smart grids. In: Proceedings of IEEE ICASSP 2016, pp 2194–2198
12. Nadendla VSS, Varshney PK (2016) Design of binary quantizers for distributed detection under secrecy constraints. IEEE Trans Signal Process 64:2636–2648
13. Li Z, Oechtering TJ (2017) Privacy-constrained parallel distributed Neyman–Pearson test. IEEE Trans Signal Inf Process Netw 3:77–90
14. You Y, Li Z, Oechtering TJ (2018) Optimal privacy-enhancing and cost-efficient energy management strategies for smart grid consumers. Proc IEEE SSP 2018:826–830
15. Li Z, Oechtering TJ (2018) Privacy-utility management of hypothesis tests. In: Proceedings of IEEE ITW 2018, pp 1–5
16. Li Z, Oechtering TJ, Gündüz D (2019) Privacy against a hypothesis testing adversary. IEEE Trans Inf Forensics Secur 14:1567–1581
17. Giaconi G, Gündüz D, Poor HV (2018) Privacy-aware smart metering: progress and challenges. IEEE Signal Process Mag 35:59–78
18. Rajagopalan SR, Sankar L, Mohajer S, Poor HV (2011) Smart meter privacy: a utility-privacy framework. In: Proceedings of IEEE SmartGridComm 2011, pp 190–195
19. Garcia FD, Jacobs B (2010) Privacy-friendly energy-metering via homomorphic encryption. In: Proceedings of STM 2010, pp 226–238
20. Efthymiou C, Kalogridis G (2010) Smart grid privacy via anonymization of smart metering data. In: Proceedings of IEEE SmartGridComm 2010, pp 238–243
21. Gündüz D, Gómez-Vilardebó J (2013) Smart meter privacy in the presence of an alternative energy source. In: Proceedings of IEEE ICC 2013, pp 2027–2031
22. Gómez-Vilardebó J, Gündüz D (2015) Smart meter privacy for multiple users in the presence of an alternative energy source. IEEE Trans Inf Forensics Secur 10:132–141
23. Giaconi G, Gündüz D, Poor HV (2018) Smart meter privacy with renewable energy and an energy storage device. IEEE Trans Inf Forensics Secur 13:129–142
24. Yao J, Venkitasubramaniam P (2013) On the privacy-cost tradeoff of an in-home power storage mechanism. In: Proceedings of Allerton 2013, pp 115–122
25. Tan O, Gómez-Vilardebó J, Gündüz D (2017) Privacy-cost trade-offs in demand-side management with storage. IEEE Trans Inf Forensics Secur 12:1458–1469
26. Li S, Khisti A, Mahajan A (2018) Information-theoretic privacy for smart metering systems with a rechargeable battery. IEEE Trans Inf Theory 64:3679–3695
27. Ács G, Castelluccia C (2011) I have a DREAM! (DiffeRentially privatE smArt Metering). In: Proceedings of IH 2011, pp 118–132
28. Zhang Z, Qin Z, Zhu L, Weng J, Ren K (2017) Cost-friendly differential privacy for smart meters: exploiting the dual roles of the noise. IEEE Trans Smart Grid 8:619–626
29. Eibl G, Engel D (2017) Differential privacy for real smart metering data. Comput Sci Res Dev 32:173–182
30. Csiszár I, Körner J (2011) Information theory: coding theorems for discrete memoryless systems. Cambridge University Press
31. Nikaidô H (1954) On von Neumann's minimax theorem. Pac J Math 4:65–72

32. Krishnamurthy V (2016) Partially observed markov decision processes: from filtering to controlled sensing. Cambridge University Press
33. Bellman R (1954) The theory of dynamic programming. Bull Am Math Soc 60:503–516
34. Van Erven T, Harremoës P (2014) Rényi divergence and Kullback–Leibler divergence. IEEE Trans Inf Theory 60:3797–3820
35. Kolter JZ, Johnson MJ (2011) REDD: a public data set for energy disaggregation research. In: Proceedings of the SustKDD workshop on data mining applications in sustainability, pp 1–6

Chapter 4
Statistical Parameter Privacy

Germán Bassi, Ehsan Nekouei, Mikael Skoglund and Karl H. Johansson

Abstract We investigate the problem of sharing the outcomes of a parametric source with an untrusted party while ensuring the privacy of the parameters. We propose privacy mechanisms which guarantee parameter privacy under both Bayesian statistical as well as information-theoretic privacy measures. The properties of the proposed mechanisms are investigated and the utility-privacy trade-off is analyzed.

4.1 Introduction

As the costs of digitizing, storing, and analyzing real-world data constantly decrease, more and more parts of our lives are increasingly being done through digital means. Once this information leaves our control, it can be duplicated and inspected at will. The data might be collected without our explicit knowledge and consent, e.g., our online behavior which is used to personalize the ads with see, or we might actively seek to share the information, e.g., by publishing our movie preferences on social media. In the latter case, there is usually a benefit or utility to be gained by sharing data. However, revealing sensitive information might be undesired for many people even if there is some gain involved. Moreover, if the benefit obtained is directly related to the fidelity of the shared information, a natural trade-off between utility and privacy arises.

Over the past two decades, there has been a surge of research on the problem of utility versus privacy and the design of privacy mechanisms to safely share data.

G. Bassi · M. Skoglund · K. H. Johansson
KTH Royal Institute of Technology, Stockholm, Sweden
e-mail: germanb@kth.se

M. Skoglund
e-mail: skoglund@kth.se

K. H. Johansson
e-mail: kallej@kth.se

E. Nekouei (✉)
City University of Hong Kong, Kowloon Tong, Hong Kong
e-mail: enekouei@cityu.edu.hk

© Springer Nature Singapore Pte Ltd. 2020
F. Farokhi (ed.), *Privacy in Dynamical Systems*,
https://doi.org/10.1007/978-981-15-0493-8_4

Two of the most well-known approaches for providing privacy in databases are k-anonymity [20] and differential privacy [11]. These strategies are normally implemented when data from different users are employed to estimate statistical population parameters; it is desired that the estimated values are close to the true parameters while the amount of information revealed about any particular user is low. This goal is achieved by suppressing values in the database, by clustering similar values, or by distorting the values with noise. As an example, a group of users participating in a medical survey might have their names removed from the database, their ages assigned to specific age-groups, and their weights modified by the addition of a zero-mean random variable. The effect in the inference performance of such privacy mechanisms has been addressed more recently; for instance, minimax risk bounds and minimax optimal estimation procedures for several canonical families of problems are studied in [3, 10] (see also the references therein).

A complementary problem to the one just described appears when the shared data should closely mirror the real one but must prevent an observer from learning some specific patterns or sensitive statistics of the raw data. This is commonly the case for users who share a stream of data with a third party in order to obtain a service; the more faithful the data is, the better the service provided but the easier the analysis of hidden information. As an example, a user may submit a scanned document to an online service with the goal of performing optical character recognition, however, the user might not want the online service to infer any personality traits in the handwriting or the author's identity with respect to previously submitted documents. In this work, we focus on this second type of problems.

To the best of our knowledge, this is a less explored direction of research in privacy. The reason might be strictly technological since it was not until recently that large amounts of data from a single user could be collected; on the other hand, tiny bits of information from vast numbers of users have been compiled in databases for many years now. The design of this type of privacy filter is also inherently more complex since each entry in the sequence cannot be processed independently and the statistics of the whole sequence must be taken into account. Recently, the authors in [12] define a privacy-preserving strategy that minimizes the Fisher information about the private parameters in the released data. The Cramér–Rao lower bound [9, Theorem 11.10.1] establishes that this strategy maximizes a lower bound on the mean square error of any unbiased estimator of the parameters. The interested reader is referred to [5] for a broader study on the performance degradation of *any* parameter estimator due to a privacy filter. In particular, the authors investigate the relationship between the length n of the data sequence and the inference performance of the parameters. The main result from this work, a privacy filter that hinders the estimation of sensitive parameters, is reproduced in the present manuscript.

Several different works have studied the general relationship between utility and privacy. The authors of [6] introduce a general framework of utility versus privacy where the former is defined as a bounded distortion and the latter as a log-loss cost, which yields a trade-off similar to the rate-distortion function. If both utility and privacy are measured using entropy or mutual information, a different utility-privacy region is provided in [7]; however, only the two extreme cases of perfect privacy and perfect utility are properly characterized therein. The concept of maximal correlation as a measure of privacy is introduced in [1] and it is shown in [2] that this measure is equivalent to maximizing the MMSE of the private data given the shared data. The authors of [13] also define a problem similar to rate-distortion, where privacy is determined by mutual information, and they characterize optimal asymptotic leakages for i.i.d. and general privacy mechanisms. Finally, the use of a secret key to hinder the success of an eavesdropper is addressed in [23]; the authors argue for the use of distortion-based privacy metrics instead of stronger information-theoretic ones to reduce the size of the secret key.

Information privacy and the design of privacy filters have also been studied in dynamic settings. The authors of [22] study the design of privacy-preserving filters in a cloud-based control problem using the notion of directed information as the privacy metric. Le Ny and Pappas in [15] propose privacy-preserving filtering algorithms for ensuring the privacy of states or measurements of dynamical systems, based on differential privacy. The authors of [21] study the state estimation problem in a distribution power network subject to differential privacy constraints for the consumers. Wang et al. in [24] propose privacy-preserving mechanisms for ensuring the privacy of the initial states and the preferred target way-points in a distributed multi-agent control system. Privacy-preserving average consensus algorithms, for preserving the privacy of initial states, are addressed in [17, 18].

4.1.1 Organization

After the comprehensive introduction into the problem of parameter privacy, the rest of this chapter is devoted to the analysis of two different privacy-preserving filters. To facilitate the reading, we have organized the work as follows.

In Sect. 4.2, we present the system model for the problem of parameter privacy and some important definitions. In particular, we link the performance of the parameter estimation to the mutual information between the parameter and the released data. The section ends with the overview of the two different privacy filters described in this work. The first of these filters is introduced in Sect. 4.3. An achievable scheme that distorts the shared data is outlined; the proposed privacy mechanism seeks to confuse the adversary by introducing an auxiliary parameter that behaves as the true one. It is shown that the filter limits the amount of information released to the eavesdropper. A Gaussian example is used to illustrate the trade-off between distortion and privacy. In Sect. 4.4, the structure of the second privacy filter is studied. In this scheme, the privacy filter design problem is posed as a convex optimization problem which achieves the Pareto boundary of the distortion-privacy region. An upper bound on the leakage of private information under this scheme is obtained. The distortion-privacy trade-off for this filter is studied with a numerical example.

4.1.2 Notation

In the rest of this chapter, lowercase letters such as x and y are mainly used to represent constants or realizations of random variables, capital letters such as X and Y stand for the random variables in itself, and calligraphic letters such as $\mathcal{X}$ and $\mathcal{Y}$ are reserved for sets. In the case of Greek letters, we use Θ, θ, and $\boldsymbol{\Theta}$ to denote a random variable, its realization, and its support set, respectively.

We use X^n to denote the sequence of independent and identically distributed (i.i.d.) random variables $\{X_k\}_{k=1}^n$. Given three random variables X, Y, and Z, if its joint probability distribution can be decomposed as $p(xyz) = p(x)p(y|x)p(z|y)$, then they form a Markov chain, denoted by $X \rightarrow Y \rightarrow Z$.

Entropy is denoted by $H(\cdot)$ and mutual information, $I(\cdot;\cdot)$. Throughout the work and unless stated otherwise, log refers to logarithm in base 2.

4.2 System Model and Overview of Results

In this section, we first introduce the general model studied in this work and some useful definitions. We then present an overview of results for two particular privacy-preserving filters.

4.2.1 System Model

Consider the three-user problem depicted in Fig. 4.1, where Alice wants to share with Bob the outcomes of a random parametric source she observes. The value of the parameter, which in this work constitutes private information that Alice does not want to disclose, might even be unknown to her. For example, the observation might be a handwritten note (a sequence of characters) by Alice while the parameter represents her personality traits.

In the absence of any constraint on the rate of information between the users, Alice may choose to directly send the observed sequence of values. However, the communication is overheard by Eve, who is interested in characterizing the statistical properties of the random parametric source, i.e., estimate the parameter. In order to protect her privacy, Alice needs to share a distorted version of the source, but one

Fig. 4.1 General system model

$$X^n \sim p_{\theta_0}(x^n) \rightarrow \boxed{\text{Alice}} \xrightarrow{Y^n} \boxed{\text{Bob}} \xrightarrow{Y^n} \mathbb{E}[d(X^n, Y^n)] \leq D$$

$$\xrightarrow{} \boxed{\text{Eve}} \xrightarrow{\hat{\theta}_n} \mathbb{E}[\ell(\Theta, \hat{\theta}_n(Y^n))] \geq \varepsilon$$

that it is still useful for Bob. In our previous example with the handwritten note, the font style may change as long as Bob is able to correctly read the text.

More precisely, we assume that Alice observes n samples of the random variable $X \in \mathcal{X}$ where the samples are i.i.d. with respect to the distribution P_{θ_0}. The probability measure P_{θ_0} is a member of a parameterized family of distributions $\mathcal{P}_\Theta = \{P_\theta : \theta \in \boldsymbol{\Theta}\}$ on a measurable space, where θ_0 is a point in the interior of $\boldsymbol{\Theta}$. Moreover, $p_{\theta_0}(x)$ is the probability density function (PDF) of P_{θ_0} with respect to a fixed σ-finite measure $\mu(dx)$; it is assumed that $p_{\theta_0}(x)$ is non-zero almost everywhere on $\mathcal{X}$ and the corresponding probability measure, P_{θ_0}, is absolutely continuous with respect to Lebesgue measure.

The value of the parameter is chosen randomly by nature according to the known prior distribution $p(\theta)$ with respect to Lebesgue measure; thus, the parameter is regarded as a random variable, which we denote Θ.[1] As previously mentioned, Alice produces a distorted sequence Y^n which is based on the observed sequence X^n and shares it with Bob. The channel between Alice and Bob has no rate limitation, and the purpose of distorting the sequence is to prevent Eve from increasing her knowledge about the unknown parameter Θ beyond what is specified by the prior distribution.

We further assume that Bob has no advantage over Eve. The communication is received by both users with the same level of quality and they are all aware of the strategy employed by Alice to distort the observed sequence. If a certain stochastic transformation is used to increase the privacy, the particular realization of the mapping is unknown to both Bob and Eve.

4.2.2   Useful Definitions and Preliminary Result

We present here some definitions needed to characterize the loss in the fidelity of the sequence Y^n with respect to X^n and the increase in privacy. We start with some general notions.

Definition 4.1 Let X denote an absolutely continuous random variable with probability density function $p_X(x)$. Then, the *differential entropy* of X is defined as

$$h(X) = - \int_{\mathcal{X}} p_X(x) \log p_X(x) \, dx. \tag{4.1}$$

Definition 4.2 Consider the random variable X distributed according to the probability density function $p_\theta(x)$, where θ is a parameter taking values in $\mathbb{R}^d$. Then, the *Fisher information matrix* about θ contained in $X|_{\Theta=\theta}$ is defined as a $d \times d$ matrix with the (i, j)th entry given by

---

[1]Note that Θ stands for the parameter taken as a random variable, whereas $\boldsymbol{\Theta}$ corresponds to the parameter space.

$$\left[\mathbf{I}_X(\theta)\right]_{i,j} = \mathbb{E}\left[\left(\frac{\partial}{\partial \theta_i} \log p_\theta(x)\right)\left(\frac{\partial}{\partial \theta_j} \log p_\theta(x)\right) \,\Big|\, \Theta = \theta\right] \quad 1 \le i, j \le d. \quad (4.2)$$

Definition 4.3 The *distortion* between the sequences x^n and y^n is defined as

$$d(x^n, y^n) \triangleq \frac{1}{n}\sum_{i=1}^{n} d(x_i, y_i), \quad (4.3)$$

where the distortion function d is a mapping $\mathcal{X} \times \mathcal{Y} \to \mathbb{R}^+$.

Definition 4.4 The *privacy* (distortion) between the parameter θ and an estimate $\tilde{\theta}$, i.e., $\ell(\theta, \tilde{\theta})$, is given by the mapping $\ell : \boldsymbol{\Theta} \times \boldsymbol{\Theta} \to \mathbb{R}^+$, where we assume that $\inf_{\tilde{\theta}} \ell(\theta, \tilde{\theta}) = 0$ for all θ.

Definition 4.5 A distortion-privacy pair (D, ϵ) is *achievable* in this problem if there exists $N > 0$ and a privacy-preserving mapping (stochastic kernel) $f_n : \mathcal{X}^n \to \mathcal{Y}^n$ such that

$$\mathbb{E}[d(X^n, Y^n)] \le D, \quad (4.4)$$

$$\inf_{\hat{\theta}_n} \mathbb{E}[\ell(\Theta, \hat{\theta}_n(Y^n))] \ge \epsilon, \quad (4.5)$$

for $n > N$, where $Y^n = f_n(X^n)$ and the infimum is taken over all measurable functions $\hat{\theta}_n : \mathcal{Y}^n \to \boldsymbol{\Theta}$ that are possible estimators of the parameter Θ.

An important information-theoretic function is the rate-distortion (RD) function. We introduce it here for completeness.

Definition 4.6 ([9, Sect. 10.2]) The *(information) rate-distortion function* for a random variable Θ with distortion measure $\ell(\cdot, \cdot)$ is defined as

$$R_{\Theta,\ell}(D) \triangleq \min_{p(\tilde{\theta}|\theta): \mathbb{E}[\ell(\Theta, \tilde{\Theta})] \le D} I(\Theta; \tilde{\Theta}). \quad (4.6)$$

We assume that there exists $D \ge 0$ such that $R_{\Theta,\ell}(D)$ is finite.

We note that the RD function has the following properties:

- The infimum over $D \ge 0$ such that $R_{\Theta,\ell}(D)$ is finite is denoted $D_{\min}$; the corresponding rate is $R_{\max} \triangleq \lim_{D \to D_{\min}^+} R_{\Theta,\ell}(D)$.
- The RD function $R_{\Theta,\ell}(D)$ is a non-increasing convex function of D on the interval $(D_{\min}, \infty)$. It is monotonically decreasing on the interval $(D_{\min}, D_{\max})$ and constant with $R_{\Theta,\ell}(D) = R_{\min}$ on $[D_{\max}, \infty)$.
- The inverse function $R_{\Theta,\ell}^{-1}(r)$ is well defined on $(R_{\min}, R_{\max})$ and monotonically decreasing. This function is known as the *distortion-rate* (DR) *function*.

Before proceeding with the overview of results in the next subsection, we present an important lemma that relates Eve's performance in estimating the random parameter Θ with the sequence transmitted by Alice. From Eve's point of view, the setting of Fig. 4.1 is a statistical inference problem of a random quantity Θ that cannot be directly observed; only an indirect measurement Y^n is obtained. However, the pair (Θ, Y^n) has a given joint probability distribution, and thus Eve may calculate an estimate $\hat{\theta}_n(Y^n)$ of the parameter Θ [14].

Lemma 4.1 ([5, Lemma 2]) *For any estimator $\hat{\theta}_n$ and any distortion function $\ell(\cdot, \cdot)$, the expected privacy (the Bayes risk of the estimator) is bounded from below by:*

$$\mathbb{E}[\ell(\Theta, \hat{\theta}_n)] \geq R_{\Theta,\ell}^{-1}(I(\Theta; Y^n)), \tag{4.7}$$

where $R_{\Theta,\ell}^{-1}(\cdot)$ is the DR function of the random variable Θ.

Proof The proof follows from the data-processing inequality [9, Sect. 2.8], the Markov chain

$$\Theta \to Y^n \to \hat{\theta}_n(Y^n), \tag{4.8}$$

and the definition of the RD function in (4.6). Please refer to [5] for more details. $\square$

The preceding lemma is quite powerful in that it allows us to bound the performance of *any* estimator $\hat{\theta}_n$ of Θ based on Y^n without knowing how that estimator is calculated. Analytical solutions for the DR function in (4.7), on the other hand, are only known for a handful of random variables and distortion measures. In many situations, we may need to employ the looser Shannon lower bound or compute the RD function numerically using the Blahut–Arimoto algorithm [9, Chap. 10].

Gaussian Example

Assume that Alice observes n i.i.d. samples of the process $X|_{\Theta=\theta} \sim \mathcal{N}(\theta, \sigma_X^2)$, where the mean is fixed throughout the process but it has an unknown value. Additionally, assume that $\Theta \sim \mathcal{N}(0, \sigma_\Theta^2)$, and consider the square error distortion function for the estimation at Eve, i.e., $\ell(\theta, \hat{\theta}) \triangleq (\theta - \hat{\theta})^2$.

In this setting, the expected distortion is the mean square error, and the DR function is [9, Theorem 10.3.2]

$$R_{\Theta,\ell}^{-1}(r) = \sigma_\Theta^2 2^{-2r}. \tag{4.9}$$

If Alice sends the sequence X^n without any distortion, i.e., $Y = X$, then

$$I(\Theta; X^n) = \frac{1}{2} \log\left(\frac{\sigma_X^2 + n\sigma_\Theta^2}{\sigma_X^2}\right) = \frac{1}{2} \log \frac{n\sigma_\Theta^2}{\sigma_X^2} + o(1), \tag{4.10}$$

which according to Lemma 4.1 yields

$$\text{MSE}_{\hat{\theta}_n} \geq \frac{\sigma_X^2 \sigma_\Theta^2}{\sigma_X^2 + n\sigma_\Theta^2} \simeq \frac{\sigma_X^2}{n}, \tag{4.11}$$

where the approximation is valid for large n such that $\sigma_X^2 \ll n\sigma_\Theta^2$. We see that the lower bound approaches 0 as $n \to \infty$; hence, there might exist an estimator that attains a vanishing mean square error.

We note that in this particular example the general lower bound in Lemma 4.1 is tight. The right-hand side of (4.11) is known to be the minimum mean square error for the considered estimation problem; thus, it can be attained by a specific estimator: the MMSE estimator [14, Sect. 11.4].

We conclude this part with an important remark. Lemma 4.1 states that the expected distortion of the unknown estimator $\hat{\theta}_n$ is bounded from below by a monotonically decreasing function (the DR function) of the mutual information between Θ and Y^n. Consequently, the mutual information $I(\Theta; Y^n)$ should be *minimized* in order to hinder Eve's estimation performance.

4.2.3 Overview of Results

Two different privacy-preserving filters are presented in this work. The following is an overview of their characteristics and performance; both filters are analyzed in more detail in Sects. 4.3 and 4.4.

4.2.3.1 First Scheme

For smooth parametric families $\mathcal{P}_\Theta$ with a *continuous* parameter $\Theta \in \boldsymbol{\Theta} \subset \mathbb{R}^d$, it is a well-known fact that $I(\Theta; X^n) \propto \frac{d}{2} \log n$ (see e.g., [8, 19]). Therefore, without properly distorting X^n, the amount of information about Θ gathered by the eavesdropper increases with n. According to Lemma 4.1, Eve might thus be able to estimate the parameter with arbitrarily high precision.

In Sect. 4.3, we analyze a privacy-preserving filter like the one depicted in Fig. 4.2, where the auxiliary random parameter Φ is added to prevent an accurate estimation of Θ. Eve is only able to estimate a function of the true and auxiliary parameters, i.e., only $\Psi = \psi(\Theta, \Phi)$ may be estimated, where the function $\psi(\cdot)$ depends on the statistics of the source and the filter. Namely, the randomness in the auxiliary parameter acts as noise of a virtual noisy channel for the inference problem: $I(\Theta; Y^n) = I(\Theta; \Psi) + O(1)$. Consequently, we prevent Eve from collecting unlimited amount of information about Θ as n increases.

Fig. 4.2 Stochastic filter.
Both the mapping $\mathcal{X} \to \mathcal{Y}$
and the choice of the
mapping are random

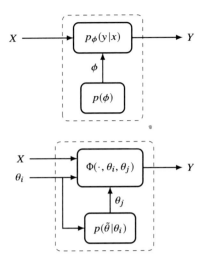

Fig. 4.3 Deterministic filter
for $\Theta = \theta_i$ and $\tilde{\Theta} = \theta_j$. The
mapping $\Phi\left(\cdot, \theta_i, \theta_j\right)$ is
deterministic but the choice
of $\tilde{\Theta}$ is stochastic

4.2.3.2 Second Scheme

Under the second scheme, we assume that $\mathcal{X} = \mathbb{R}^d$ and that the parameter set $\boldsymbol{\Theta}$ has a finite cardinality. In the proposed filter and at each time k, Alice generates the random variable Y_k according to the mapping

$$Y_k = \Phi\left(X_k, \theta, \tilde{\theta}\right), \tag{4.12}$$

where $\tilde{\theta}$ belongs to $\boldsymbol{\Theta}$ and $\Phi\left(\cdot, \theta, \tilde{\theta}\right)$ is a map from $\mathbb{R}^d$ to $\mathbb{R}^d$. The map $\Phi\left(\cdot, \theta, \tilde{\theta}\right)$ is designed such that the common PDF of $\{Y_1, \ldots, Y_n\}$ is equal to $p_{\tilde{\theta}}(x)$. We assume that θ is known by Alice and $\tilde{\theta}$ is selected, by the privacy filter, to simultaneously ensure the privacy of θ and accuracy of the revealed information to Bob. The construction of the map $\Phi\left(\cdot, \theta, \tilde{\theta}\right)$ and the generation of $\tilde{\theta}$ are discussed in Sect. 4.4. Figure 4.3 shows a pictorial representation of the privacy filter under the second scheme.

4.3 First Scheme

In this section, we focus on parameter sets $\boldsymbol{\Theta} \subset \mathbb{R}^d$. The unfavorable result obtained in the Gaussian example in the preceding section, in particular (4.10) and (4.11), is not an isolated case but rather the norm for most (well-behaved) parametric sources. Given that the parameter $\Theta \in \mathbb{R}^d$, i.e., it has an infinite precision, it is expected that the mutual information (4.10) grows unboundedly; hence, an observer is able to estimate the parameter Θ with arbitrarily low error as n increases.

The asymptotic behavior in (4.10) is a special case of a much larger set of results. Specifically, if the density $p_\theta(x)$ satisfies suitable smoothness conditions, it is shown by Clarke and Barron [8] that

$$I(\Theta; X^n) = \frac{d}{2} \log \frac{n}{2\pi e} + h(\Theta) + \frac{1}{2} \mathbb{E}\big[\log |\mathbf{I}_X(\Theta)|\big] + o(1), \qquad (4.13)$$

where d is the dimension of the parameter space, $h(\Theta)$ is the differential entropy of the parameter, and $\mathbf{I}_X(\theta)$ is the Fisher information matrix about θ contained in $X|_{\Theta=\theta}$. In the aforementioned Gaussian example, we obtain (4.10) from (4.13) by noting that $d = 1$, $h(\Theta) = \frac{1}{2}\log 2\pi e\sigma_\Theta^2$, and $\mathbf{I}_X(\theta) = 1/\sigma_X^2 \ \forall \theta \in \mathbb{R}$. For all the parametric sources where (4.13) holds, e.g., all the exponential families, Eve may attain good inference performance if Alice reveals her observation X^n without distortion.

4.3.1 A Simple Privacy-Preserving Strategy

Let us define the privacy filter as a conditional distribution of Y given X belonging to a parametric family of distributions $\mathcal{P}_\Phi = \{P_\phi : \phi \in \boldsymbol{\Phi}\}$, where $\boldsymbol{\Phi} \subset \mathbb{R}^{d'}$, and where the auxiliary random parameter Φ is distributed according to the prior distribution $p(\phi)$. Therefore, $\{\mathcal{P}_\Phi, p(\phi)\}$ determines the privacy filter.

In a well-designed privacy filter, the auxiliary parameter Φ combines with Θ in a way that the sequence Y^n is consistent with the observation of a parametric family of distributions $\mathcal{P}_\Psi = \{P_\psi : \psi \in \boldsymbol{\Psi}\}$, where $\psi = \psi(\theta, \phi)$ and $\boldsymbol{\Psi} \subset \mathbb{R}^{d'}$. In other words, Ψ is a *sufficient statistic* for Y [9, 16]. Eve may thus be able to estimate ψ with arbitrarily low error as n increases but she has a non-vanishing uncertainty about θ given by the randomness in ϕ.

Theorem 4.1 *If the privacy filter $\{\mathcal{P}_\Phi, p(\phi)\}$ satisfies some suitable smoothness conditions (defined in the proof), it achieves according to Def. 4.5 all distortion-privacy pairs (D, ϵ) such that $\mathbb{E}[d(X^n, Y^n)] \leq D$ and $\epsilon \leq R_{\Theta,\ell}^{-1}(I(\Theta; Y^n))$, where*

$$I(\Theta; Y^n) = h(\Psi) - h(\Phi) + \frac{1}{2} \mathbb{E}\left[\log \frac{|\mathbf{I}_Y(\Psi)|}{|\mathbf{I}_Y(\Phi)|}\right] + o(1). \qquad (4.14)$$

Therefore, the privacy level ϵ remains asymptotically bounded away from zero.

Remark 4.1 Since Φ is independent of Θ, we have that $h(\Phi) = h(\Psi|\Theta)$, and thus (4.14) is equivalently

$$I(\Theta; Y^n) = I(\Theta; \Psi) + O(1). \qquad (4.15)$$

In other words, the privacy filter creates a noisy channel for the parameter.

Before presenting the proof of the theorem, we revisit the Gaussian example.

Fig. 4.4 Trade-off between the distortion level D and the corresponding maximum privacy level $\epsilon = \inf_{\hat{\theta}_n} \mathrm{MSE}_{\hat{\theta}_n}$. The relation between the curves for $I(\Theta; Y^n)$ and $\mathrm{MSE}_{\hat{\theta}_n}$ is given by the DR function (4.9). These curves are calculated assuming $\sigma_{\Theta}^2 = 1$ and $\sigma_Z^2 = 10^{-3}$

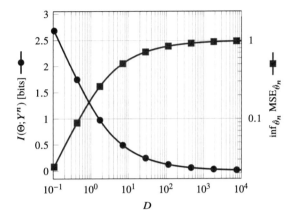

Gaussian Example (cont.)

Let us continue with the Gaussian example where we now consider the square error distortion function for the reconstruction at Bob, i.e., $d(x, y) \triangleq (x - y)^2$.

The privacy filter $\{\mathcal{P}_{\Phi}, p(\phi)\}$ is chosen to mimic the parametric source that Alice tries to protect. In particular, for a fixed $\sigma_Z^2 < D$, the auxiliary parameter Φ is drawn uniformly at random from the interval $[D - \sigma_Z^2, D - \sigma_Z^2]$ and the filter's output is constructed in an i.i.d. manner: for each time $i \in [1 : n]$, $Y_i = X_i + Z_i$, where $Z_i|_{\Phi = \phi} \sim \mathcal{N}(\phi, \sigma_Z^2)$ and is independent of X_i. This choice of filter satisfies $\mathbb{E}[d(X^n, Y^n)] \leq D$ and the smoothness conditions needed for Theorem 4.1.

With this privacy-preserving strategy, the eavesdropper observes n i.i.d. samples of the process $Y|_{\Psi = \psi} \sim \mathcal{N}(\psi, \sigma_X^2 + \sigma_Z^2)$, where the mean $\Psi = \Phi + \Theta$ is distributed according to

$$p(\psi) = \frac{1}{2\sqrt{D}} \left[F\left(\frac{\psi + \sqrt{D}}{\sigma_{\Theta}} \right) - F\left(\frac{\psi - \sqrt{D}}{\sigma_{\Theta}} \right) \right] \tag{4.16}$$

and $F(\cdot)$ is the cumulative distribution function of the standard normal distribution. We may easily obtain the Fisher information

$$\mathbf{I}_Y(\psi) = \mathbf{I}_Y(\phi) = (\sigma_X^2 + \sigma_Z^2)^{-1}, \tag{4.17}$$

which holds for all values of ϕ and ψ. Therefore, using (4.14) we have that

$$I(\Theta; Y^n) = h(\Psi) - \log 2\sqrt{D} + o(1), \tag{4.18}$$

where $h(\Psi)$ does not have a closed-form expression and has to be calculated numerically using (4.16).

The mutual information (4.18) and the corresponding lower bound from Lemma 4.1 are plotted in Fig. 4.4 for different values of D; we note that the DR function is found in (4.9). The curves show the effect of the privacy-preserving strategy and the trade-off between the distortion level D and the loss in Eve's inference performance.

4.3.2 Proof of Theorem 4.1

In order to protect her privacy, Alice needs to distort the observed sequence x^n with some randomness that behaves like the parameter θ from the point of view of Eve.

Given the privacy filter $\{\mathcal{P}_\Phi, p(\phi)\}$, Alice selects a distribution P_ϕ, whose probability density function with respect to a fixed σ-finite measure $\mu(dy)$ is $p_\phi(y|x)$, according to the prior $p(\phi)$. Then, given the original sequence x^n and the specific distribution P_ϕ, she transmits the distorted symbols $Y_i \sim p_\phi(y|x_i)$ for $i \in [1:n]$. Therefore, the joint density of the sequences is:

$$p_{\theta,\phi}(x^n, y^n) = \prod_{i=1}^{n} p_\phi(y_i|x_i) p_\theta(x_i) . \tag{4.19}$$

Due to the i.i.d. nature of the source and the privacy filter, and conditioned on the true value of the parameters, the sequence observed by Eve is distributed according to

$$p_{\theta,\phi}(y) = \int_{\mathcal{X}} p_\phi(y|x) p_\theta(x) \mu(dx) , \tag{4.20}$$

where both θ and ϕ are unknown to her. As previously mentioned, privacy is possible if, in the marginal density (4.20), the auxiliary parameter ϕ combines with θ such that $\psi = \psi(\theta, \phi)$ is a sufficient statistic for the parametric family of distributions $\mathcal{P}_\Psi$.

In this case, we may expand the quantity of interest as follows

$$\begin{aligned} I(\Theta; Y^n) &= I(\Theta, \Phi; Y^n) - I(\Phi; Y^n|\Theta) \\ &= I(\Psi; Y^n) - I(\Phi; Y^n|\Theta) , \end{aligned} \tag{4.21}$$

where the last equality is due to Ψ being a sufficient statistic for (θ, ϕ), i.e., the Markov chain $(\Theta, \Phi) \to \Psi \to Y^n$ holds. Assuming all the probability distributions involved in (4.21) satisfy suitable smoothness conditions, we may characterize the asymptotic behavior of both terms similarly to (4.13). In broad terms, these conditions are:

- the densities $p_\theta(x)$ and $p_\phi(y|x)$ are twice continuously differentiable almost everywhere, and the first and second derivatives of $\log p_\theta(x)$ and $\log p_\phi(y|x)$ are square-integrable;
- the priors are continuous and positive almost everywhere;
- the appropriate Fisher information matrices are positive definite; and,
- the posterior distribution of the parameters concentrates around the true value as n increases.

We refer the reader to [5] for more details on these conditions. If these conditions are satisfied, the first term on the right-hand side of (4.21) may be written as

$$I(\Psi; Y^n) = \frac{d'}{2} \log \frac{n}{2\pi e} + h(\Psi) + \frac{1}{2} \mathbb{E}\big[\log |\mathbf{I}_Y(\Psi)|\big] + o(1). \qquad (4.22)$$

On the other hand, the second term is an expectation on Θ:

$$I(\Phi; Y^n|\Theta) = \int_{\Theta} I(\Phi; Y^n|\theta) p(\theta) d\theta, \qquad (4.23)$$

where $I(\Phi; Y^n|\theta)$ implies that the parameter is now fixed and known. Then, $I(\Phi; Y^n|\theta)$ is equal to

$$I(\Phi; Y^n|\theta) = \frac{d'}{2} \log \frac{n}{2\pi e} + h(\Phi) + \frac{1}{2} \mathbb{E}\big[\log |\mathbf{I}_Y(\Phi)|\big] + o(1). \qquad (4.24)$$

Joining these results, we obtain the expression (4.14), which concludes the proof of Theorem 4.1. $\qquad\square$

4.4   Second Scheme

For the rest of the chapter, we assume that each random variable $X_k = [X_k^1, \ldots, X_k^d]^\top$ takes values in $\mathbb{R}^d$. Furthermore, X_k^l denotes the lth entry of the random variable X_k while $X_k^{1:l-1}$ denotes the collection of the first $l-1$ entries of X_k. In this section, we also assume that the parameter set Θ consists of m elements, i.e., $\Theta = \{\theta_1, \ldots, \theta_m\}$.

The conditional cumulative distribution function (CDF) of X^l given $X^{1:l-1} = x^{1:l-1}$ and $\Theta = \theta_i$ is defined as

$$F_{l,\theta_i}\left(z \,|x^{1:l-1}\right) = \int_{-\infty}^{z} p_{\theta_i}\left(x \,|x^{1:l-1}\right) dx, \qquad (4.25)$$

where $p_{\theta_i}\left(x \,|x^{1:l-1}\right)$ is the conditional PDF of X^l given $(X^{1:l-1} = x^{1:l-1}, \Theta = \theta_i)$ which is computed by Bayes' rule and the marginalization of $p_\theta(x)$. We use the convention that $p_{\theta_i}\left(x \,|x^{1:0}\right) = p_{\theta_i}(x)$. Note that $F_{l,\theta_i}(\cdot|\cdot)$ is a map from $\mathbb{R}^l$ to $[0, 1]$ and is non-decreasing in the first argument when the second argument is fixed. We use $F_{l,\theta_i}^{-1}\left(\cdot \,|x^{1:l-1}\right)$ to denote the inverse of the function $F_{l,\theta_i}\left(\cdot \,|x^{1:l-1}\right)$ for $1 \le l \le d$.

The second privacy filter comprises the map $\Phi\left(\cdot, \theta, \tilde{\theta}\right)$ and a stochastic kernel for generating $\tilde{\theta}$. We first describe the structure of $\Phi\left(\cdot, \theta, \tilde{\theta}\right)$. Given $\Theta = \theta_i$, $\tilde{\Theta} = \theta_j$ and the observation at time k, i.e., $X_k = [X_k^1, \ldots, X_k^d]^\top$, Alice sequentially generates the entries of $Y_k = [Y_k^1, \ldots, Y_k^d]^\top$ as follows. For $1 \le l \le d$, the lth entry of Y_k, i.e., Y_k^l, is generated according to

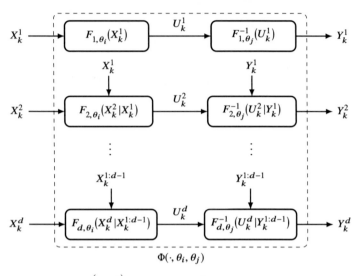

Fig. 4.5 The structure of $\Phi\left(\cdot, \theta, \tilde{\theta}\right)$ for $\Theta = \theta_i$ and $\tilde{\Theta} = \theta_j$

$$Y_k^l = \phi^l\left(X_k^{1:l}, \theta_i, \theta_j\right),\tag{4.26}$$

where $\phi^l\left(X_k^{1:l}, \theta_i, \theta_j\right) = F_{l,\theta_j}^{-1}\left(U_k^l \left| Y_k^{1:l-1}\right.\right)$ and $U_k^l = F_{l,\theta_i}\left(X_k^l \left| X_k^{1:l-1}\right.\right)$. The non-linear map $\Phi\left(\cdot, \theta_i, \theta_j\right)$ can be written as $\Phi\left(\cdot, \theta_i, \theta_j\right) = [\phi^1\left(\cdot, \theta_i, \theta_j\right), \ldots, \phi^d\left(\cdot, \theta_i, \theta_j\right)]^\top$. Figure 4.5 shows the structure of $\Phi\left(\cdot, \theta_i, \theta_j\right)$ for $\Theta = \theta_i$ and $\tilde{\Theta} = \theta_j$.

The following lemma studies the statistical properties of the output of this privacy-preserving filter.

Lemma 4.2 *Consider the construction above and assume that $\Theta = \theta_i$ and $\tilde{\Theta} = \theta_j$ for $1 \leq i, j \leq m$. Then, the sequence of random variables $\{Y_k\}_k$ are jointly independent and distributed according to $p_{\theta_j}(x)$.*

Proof See [4]. □

We next discuss the optimal generation of $\tilde{\Theta}$. The parameter $\tilde{\Theta}$ is selected from the set $\boldsymbol{\Theta}$ using a stochastic kernel. More precisely, given $\Theta = \theta_i$, the value of $\tilde{\Theta}$ is generated according to the following stochastic kernel

$$\tilde{\Theta} = \theta_j \ \text{w.p.} \ P_{ji} = \Pr\left(\tilde{\Theta} = \theta_j \middle| \Theta = \theta_i\right),\tag{4.27}$$

where $\sum_j P_{ji} = 1$ for all i and w.p. stands for *with probability*. The randomization probabilities are designed such that the accuracy of the output of the privacy filter is maximized while a certain privacy level for the parameter Θ is achieved.

To discuss the design of the randomization probabilities, we first define the privacy metric as follows. The privacy level of the parameter Θ is captured by the mutual information between the Θ and $\tilde{\Theta}$ which is defined as

$$I(\Theta; \tilde{\Theta}) = \sum_{i,j} \Pr\left(\Theta = \theta_i, \tilde{\Theta} = \theta_j\right) \log \frac{\Pr\left(\Theta = \theta_i, \tilde{\Theta} = \theta_j\right)}{\Pr(\Theta = \theta_i) \Pr\left(\tilde{\Theta} = \theta_j\right)} \qquad (4.28)$$

Note that when the privacy metric equal to zero, $\tilde{\Theta}$ contains no information about Θ and the maximum privacy level is achieved.

The optimal randomization probabilities are obtained, by minimizing the average distortion between X^n and Y^n subject to a privacy level of Θ, using the following optimization problem

$$\begin{aligned}
\underset{\{P_{ji}\}_{j,i}}{\text{minimize}} \quad & \mathbb{E}[d(X^n, Y^n)] \\
& P_{ji} \geq 0, \forall i, j \\
& \sum_j P_{ji} = 1, \quad \forall i \\
& I(\Theta; \tilde{\Theta}) \leq I_0
\end{aligned} \qquad (4.29)$$

The next theorem states that the optimization problem above is a convex optimization problem. Hence, the optimal randomization probabilities can be computed efficiently.

Theorem 4.2 *The optimal privacy filter design problem in* (4.29) *is a convex optimization problem.*

Proof See [4]. □

To study the privacy level of Θ under the proposed scheme, consider an estimator of Θ based on Y^n, denoted by $\hat{\Theta}(Y^n)$. Using Fano's inequality [9], the error probability of any estimator of Θ based on Y^n, can be bounded from below as

$$\Pr\left(\Theta \neq \hat{\Theta}(Y^n)\right) \geq \frac{H(\Theta \mid Y^n) - 1}{\log |\Theta|}, \qquad (4.30)$$

where $H(\Theta \mid Y^n)$ denotes the conditional entropy of Θ given Y^n. Using the definition of mutual information, we have that

$$H(\Theta \mid Y^n) = H(\Theta) - I(\Theta; Y^n). \qquad (4.31)$$

Notice that the following Markov chain holds

$$\Theta \rightarrow \tilde{\Theta} \rightarrow Y^n \rightarrow \hat{\Theta}(Y^n). \qquad (4.32)$$

Thus, according to the data-processing inequality, we have that

$$I(\Theta; \hat{\Theta}(Y^n)) \leq I(\Theta; Y^n) \leq I(\Theta; \tilde{\Theta}). \qquad (4.33)$$

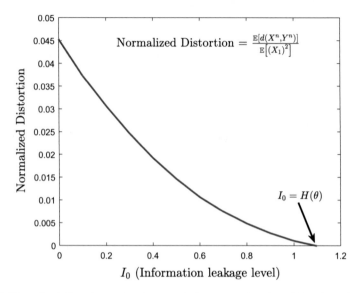

Fig. 4.6 Normalized distortion, under the second scheme, as a function of the privacy level

Hence, the upper bound on the mutual information between Θ and $\tilde{\Theta}$ in (4.29) ensures the privacy of Θ. That is, according to Fano's inequality, the privacy constraint imposes a lower bound on the performance of Bob in recovering Θ using the output of the privacy filter.

Gaussian Example

In this section, the distortion-privacy trade-off is numerically studied for a Gaussian information source under the second scheme. In our numerical result, X^n is modeled as a sequence of i.i.d. Gaussian random variables with zero mean and variance $\Theta \in \boldsymbol{\Theta} = \{1, 2, 3\}$. It is assumed that Θ is uniformly distributed over $\boldsymbol{\Theta}$.

Figure 4.6 shows the optimal level of the normalized distortion between the input and the output of the privacy filter as a function of the privacy level I_0. According to this figure, the minimum distortion level is achieved when I_0 is equal to $H(\Theta)$. In this example, the optimal randomization probabilities are computed using the fmincon solver in MATLAB®. Note that, when $\Theta = \tilde{\Theta}$, the input and output of the privacy filter are the same. Thus, the distortion is zero in this case and the leakage of the private information is at its maximum level.

Moreover, the distortion level increases as the leakage level of private information becomes small, since the mutual information between Θ and $\tilde{\Theta}$ decreases. The maximum distortion is attained when Θ and $\tilde{\Theta}$ are statistically independent. In this case, perfect privacy is achieved since the leakage level of private information is equal to zero.

4.5 Final Remarks

In this chapter, we studied the problem of statistical parameter privacy wherein the outputs of a parametric source are shared with an untrusted party. The objective is to design privacy filters which ensure the accuracy of the shared information while guaranteeing the privacy of the parameters.

Two different schemes were proposed for the statistical parameter privacy problem, where the mutual information was used as the privacy measure. In the first scheme, it was assumed that the parameter belonged to a continuous alphabet and the mutual information was exploited as a proxy for a Bayesian statistical metric of privacy. On the other hand, the parameter was assumed to belong to a finite set of possibilities under the second scheme, and the mutual information was used directly as the privacy measure via Fano's inequality. The optimal distortion-privacy trade-off was analyzed for this scheme.

Acknowledgements This work was supported by the Knut and Alice Wallenberg foundation and the Swedish Foundation for Strategic Research.

References

1. Asoodeh S, Alajaji F, Linder T (2015) On maximal correlation, mutual information and data privacy. In: 2015 IEEE 14th Canadian workshop on information theory (CWIT), pp 27–31
2. Asoodeh S, Alajaji F, Linder T (2016) Privacy-aware MMSE estimation. In: 2016 IEEE international symposium on information theory (ISIT), pp 1989–1993
3. Barber RF, Duchi JC (2014) Privacy and statistical risk: formalisms and minimax bounds. arXiv:1412.4451 [cs, math, stat]
4. Bassi G, Nekouei E, Skoglund M, Johansson KH (2019) Statistical parameter privacy. Technical report, KTH Royal Institute of Technology, Stockholm, Sweden. https://www.dropbox.com/s/zx49o4oxn95k3c9/main.pdf?dl=0
5. Bassi G, Piantanida P, Skoglund M (2018) Lossy communication subject to statistical parameter privacy. In: 2018 IEEE international symposium on information theory (ISIT), pp 1031–1035
6. du Calmon FP, Fawaz N (2012) Privacy against statistical inference. In: 2012 50th annual Allerton conference on communication, control, and computing (Allerton), pp 1401–1408
7. Chakraborty S, Bitouzé N, Srivastava M, Dolecek L (2013) Protecting data against unwanted inferences. In: 2013 IEEE information theory workshop (ITW), pp 1–5
8. Clarke BS, Barron AR (1990) Information-theoretic asymptotics of Bayes methods. IEEE Trans Inf Theory 36(3):453–471
9. Cover TM, Thomas JA (2006) Elements of information theory, 2nd edn. Wiley, New York, NY
10. Duchi JC, Jordan MI, Wainwright MJ (2018) Minimax optimal procedures for locally private estimation. J Am Stat Assoc 113(521):182–201
11. Dwork C, McSherry F, Nissim K, Smith A (2006) Calibrating noise to sensitivity in private data analysis. In: Halevi S, Rabin T (eds) Theory of cryptography. Springer, Berlin, pp 265–284
12. Farokhi F, Sandberg H (2018) Fisher information as a measure of privacy: preserving privacy of households with smart meters using batteries. IEEE Trans Smart Grid 9(5):4726–4734
13. Kalantari K, Sankar L, Kosut O (2017) On information-theoretic privacy with general distortion cost functions. In: 2017 IEEE international symposium on information theory, pp 2865–2869
14. Kay SM (1993) Fundamentals of statistical signal processing: estimation theory. Prentice-Hall, Upper Saddle River, NJ, USA

15. Le Ny J, Pappas GJ (2014) Differentially private filtering. IEEE Trans Autom Control 59(2):341–354
16. Lehmann EL, Casella G (1998) Theory of point estimation, 2nd edn. Springer, New York, NY
17. Mo Y, Murray RM (2017) Privacy preserving average consensus. IEEE Trans Autom Control 62(2):753–765
18. Nozari E, Tallapragada P, Cortés J (2017) Differentially private average consensus: obstructions, trade-offs, and optimal algorithm design. Automatica 81:221–231
19. Rissanen J (1986) Stochastic complexity and modeling. Ann Stat 14(3):1080–1100
20. Samarati P, Sweeney L (1998) Protecting privacy when disclosing information: k-anonymity and its enforcement through generalization and suppression. Technical report, Computer Science Laboratory, SRI International
21. Sandberg H, Dán G, Thobaben R (2015) Differentially private state estimation in distribution networks with smart meters. In: 2015 54th IEEE conference on decision and control (CDC), pp 4492–4498
22. Tanaka T, Skoglund M, Sandberg H, Johansson K (2017) Directed information as privacy measure in cloud-based control. Technical report, KTH Royal Institute of Technology, Sweden. https://arxiv.org/abs/1705.02802
23. Tsai CY, Agarwal GK, Fragouli C, Diggavi S (2017) A distortion based approach for protecting inferences. In: 2017 IEEE international symposium on information theory (ISIT), pp 1913–1917
24. Wang Y, Huang Z, Mitra S, Dullerud GE (2017) Differential privacy in linear distributed control systems: entropy minimizing mechanisms and performance tradeoffs. IEEE Trans Control Netw Syst 4(1):118–130

Chapter 5
Privacy Verification and Enforcement via Belief Manipulation

Bo Wu, Hai Lin and Ufuk Topcu

Abstract We study the problem of verifying and enforcing the privacy of a partially observable stochastic system in the presence of an eavesdropping intruder. The intruder has the knowledge of the system model. It also has partial observation to system states and actions based on which it updates the belief which is a probability distribution over the system states. The intruder's objective is to learn the system's secrets in terms of whether the system is currently in some critical or sensitive states. We model the system as a partially observable Markov decision process and propose a notion of privacy with respect to the intruder 's belief. Our key observation is that the evolution of the intruder's belief can be represented as a discrete-time switched system. The privacy verification and enforcement can thus be cast into a reachability problem with respect to the unsafe subset of beliefs defined by the privacy requirement. We show how to apply control theory to verify and enforce the system's privacy with two proposed approaches.

5.1 Introduction

Privacy is becoming one of the most critical concerns in many practical systems [9, 14, 22, 33]. The vulnerabilities to information leaking pose significant challenges in systems that may have a huge social or economic impact if their privacy is compro-

B. Wu (✉)
Oden Institute for Computational Engineering and Sciences, University of Texas at Austin,
201 E 24th St, Austin, TX 78712, USA
e-mail: bwu3@utexas.edu

H. Lin
Department of Electrical Engineering, University of Notre Dame, Notre Dame, IN 46556, USA
e-mail: hlin1@nd.edu

U. Topcu
Department of Aerospace Engineering and Engineering Mechanics, Oden Institute for
Computational Engineering and Sciences, University of Texas at Austin,
201 E 24th St, Austin, TX 78712, USA
e-mail: utopcu@utexas.edu

© Springer Nature Singapore Pte Ltd. 2020
F. Farokhi (ed.), *Privacy in Dynamical Systems*,
https://doi.org/10.1007/978-981-15-0493-8_5

mised. Examples of such systems include automobiles, healthcare systems, robotic systems, power grid and so on. Development and adaptation of privacy preserving planning framework is receiving increasing interest in both academia and industry due to its great potential to significantly improve the system reliability.

In the recent years, a notion called "opacity" is receiving an increasing interest in privacy analysis and enforcement. Opacity is a confidentiality property that characterizes a system's capability to hide its "secret" information from being inferred by outside passive observers with possibly malicious intentions (termed as intruders in the sequel). The intruder is assumed to know the system's model and has (partial) access to the system's outputs but cannot observe the system states. The system is opaque if the intruder never decides that the secret happens with absolute certainty.

The opacity problem was first introduced in the computer science community [21] and quickly spread to Petri-net and discrete event system (DES) researchers. Various notions of opacity have been proposed in both deterministic and stochastic models. Interested readers are referred to [15] for a comprehensive review. In this chapter, we are interested in the current-state opacity (CSO), where the secret information is whether or not the current state of the system is a secret state. There are essentially two main directions in the opacity research—verification and enforcement. Algorithms are designed to verify if the system is opaque from the intruders [19]. And to enforce the opacity, the proposed approaches include supervisor synthesis [27], insertion functions [16, 35, 37, 38] or edit functions [17, 39] to control or manipulate the observed behavior.

The current definition of the current-state opacity relies on the absolute certainty that the current state belongs to the secret states. However, with a probabilistic model, in some cases, the intruder may just be able to maintain a belief distribution over the system states. In other words, the intruder may only infer that the current state is a secret state with a certain probability based on the observation history. As mentioned in [28], such scenario may not be characterized as a CSO violation by its definition, but still may potentially pose a security threat if the intruder deems the current state being a secret state with a high confidence.

Thus, in this chapter we are motivated to introduce a new opacity notion where the system is considered opaque if the intruder's confidence that the current state is a secret state never exceeds a given threshold.

Typically, the intruder updates its belief by the Bayesian rule. Such update depends on the action executed by the system and the observation available to the intruder. Therefore, a key observation that we make is that the belief update can be seen as a discrete-time switched system where the continuous belief dynamics is governed by the action observed. The privacy verification problem can be equivalently cast into verifying whether the solutions of the belief switched system avoid a privacy unsafe subset of the belief space, where the privacy specification is violated.

Such a safety verification problem is a familiar subject to the control community [3, 11, 12, 24, 31]. Existing approaches are generally categorized into abstraction and deductive verification [5]. In this chapter, we introduce two approaches, one in each category.

In the first approach [36], we abstract the continuous belief space into a finite set of grids where similar idea have been applied to abstract the timed automata into zones and regions [6] for the ease of analysis. By proving that the belief dynamics is mixed monotone, we could efficiently obtain the abstraction that serves as an over-approximation of the underlying continuous dynamic [10]. The belief abstraction idea has also been proposed in [8], but their belief space is the power set of the state space, which is discrete and finite.

With the abstracted finite belief transition system, we introduce two different approaches to enforce the privacy requirement. The first approach identifies the actions are guaranteed to preserve privacy. The second approach is inspired by the edit function [39] and directly manipulates the observations to the intruder.

The second approach [2] to safety verification relies on the construction of a function of the states, called the *barrier certificate* that satisfies a Lyapunov-like inequality [24]. The barrier certificates have shown to be useful in several system analysis and control problems running the gamut of bounding moment functional of stochastic systems [1] to collision avoidance of multi-robot systems [32]. It was also shown in [34] that any safe dynamical system admits a barrier certificate. We propose conditions for privacy verification of POMDPs using barrier certificates. From a computational stand point, we formulate a set of sum-of-squares programs (SOSP) to verify the privacy requirement.

5.1.1 Organization

The rest of this book chapter is organized as follows. Section 5.2 provides the necessary preliminaries to introduce the modeling framework and some key definitions. Section 5.3 deals with the efficient abstraction of the belief space based on the mixed monotone property. Then based on abstractions, we show how to enforce privacy. In Sect. 5.4, we apply the method based on barrier certificates to the privacy verification problem of POMDPs, and present a set of SOSP sufficient conditions, respectively.

5.1.2 Notation

The notations employed in this chapter are relatively straightforward. $\mathbb{R}_{\geq 0}$ denotes the set $[0, \infty)$ and $\mathbb{Z}_{\geq 0}$ denotes the set of integers $\{0, 1, 2, \ldots\}$. For a finite set A, we denote by $|A|$ the cardinality of the set A, A^* represents the set of all finite strings over A plus the empty string ϵ. Given a matrix Q, we denote by Q^T the transpose of Q. The notation $0_{n \times m}$ is the $n \times m$ matrix with zero entries. A probability simplex $\mathcal{B}$ of a dimension n is the set $\{b | b \in \mathbb{R}^n, b \geq 0, \sum_{i=1}^n b(i) = 1\}$ where $b(i)$ is the i-th element of the vector b. For two vectors, a and b with the same size, $a \succeq b$ implies entry-wise inequality. An interval $I = [x_1, x_2]$ of a dimension n is defined as $\{x \in \mathbb{R}^n | x_1 \leq x \leq x_2\}$. $int(I)$ denotes the interior of I, $\mathcal{R}[x]$ accounts for the set of polynomial functions with real coefficients in $x \in \mathbb{R}^n$, $p : \mathbb{R}^n \to \mathbb{R}$ and $\Sigma \subset \mathcal{R}$ is the subset of polynomials with an SOS decomposition; i.e., $p \in \Sigma[x]$ if and only if there are $p_i \in \mathcal{R}[x]$, $i \in \{1, \ldots, k\}$ such that $p = p_i^2 + \cdots + p_k^2$.

5.2 System Model and Definitions

In this section, we first introduce our modeling framework and then present a formal definition of our notion of privacy.

5.2.1 A General Model

Consider a dynamical system as show in Fig. 5.1. The system has a finite set of states. The state evolution is triggered by a finite set of actions. There is a potentially malicious intruder that can (imperfectly) observe the current states and the actions of the system. The intruder also has the knowledge of the system model. The system has a subset of secret states. Such secret states could refer to sensitive locations or critical status that should be kept private when the system is at such a state. The intruder's objective is to determine with a high certainty if the system is currently in one of the secret states based on its history of state and action observations. In order to protect the privacy of the system, we need to make sure that the intruder may never achieve its objective.

5.2.2 System Model

The system we consider in this chapter is modeled as a partially observable Markov decision process (POMDP) from the intruder's perspective. As a comprehensive model for planning in partially observable stochastic environments, POMDPs

Fig. 5.1 A general system model with an intruder. The shaded state is a secret state. The intruder tries to figure out whether the system is in the secret state based on its (partial) observations of the system's states and actions

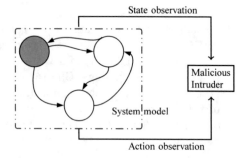

characterize the uncertainties in the state evolution due to actuation imperfections or interactions with a stochastic environment as well as the observation noises.

Formally, a POMDP [18] is a tuple $\mathcal{P} = (Q, \iota_{init}, A, T, Z, O)$ where

- Q is a finite set of states;
- $\iota_{init} : Q \rightarrow [0, 1]$ defines the distribution of the initial states, i.e., $\iota_{init}(q)$ denotes the probability of starting at $q \in Q$;
- A is a finite set of actions;
- $T : Q \times A \times Q \rightarrow [0, 1]$ is the probabilistic transition function where $T(q, a, q') := P(q_t = q'|q_{t-1} = q, a_{t-1} = a)$, for any $t \in \mathbb{Z}_{\geq 1}, q, q' \in Q, a \in A$;
- Z is the set of all possible observations;
- $O : Q \times A \times Z \rightarrow [0, 1]$ is the observation function where $O(q, a, z) := P(z_t = z|q_t = q, a_{t-1} = a)$, for any $t \in \mathbb{Z}_{\geq 1}, q \in Q, a \in A$ and $z \in Z$.

Furthermore, we assume that there is a set of secret states $Q_s \subset Q$. We want to conceal the information that the system is currently in some secret state $q \in Q_s$.

From the system's perspective, we assume it has full observation of its states, i.e. $O(q, a, z) = 1$ if $q = z$ and $O(q, a, z) = 0$ otherwise. Therefore, the system reduces to a Markov decision process (MDP) [26] which is a tuple $\mathcal{M} = (Q, \iota_{init}, A, T)$.

5.2.3 Belief Update Equations as Discrete-Time Switched Systems

For a system modeled as a POMDP, the intruder may have access to its history of observations and the executed actions. It is then possible for the intruder to maintain a belief $b : Q \rightarrow [0, 1]$ and $\sum_q b(q) = 1$. By definition, a belief is a probability distribution over the states that the system could be in. Therefore, a belief state belongs to the probability simplex $\mathcal{B}$ with a dimension $|Q|$.

The initial belief $b_0 = \iota_{init}$ and at any time step t, it can be updated as follows based on the belief state b_{t-1}, the action $a = a_{t-1}$ and the observation z_t.

$$b_t = f_a(b_{t-1}, z_t), \tag{5.1}$$

where $b_t(q') = f_a^{q'}(b_{t-1}, z_t)$ is computed as

$$
\begin{aligned}
f_a^{q'}(b_{t-1}, z_t) &= P(q'|z_t, a_{t-1}, b_{t-1}) \\
&= \frac{P(z_t|q', a_{t-1}, b_{t-1})P(q'|a_{t-1}, b_{t-1})}{P(z_t|a_{t-1}, b_{t-1})} \\
&= \frac{P(z_t|q', a_{t-1}, b_{t-1})\sum_{q \in Q} P(q'|a_{t-1}, b_{t-1}, q)P(q|a_{t-1}, b_{t-1})}{P(z_t|a_{t-1}, b_{t-1})} \\
&= \frac{O(q', a_{t-1}, z_t)\sum_{q \in Q} T(q, a_{t-1}, q')b_{t-1}(q)}{\sum_{q' \in Q} O(q', a_{t-1}, z_t)\sum_{q \in Q} T(q, a_{t-1}, q')b_{t-1}(q)}.
\end{aligned}
\tag{5.2}
$$

Equation (5.1) defines a discrete-time switched system. Each mode in this switched system corresponds to an action $a \in A$. The dynamics in each mode is determined by its previous belief state b_{t-1}, observation z_t as the input and the action a.

5.2.4 Privacy in Belief Space

Our notion of privacy is defined with respect to the belief of the intruder. We require that the intruder, even with access to the system's actions and observations since $t = 0$, is never confident that the system is currently in a secret state with a probability larger than or equal to a constant $\lambda \in [0, 1]$ at any time t. We formalize the definition as follows.

Definition 5.1 Given a POMDP $\mathcal{P} = (Q, \iota_{init}, A, T, Z, O)$ and a set of secret states $Q_s \subset Q$, $\mathcal{P}$ is current-state opaque (λ-CSO) under all possible evolutions of the switched system in (5.1) if

$$\sum_{q \in Q_s} b_t(q) \le \lambda, \forall t. \tag{5.3}$$

Intuitively, $\sum_{q \in Q_s} b_t(q)$ denotes the probability the system being in some secret state at time t based on intruder's observations of the system. As long as (5.3) holds, the intruder is never confident that the system is in a secret state.

Inventory Management Example

Suppose there is an MDP $\mathcal{M}$ that models the evolution of inventory levels of a company, which has three states, where q_1 and q_2 represents low and high inventory level and q_3 represents the medium inventory level as seen in Fig. 5.2. The company would like to keep the current inventory level being high or low as a secret, because the intruders, either suppliers or competitors, may leverage such information to adjust the price of the goods for their own benefits. Therefore $q_1, q_2 \in Q_s$ and q_3 is a non-secret state. $A = \{\sigma_1, \sigma_2\}$ represents two different purchase quantities. The transition probabilities are because of random demands. Our objective is to verify if this inventory management system is λ-CSO with respect to a given λ.

Fig. 5.2 The MDP model
for the inventory
management system. The
transition probabilities are
omitted for clarity

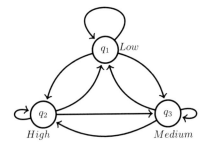

We conclude this section with the following important remark. From Definition 5.1, we can define a privacy unsafe set

$$B_u = \{b | \sum_{q \in Q_s} b(q) > \lambda\} \subset B \tag{5.4}$$

that the belief state should never enter. To make sure the system is private, we need to verify λ-CSO by analyzing whether all the belief evolutions in (5.1) can avoid a given B_u. Such a verification problem is equivalent to the reachability problem in a switched system, which is conjectured to be undecidable [30]. Therefore, in the following sections, we introduce two different approaches that serve as sufficient conditions to verify λ-CSO.

5.3 Abstraction-Based Verification and Enforcement of λ-CSO

In this section, we consider the following scenario. Given a system modeled as an MDP $\mathcal{M} = (Q, \iota_{init}, A, T)$, there is an intruder that has the knowledge of $\mathcal{M}$ and is capable of observing all the actions but not the actual states. Note that the state is fully observable for the system. This is equivalent to a POMDP $\mathcal{P} = (Q, \iota_{init}, A, T, O, Z)$ with $Z = \{null\}$ and $O(q, a, z) = 1$ for any q and a. We also assume that $A(q) = A$ for all $q \in Q$. That is, all the actions are defined at each state of the MDP. With these assumptions, the belief dynamics become

$$b_{t+1} = P_a b_t, \tag{5.5}$$

where $a \in A$ and $P_a(i, j) = T(q_j, a, q_i)$. The dynamics given any action a is linear. The belief update is therefore characterized as a switched linear system.

5.3.1 Mixed Monotonic Belief Updates

We first show that (5.5) is actually mixed monotone where efficient abstraction method is available [10].

Definition 5.2 A system

$$x = F(x), \tag{5.6}$$

is mixed monotone, where $x \in X \subset \mathbb{R}^n$ and $F : X \to X$ is a continuous map, if there exists a decomposition function $f : X \times X \to X$ such that (1) $F(x) = f(x, x), \forall x \in X$, (2) $x_1 \le x_2 \Rightarrow f(x_1, y) \le f(x_2, y), \forall x_1, x_2, y \in X$, (3) $y_1 \ge y_2 \Rightarrow f(x, y_1) \le f(x, y_2), \forall x, y_1, y_2 \in X$, where $\le$ denotes the element-wise inequality. A switched system is mixed monotone if it is mixed monotone for each mode (action) $a \in A$.

Since $\sum_{q \in Q} b_t(q) = 1$ and we denote $N = |Q|$, (5.5) can be equivalently written as an $N - 1$-dimension dynamical system

$$b_{t+1}^{[1,N-1]} = F_a(b_t^{[1,N-1]}), \tag{5.7}$$

where $b_t^{[1,N-1]} = [b_{t,1}, ..., b_{t,N-1}]^T$ and the function mapping F_a will be shown in the following lemma which proves that (5.7) is indeed mixed monotone.

Lemma 5.1 *The switched system (5.7) is mixed monotone.*

Proof From (5.5), for $a \in A$ and at time $t + 1$ we have

$$b_{t+1} = P_a b_t = \begin{bmatrix} p_{1,1}^a & \cdots & p_{1,N}^a \\ p_{2,1}^a & \cdots & p_{2,N}^a \\ \vdots & \cdots & \vdots \\ p_{N,1}^a & \cdots & p_{N,N}^a \end{bmatrix} \begin{bmatrix} b_{t,1} \\ b_{t,2} \\ \vdots \\ b_{t,N} \end{bmatrix}. \tag{5.8}$$

Since the probabilities have to sum to one, we have $\sum_{i=1}^{N} p_{i,j}^a = 1, \forall j \in \{1, ..., N\}$ and $\sum_{i=1}^{N} b_{t,i} = 1$. Therefore, (5.8) can be rewritten as

$$\begin{bmatrix} b_{t+1,1} \\ b_{t+1,2} \\ \vdots \\ b_{t+1,N} \end{bmatrix} = \begin{bmatrix} p_{1,1}^a & \cdots & p_{1,N}^a \\ p_{2,1}^a & \cdots & p_{2,N}^a \\ \vdots & \cdots & \vdots \\ 1 - \sum_{i=1}^{N-1} p_{i,1}^a & \cdots & 1 - \sum_{i=1}^{N-1} p_{i,N}^a \end{bmatrix} \begin{bmatrix} b_{t,1} \\ b_{t,2} \\ \vdots \\ 1 - \sum_{i=1}^{N-1} b_{t,i} \end{bmatrix}. \tag{5.9}$$

Since $b_{t+1,N} = 1 - \sum_{i=1}^{N-1} b_{t+1,i}$, from (5.9) we have the following equation on the $(N-1)$-dimensional system

$$
\begin{bmatrix} b_{t+1,1} \\ b_{t+1,2} \\ \vdots \\ b_{t+1,N-1} \end{bmatrix} = \begin{bmatrix} p^a_{1,1} & \cdots & p^a_{1,N-1} \\ p^a_{2,1} & \cdots & p^a_{2,N-1} \\ \vdots & \cdots & \vdots \\ p^a_{N-1,1} & \cdots & p^a_{N-1,N-1} \end{bmatrix} \begin{bmatrix} b_{t,1} \\ b_{t,2} \\ \vdots \\ b_{t,N-1} \end{bmatrix}
$$
$$
- \begin{bmatrix} p^a_{1,N} & \cdots & p^a_{1,N} \\ p^a_{2,N} & \cdots & p^a_{2,N} \\ \vdots & \cdots & \vdots \\ p^a_{N-1,N} & \cdots & p^a_{N-1,N} \end{bmatrix} \begin{bmatrix} b_{t,1} \\ b_{t,2} \\ \vdots \\ b_{t,N-1} \end{bmatrix} + \begin{bmatrix} p^a_{1,N} \\ p^a_{2,N} \\ \vdots \\ p^a_{N-1,N} \end{bmatrix} \tag{5.10}
$$

Equivalently, we have

$$
b^{[1,N-1]}_{t+1} = F_a(b^{[1,N-1]}_t) = A_{a,1} b^{[1,N-1]}_t - A_{a,2} b^{[1,N-1]}_t + B_a \tag{5.11}
$$

where $A_{a,k}(i,j) \geq 0, \forall i, j \in \{1, ..., N-1\}, k \in \{1,2\}, a \in A$ and $B_a(i) \geq 0$ for all $i \in \{1, ..., N-1\}$. We define

$$
f_a(x, y) = A_{a,1} x - A_{a,2} y + B_a, \tag{5.12}
$$

from Definition 5.2, it is not hard to find that all the three conditions are satisfied. Since it holds for arbitrary $a \in A$, by definition, the switched system (5.7) is mixed monotone. $\qquad \square$

Mixed monotone systems admit efficient over-approximation of the reachable set by evaluating the function f at two points as proven in Theorem 5.1.

Theorem 5.1 ([10]) *Given a mixed monotone system as defined in (5.6) with decomposition function $f(x, y)$, given $x_1, x_2 \in X \subset \mathbb{R}$ with $x_1 \leq x_2$, we have*

$$
f(x_1, x_2) \leq F(x) \leq f(x_2, x_1), \forall x \in [x_1, x_2] \tag{5.13}
$$

This theorem is a direct result of the mixed monotone property and is the key to the efficient abstraction, which can be seen more clearly from the following formula.

$$
F([x_1, x_2]) \subseteq [f(x_1, x_2), f(x_2, x_1)] \tag{5.14}
$$

where $F(X') = \{F(x)|x \in X'\}$ and $X' \in X$ is called the one-step reachable set from X' [10]. We have $x \in [x_1, x_2]$ if and only if $x_1 \leq x \leq x_2$. It can be observed from (5.14) that it is sufficient to evaluate the decomposition function f at two points x_1 and x_2 to compute an over-approximation of the one-step reachable set where the bounding has been shown to be tight [10].

5.3.2 Verification of λ-CSO

Now we are ready to abstract the belief space dynamics as defined in (5.7) to a nondeterministic finite automaton (NFA).

Definition 5.3 ([7]) An NFA is a tuple $\mathcal{T} = (S, \Sigma, \delta, I)$ where

- S is a finite set of states;
- Σ is a finite set of actions;
- $\delta : S \times \Sigma \to 2^S$ is the transition function;
- $I \subseteq S$ is a set of initial states.

The transition function δ can be extended to $S \times \Sigma^*$ in a natural way. We say that a state s is reachable if there exists an $\omega \in \Sigma^*$ and $s_0 \in I$ such that $\delta(s_0, \omega)$ is defined. Given the initial set $I \subseteq S$ of states, the language generated by $\mathcal{T}$ is defined by $\mathcal{L}(\mathcal{T}) = \{\omega \in \Sigma^* | \exists s \in I, \delta(s, \omega) \text{ is defined}\}$.

The first step is to partition the probability simplex $\mathcal{B}$ into a finite set of intervals $\{I_s\}, s \in S$, where $I_s = [x_1^s, x_2^s], x_1^s \leq x_2^s, \mathcal{B} \subseteq \bigcup_{s \in S} I_s, int(I_s) \cap int(I_{s'}) = \emptyset, int(I_s) \cap \mathcal{B} \neq \emptyset$, for all $s, s' \in S$ where $s \neq s'$.

Recall the opacity requirement (5.3), which basically defines an unsafe set $\mathcal{B}_u$ in (5.4) that the intruder's belief should never enter. The following lemma then shows that there exists a simple algorithm to determine whether a partition I_s has an overlap with $\mathcal{B}_u$.

Lemma 5.2 *Given an interval* $I_s = [x_1, x_2], x_1 \leq x_2$ *and the set* $\mathcal{B}_u$, *then* $I_s \cap \mathcal{B}_u \neq \emptyset$ *if* $\sum_{q \in Q_s} x_2(q) > \lambda$.

From Lemma 5.2, we can define if a partition I_s is bad.

Definition 5.4 $I_s \in S$ is bad if $I_s \cap \mathcal{B}_u \neq \emptyset$. We assume that the initial belief state is always outside of $\mathcal{B}_u$. If the grid that contains the initial belief state is bad due to the overlapping, we may re-partition this grid into two smaller grids such that the initial belief state is no longer in a bad partition.

Inventory Management Example (cont.)

For our inventory management example, suppose the initial condition is that $b_0(q_1) = 0.3$ and $b_0(q_2) = 0.1$. The privacy requirement is that $b(q_1) + b(q_2) \leq 0.8$ all the time. The transition probabilities are

$$P_{\sigma_1} = \begin{bmatrix} 0.2 & 0 & 0.1 \\ 0.4 & 0.3 & 0.2 \\ 0.4 & 0.7 & 0.7 \end{bmatrix}, P_{\sigma_2} = \begin{bmatrix} 0.4 & 0.65 & 0.3 \\ 0.2 & 0 & 0.2 \\ 0.4 & 0.35 & 0.5 \end{bmatrix}. \tag{5.15}$$

The probabilistic simplex is partitioned by squares of width 0.2 as shown in Fig. 5.3a. Note that the partitioned grids can have arbitrary sizes and need not to be equal. Here

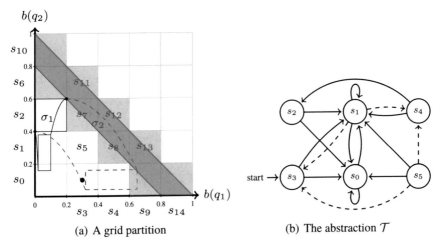

(a) A grid partition   (b) The abstraction $\mathcal{T}$

Fig. 5.3 In Fig. 5.3a, the big dot denotes the initial belief. The blue shaded area denotes $\mathcal{B}_u$ and all the grey shaded grids are bad partitions. In Fig. 5.3b, lines denote the transitions induced by σ_1 and dashed lines for σ_2. At each state, the action that may lead to $\mathcal{B}_u$ is omitted

we use the equal size grids just for simplicity. As shown in Fig. 5.3a, there are 15 grids in total.

The second step is to construct the NFA $\mathcal{T} = (S, \Sigma, \delta, I)$ given the MDP model $\mathcal{M} = (Q, \iota_{init}, A, T)$ and the partition $\{I_s\}$, where $\Sigma = A$. To determine the transition relation in $\mathcal{T}$ for every $s \in S$ and $\sigma \in \Sigma$, if $I_s = [x_1, x_2]$ and $[f_\sigma(x_1, x_2), f_\sigma(x_2, x_1)] \cap I_{s'} \neq \emptyset$, where $f_\sigma(., .)$ is defined as in (5.12), we will have $s' \in \delta(s, \sigma)$. That is, if our over-approximated one-step reachable set for I_s has a non-empty intersection with the partitioned region $I_{s'}'$ given the action σ, there will be a transition relation $s' \in \delta(s, \sigma)$ in the abstraction system $\mathcal{T}$.

Theorem 5.2 *If no bad partition is reachable in $\mathcal{T}$, POMDP $\mathcal{P}$ is λ-CSO.*

Inventory Management Example (cont.)

In Fig. 5.3a, all the shaded grids are bad partitions and there are 6 states (correspondingly 6 partitions) of interest in $\mathcal{T}$.

Let's take s_2 as an example to determine the transitions in $\mathcal{T}$. By mixed monotone property, we only have to evaluate two points, namely $s_1 = (0, 0.4)$ and $s_2 = (0.2, 0.6)$. In Fig. 5.3a, from (5.14), the one-step reachable sets from s_2 by executing σ_1 and σ_2 are bounded by the solid and dashed rectangles that are connected to s_2 with solid and dashed curves labeled by σ_1 and σ_2 respectively. With action $\sigma_1 \in \Sigma$, the over-approximation reachable set overlaps with s_0 and s_1. Therefore, we have $s_0 \in \delta(s_2, \sigma_1)$ and $s_1 \in \delta(s_2, \sigma_1)$ where the corresponding transitions can be found in Fig. 5.3b. Similarly, we have $s_3 \in \delta(s_2, \sigma_2)$, $s_4 \in \delta(s_2, \sigma_2)$ and $s_9 \in \delta(s_2, \sigma_2)$ where s_9 is a bad partition.

If $A = \{\sigma_1\}$, $\mathcal{T}$ will be the same in Fig. 5.3b with the dashed transitions ignored since they are for σ_2. The model is guaranteed to be λ-CSO with $\lambda = 0.8$, since all the transitions are among the six partitions that are not shaded in Fig. 5.3a.

It should be noted that if a bad partition is reachable, it does not necessarily mean that the system is not λ-CSO. The reason is that such abstraction is an over-approximation that could produce spurious trajectories that do not actually exist in (5.7). This is generally unavoidable in the partition-based approaches. However, since we are only interested in the safety property in the belief space (if a bad belief state is reachable), such spuriousness may make the results more conservative. But it does not affect its correctness, as all the transitions that are possible to happen in the concrete system (5.7) are included in the abstraction system.

5.3.3 λ-CSO Enforcement

If a bad partition that violates the opacity requirement is reachable, like s_9 is reachable from s_2 in our example, it means that the system is potentially not λ-CSO. In such a case, we would like to enforce the privacy.

All bad partitiond should be deleted from $\mathcal{T}$. In our example, as shown in Fig. 5.3a, we are only concerned with the six non-shaded grids. Furthermore, any outgoing transition (s, σ, s') where s' is a bad partition should be deleted from q. To do this, we directly disable the action σ from s, as the transitions are probabilistic. For example, in Fig. 5.3a, since we have $s_9 \in \delta(s_2, \sigma_2)$ where s_9 is a bad partition, action σ_2 will be disabled in s_2. If such pruning results in any state q' blocking, that is, all its outing transitions for all actions are pruned, then q' and all its incoming and outgoing transitions are deleted. Such process continues until no more states are pruned from $\mathcal{T}$ or the initial state of $\mathcal{T}$ is pruned. If the latter situation happens, it implies that the current partition may be too coarse so that the over-approximation is too conservative, which we may need to find a finer partition scheme, for example, by splitting the grids into two equally sized grids.

Abstraction and refinement is an active research area but is out of the scope of this chapter. It could also be the case that the belief dynamics (5.7) will eventually drive the belief state to $\mathcal{B}_u$ under arbitrary switching. If this is the case, there is no hope to find a non-empty $\mathcal{T}$ after pruning, regardless of how the belief space is partitioned. Determining whether it is true relies on the reachability analysis of the underlying switched linear systems, which is conjectured to be undecidable. The resulting NFA from Fig. 5.3a is shown in Fig. 5.3b.

Once we obtain the abstracted and refined belief model T, it is then possible to enforce the privacy requirement with two different solutions based on different capabilities of the intruders. The current intruder's belief can be mapped to the discrete states in T after pruning, from which we know the next available actions that are guaranteed to be safe. Then we select an action and update the belief following (5.1) and the discrete transition in T.

5.3.3.1 Direct Synthesis

For our inventory example, if the intruder is the supplier that can observe the purchasing actions since the purchase has to go through it, we need a purchasing strategy such that the supplier may never be sure with high confidence that the company's inventory is running low or high.

Our proposed approach is to restrict the purchasing actions to always keep the intruder's belief away from $\mathcal{B}_u$. From Theorem 5.2, it implies that at any time, we select the actions based on the current intruder's belief and the abstraction T.

5.3.3.2 Edit Function

If the intruder is the competitor, it is then possible to manipulate the reported purchase activity, such that the competitor may never infer with a high confidence of the company's inventory level being low or high. Unlike suppliers, the competitor may not distinguish between the real or the reported purchase. This approach is inspired by the edit function synthesis in [39] where the system has the capability to modify the observations of the intruder, such that the observed behavior is consistent with the model's behavior and at the same time, the intruder may never determine with certainty that the current state is a secret state.

Formally, given an MDP $\mathcal{M} = (Q, \iota_{init}, A, T)$ and its corresponding NFA $T_{\mathcal{M}} = (S, \Sigma, \delta, I)$ where $Q = S$, $\Sigma = A$, $q \in I$ if and only if $\iota_{init} > 0$, $q' \in \delta(q, a)$ if and only if $P(q, a, q') > 0$, we are looking for an edit function $f_e : \Sigma^* \rightarrow \Sigma^*$, such that

1. $\forall \omega \in \mathcal{L}(T_{\mathcal{M}})$, $f_e(\omega)$ is defined,
2. $\forall \omega \in \mathcal{L}(T_{\mathcal{M}})$, $\exists s \in I$, $\delta(s, f_e(\omega))$ is defined,
3. and $\forall \omega \in \mathcal{L}(T_{\mathcal{M}})$, after executing $f_e(\omega)$, the switched system in (5.5) satisfies (5.3).

Intuitively, the first item requires that the edit function f_e should be defined for all the possible system behaviors. The second item requires that the output of the edit function, which is observed by the intruder, should also be a valid behavior of the system. The third item requires that the output behavior of the edit function should satisfy the opacity specification. Note that from this definition, f_e may not be unique.

Given a system modeled as an MDP $\mathcal{M} = (Q, \iota_{init}, A, T)$, f_e can be implemented as a (potentially) infinite-state edit automaton $\mathcal{T}_f = (S_f, \Sigma, \delta_f, I_f)$, where $\Sigma = A$, $\delta_f \subseteq S_f \times \Sigma \times \Sigma^* \times S_f$. Therefore, each transition (s, σ, o, s') in $\mathcal{T}_f$ denotes that from state s, when $\sigma \in \Sigma$ actually happens in the system, it is modified to become $o \in \Sigma^*$ which is observed by the intruder, and then the edit automaton transits to some s'. Intuitively, if we edit every possible execution to be the empty string ϵ, the intruder will observe nothing and the system will always be opaque if it is opaque initially. However, such case may become trivial. Therefore, we restrict the transitions of the edit automaton to be of the form $\delta_f \subseteq S_f \times \Sigma \times \Sigma \times S_f$, that is, it must output one and only one event $\sigma \in \Sigma$ after an event has actually happened in the system.

In the setting of this chapter, f_e is easier to synthesize, since all actions are defined in every state, the second requirement of f_e is automatically satisfied. To guarantee the third requirement, the output of f_e can be the language generated by the abstraction $\mathcal{T} = (S, \Sigma, \delta, I)$. That is, the edit automaton $\mathcal{T}_f = (S, \Sigma, \delta_f, I)$, where given the transition $(s, o, s'), o \in \Sigma$ in $\mathcal{T}$ and given the actual event $\sigma \in \Sigma$, there is a transition (s, σ, o, s'). Therefore, the intruder observes a subset of the generated language of $\mathcal{T}$, which is guaranteed to preserve opacity. Furthermore, since we don't have any restriction on the actual event σ, the requirement 1 of the edit function is also satisfied.

Inventory Management Example (cont.)

In this example, the observation function is essentially the abstracted model $\mathcal{T}$ in Fig. 5.3b. Regardless of the real system action that is executed, starting from s_3, the edit function may select any action σ that is defined at the current belief partition s to be the observation to the intruder. Then the next abstracted belief state s' is determined by the belief dynamic (5.7). Note that such update is based on the "fake" action σ, not the real system action, which is hidden by the edit function. For example, starting from s_3 in Fig. 5.3b, the edit function may output σ_1, regardless of event σ_1 or σ_2 actually happened. If the belief state update based on the output behavior σ_1 results in s_1, from Fig. 5.3b, next time it could either output σ_1 or σ_2.

5.4 λ-CSO Verification Using Barrier Certificates

Section 5.3 introduced an abstraction-based approach to verify and enforce the privacy on a special subclass of the POMDP where no observation is ever available to any of the states. In the sequel, we develop a technique based on barrier certificates for privacy verification of belief update equations with respect to general POMDPs.

Let's revisit the belief dynamics of a POMDP as

$$b_t = f_a(b_{t-1}, z_t). \tag{5.16}$$

Verifying whether all the belief evolutions of (5.16) starting at $\mathcal{B}_0 \subset \mathcal{B}$ avoid a given privacy unsafe set $\mathcal{B}_u$ at a pre-specified time H or for all time is a cumbersome task in general and requires simulating (5.16) for all elements of the set $\mathcal{B}_0$ and for different sequences of $a \in A$ and $z \in Z$. Furthermore, POMDPs are often computationally intractable to solve exactly [13]. To surmount these challenges, we demonstrate that we can find a barrier certificate which verifies that a given privacy requirement is not violated without the need to solve the belief update equations or the POMDPs directly.

Theorem 5.3 checks whether the privacy requirement is not violated for all time.

Theorem 5.3 *Consider the belief switched dynamics (5.16). Given a set of initial conditions $\mathcal{B}_0 \subset [0, 1]^{|Q|}$, and an unsafe set $\mathcal{B}_u \subset [0, 1]^{|Q|}$ ($\mathcal{B}_0 \cap \mathcal{B}_u = \emptyset$), if there exists a function $B : [0, 1]^{|Q|} \to \mathbb{R}$ such that*

$$B(b) > 0, \quad \forall b \in \mathcal{B}_u, \tag{5.17}$$

$$B(b) < 0, \quad \forall b \in \mathcal{B}_0, \tag{5.18}$$

and

$$B\left(f_a(b_{t-1}, z_t)\right) - B(b_{t-1}) \leq 0, \forall t \in \mathbb{Z}_{\geq 1}, \ \forall a \in A \text{ and } z \in Z, \tag{5.19}$$

then there exist no solution of (5.16) such that $b_0 \in \mathcal{B}_0$ and $b_t \in \mathcal{B}_u$ for all $t \in \mathbb{Z}_{\geq 1}$ and any sequence of actions $a \in A$ and $z \in Z$. Hence, the privacy requirement is not violated.

Proof The proof is carried out by contradiction. Assume at any time instance H there exists a solution to (5.16) such that $b_0 \in \mathcal{B}_0$ and $b_H \in \mathcal{B}_u$. From inequality (5.19), we have

$$B(b_t) \leq B(b_{t-1})$$

for all $t \in \{1, 2, \ldots, H\}$ and all actions $a \in A$. Hence, $B(b_t) \leq B(b_0)$ for all $t \in \{1, 2, \ldots, H\}$. Furthermore, inequality (5.18) implies that $B(b_0) < 0$ for all $b_0 \in \mathcal{B}_0$. Since the choice of H can be arbitrary, this is a contradiction because it implies that $B(b_H) \leq B(b_0) < 0$. Therefore, there exist no solution of (5.16) such that $b_0 \in \mathcal{B}_0$ and $b_H \in \mathcal{B}_u$ for any sequence of actions $a \in A$ and $z \in Z$. $\square$

In the following, we formulate a set of conditions in terms of sum-of-squares programs (SOSPs) [23] to verify whether a given a POMDP satisfies a privacy requirement.

5.4.1 Privacy Verification for POMDPs via SOSP

The belief update equation (5.16) for a POMDP is a rational function in the belief states $b_t(q)$ where from (5.2) we know that

$$b_t(q') = \frac{S_a\left(b_{t-1}(q', z_t)\right)}{R_a\left(b_{t-1}(q', z_t)\right)}$$

$$= \frac{O(q', a_{t-1}, z_t)\sum_{q\in Q} T(q, a_{t-1}, q')b_{t-1}(q)}{\sum_{q'\in Q} O(q', a_{t-1}, z_t)\sum_{q\in Q} T(q, a_{t-1}, q')b_{t-1}(q)}. \tag{5.20}$$

Moreover, We further assume the initial belief is from a set.

$$\mathcal{B}_0 = \left\{ b_0 \in \mathbb{R}^{|Q|} \mid l_i^0(b_0) \le 0, l_i^0 \in \mathcal{R}[b], \ i = 1, 2, \ldots, n_0 \right\}, \tag{5.21}$$

where $l_i^0 \in \mathcal{R}[x]$ for all $i = 1, 2, \ldots, n_0$. At this stage, we are ready to present conditions based on SOSP to verify privacy of a given POMDP.

Theorem 5.4 *Consider the POMDP belief update dynamics* (5.20), *the privacy unsafe set* (5.4), *and the set of initial beliefs* (5.21). *If there exist polynomial functions* $B \in \mathcal{R}[b]$ *with degree* d, $p^u \in \Sigma[b]$, $p_i^0 \in \Sigma[b]$, $i = 1, 2, \ldots, n_0$, *and constants* $c_1, c_2 > 0$ *such that*

$$B(b) - p^u(b)\left(\sum_{q\in Q_s} b(q) - \lambda\right) - c_1 \in \Sigma[b], \tag{5.22}$$

$$-B(b_0) + \sum_{i=1}^{n_0} p_i^0(b_0)l_i^0(b_0) - c_2 \in \Sigma[b_0], \tag{5.23}$$

and

$$-R_a(b_{t-1})^d\left(B\left(\frac{S_a(b_{t-1})}{R_a(b_{t-1})}\right) - B(b_{t-1})\right) \in \Sigma[b_{t-1}], \tag{5.24}$$

then the privacy requirement (5.3) *is satisfied for all time.*

Proof The proof is from Theorem 5.3. SOS conditions (5.22) and (5.23) are to verify conditions (5.17) and (5.18), respectively. Furthermore, condition (5.19) for system (5.20) can be re-written as

$$B\left(\frac{S_a(b_{t-1})}{R_a(b_{t-1})}\right) - B(b_{t-1}) > 0.$$

Given the fact that $R_a(b_{t-1}(q'))$ is a positive polynomial of degree one, we can relax the above inequality into an SOS condition given by

$$-R_a(b_{t-1})^d\left(B\left(\frac{S_a(b_{t-1})}{R_a(b_{t-1})}\right) - B(b_{t-1})\right) \in \Sigma[b_{t-1}].$$

Hence, if (5.24) holds, then (5.19) is satisfied as well. Then, by Theorem 5.3, we infer the privacy requirement is satisfied at any time t.

$\square$

Inventory Management Example Revisited

Following our inventory management example, besides the purchasing actions, the intruder may also have access to the intervals between the two consecutive purchases, which suggests a POMDP $\mathcal{P}$ model that has the observation set $Z = \{z_0, z_1\}$. The observations represent a short and a long purchasing intervals respectively. The initial belief is $b(q_0) = 0.2$ and $b(q_1) = 0.2$. The transition matrix is given as

$$P_{\sigma_1} = \begin{bmatrix} 0.2, 0.2 \ 0.45 \\ 0.6, 0.5 \ 0.4 \\ 0.2, 0.3 \ 0.15 \end{bmatrix}, P_{\sigma_2} = \begin{bmatrix} 0.1, \ 0.5 \ 0.25 \\ 0.55, 0.4 \ 0.5 \\ 0.35, 0.1 \ 0.25 \end{bmatrix} \quad (5.25)$$

The observation function is defined as below where $O_\sigma(i, j) = O(q_i, \sigma, z_j)$,

$$O_{\sigma_1} = \begin{bmatrix} 0.5, \ 0.5 \\ 0.8, \ 0.2 \\ 0.7, \ 0.3 \end{bmatrix}, O_{\sigma_2} = \begin{bmatrix} 0.2, \ 0.8 \\ 0.8, \ 0.2 \\ 0.6, \ 0.4 \end{bmatrix}. \quad (5.26)$$

We are interested in finding an upper-bound of λ, i.e., the achievable privacy threshold. We check the SOSPs (5.22)–(5.24) where fix the degree d of the barrier certificate. The numerical experiments are carried out on a MacBook Pro 2.9 GHz Intel Core i5 and 8 GB of RAM. The SDPs are solved using YALMIP [20] and the SOSPs are solved using the SOSTOOLs [25] parser and solvers such as Sedumi [29].

We increase the degree of the barrier certificates from 2 to 10 and look for the smallest value of λ, for which privacy verification could be certified. Table 5.1 outlines the obtained results. As it can be observed from the table, by increasing the degree of the barrier certificate, we can find a tighter upper-bound on the best achievable privacy level. The barrier certificate of degree 2 (excluding terms smaller than 10^{-4}) constructed using Theorem 5.4 is provided below:

$$B(b) = 0.1629b(q_3)^2 - 3.9382b(q_1)^2 + 0.9280b(q_2)^2 - 0.0297b(q_1)b(q_3)$$
$$- 4.4451b(q_1)b(q_2) - 0.0027b(q_3) - 2.0452b(q_1) + 9.2633.$$

Table 5.1 Numerical results

d	2	4	6	8	10
λ^*	0.93	0.88	0.80	0.74	0.69
Computation Time (s)	5.38	8.37	12.03	18.42	27.09

5.5 Final Remarks

In this chapter, we studied the privacy verification and enforcement problem in POMDPs. We defined the notion of current state opacity (λ-CSO) with respect to the belief state of a POMDP, such that an intruder should not know with a high confidence that the system is in some secret states. Our problem was to be able to verify and enforce λ-CSO at any time.

We introduced two approaches to tackle the problem. The first approach is based on abstraction, where we showed that the belief dynamics are mix-monotone and efficient abstraction scheme exists. Based on the abstraction, we also proposed two ways to enforce our notion of privacy. Our second approach is based on barrier functions for privacy verification. We illustrated both approaches with an inventory management system example. The presentation of this chapter closely follows [4, 36].

Acknowledgements This work was supported by AFOSR FA9550-19-1-0005, NSF 1446288, NSF 1253488, NSF 1724070, NSF 1646522, NSF 1652113. and DARPA D19AP00004.

References

1. Ahmadi M, Harris AWK, Papachristodoulou A (2016) An optimization-based method for bounding state functionals of nonlinear stochastic systems. In: 2016 IEEE 55th Conference on Decision and Control (CDC). IEEE, pp 5342–5347
2. Ahmadi M, Israel A, Topcu U (2018) Controller synthesis for safety of physically-viable data-driven models. arXiv:1801.04072
3. Ahmadi M, Valmorbida G, Papachristodoulou A (2017) Safety verification for distributed parameter systems using barrier functionals. Syst Control Lett 108:33–39
4. Ahmadi M, Wu B, Lin H, Topcu U (2018) Privacy verification in pomdps via barrier certificates. In: 2018 IEEE conference on decision and control (CDC). IEEE, pp 5610–5615
5. Alur R (2011) Formal verification of hybrid systems. In: 2011 Proceedings of the international conference on embedded software (EMSOFT). IEEE, pp 273–278
6. Alur R, Dill DL (1994) A theory of timed automata. Theor Comput Sci 126(2):183–235
7. Baier C, Katoen J (2008) Principles of model checking. MIT Press
8. Bharadwaj S, Dimitrova R, Topcu U (2018) Synthesis of surveillance strategies via belief abstraction. In: 2018 IEEE conference on decision and control (CDC). IEEE, pp 4159–4166
9. Chan H, Perrig A (2003) Security and privacy in sensor networks. Computer 36(10):103–105
10. Coogan S, Arcak M (2015) Efficient finite abstraction of mixed monotone systems. In: Proceedings of the 18th international conference on hybrid systems: computation and control. ACM, pp 58–67
11. Guéguen H, Lefebvre M, Zaytoon J, Nasri O (2009) Safety verification and reachability analysis for hybrid systems. Ann Rev Control 33(1):25–36
12. Han S, Topcu U, Pappas GJ (2015) A sublinear algorithm for barrier-certificate-based data-driven model validation of dynamical systems. In: 2015 54th IEEE conference on decision and control (CDC), pp 2049–2054
13. Hauskrecht M (2000) Value-function approximations for partially observable Markov decision processes. J Artif Intell Res 13(1):33–94
14. Hubaux JP, Capkun S, Luo J (2004) The security and privacy of smart vehicles. IEEE Secur Priv 2(3):49–55

15. Jacob R, Lesage JJ, Faure JM (2016) Overview of discrete event systems opacity: models, validation, and quantification. Ann Rev Control 41:135–146
16. Ji Y, Wu YC, Lafortune S (2018) Enforcement of opacity by public and private insertion functions. Automatica 93:369–378
17. Ji Y, Yin X, Lafortune S (2019) Opacity enforcement using nondeterministic publicly-known edit functions. IEEE Trans Autom Control
18. Kaelbling LP, Littman ML, Cassandra AR (1998) Planning and acting in partially observable stochastic domains. Artif Intell 101(1):99–134
19. Lin F (2011) Opacity of discrete event systems and its applications. Automatica 47(3):496–503
20. Löfberg J (2004) Yalmip: A toolbox for modeling and optimization in MATLAB. In: Proceedings of the CACSD conference. Taipei, Taiwan. http://control.ee.ethz.ch/~joloef/yalmip.php
21. Mazaré L (2004) Using unification for opacity properties. In: Proceedings of the 4th IFIP WG1, vol 7, pp 165–176
22. McDaniel P, McLaughlin S (2009) Security and privacy challenges in the smart grid. IEEE Secur Priv 7(3)
23. Papachristodoulou A, Prajna S (2005) A tutorial on sum of squares techniques for systems analysis. In: American control conference, 2005. Proceedings of the 2005, pp 2686–2700
24. Prajna S (2006) Barrier certificates for nonlinear model validation. Automatica 42(1):117–126
25. Prajna S, Papachristodoulou A, Seiler P, Parrilo P (2013) SOSTOOLS: sum of squares optimization toolbox for MATLAB V3.00
26. Puterman ML (2014) Markov decision processes: discrete stochastic dynamic programming. Wiley
27. Saboori A, Hadjicostis CN (2012) Opacity-enforcing supervisory strategies via state estimator constructions. IEEE Trans Autom Control 57(5):1155–1165
28. Saboori A, Hadjicostis CN (2014) Current-state opacity formulations in probabilistic finite automata. IEEE Trans Autom Control 59(1):120–133
29. Sturm JF (1998) Using sedumi 1.02, a matlab toolbox for optimization over symmetric cones
30. Sun Z (2006) Switched linear systems: control and design. Springer Science & Business Media
31. Tomlin CJ, Mitchell I, Bayen AM, Oishi M (2003) Computational techniques for the verification of hybrid systems. Proc IEEE 91(7):986–1001
32. Wang L, Ames AD, Egerstedt M (2017) Safety barrier certificates for collisions-free multirobot systems. IEEE Trans Robot 33(3):661–674
33. Weber RH (2010) Internet of things-new security and privacy challenges. Comput Law Secur Rev 26(1):23–30
34. Wisniewski R, Sloth C (2016) Converse barrier certificate theorems. IEEE Trans Autom Control 61(5):1356–1361
35. Wu B, Dai J, Lin H (2018) Synthesis of insertion functions to enforce decentralized and joint opacity properties of discrete-event systems. In: 2018 annual American control conference (ACC). IEEE, pp 3026–3031
36. Wu B, Lin H (2018) Privacy verification and enforcement via belief abstraction. IEEE Control Syst Lett 2(4):815–820
37. Wu B, Liu Z, Lin H (2018) Parameter and insertion function co-synthesis for opacity enhancement in parametric stochastic discrete event systems. In: 2018 annual American control conference (ACC). IEEE, pp 3032–3037
38. Wu YC, Lafortune S (2014) Synthesis of insertion functions for enforcement of opacity security properties. Automatica 50(5):1336–1348
39. Wu YC, Raman V, Rawlings BC, Lafortune S, Seshia SA (2017) Synthesis of obfuscation policies to ensure privacy and utility. J Autom Reason 1–25

Chapter 6
Information-Theoretic Privacy Through Chaos Synchronization and Optimal Additive Noise

Carlos Murguia, Iman Shames, Farhad Farokhi and Dragan Nešić

Abstract We study the problem of maximizing privacy of data sets by adding random vectors generated via synchronized chaotics oscillators. In particular, we consider the setup where information about data sets, queries, is sent through public (unsecured) communication channels to a remote station. To hide private features (specific entries) within the data set, we corrupt the response to queries by adding random vectors. We send the distorted query (the sum of the requested query and the random vector) through the public channel. The distribution of the additive random vector is designed to minimize the mutual information (our privacy metric) between private entries of the data set and the distorted query. We cast the synthesis of this distribution as a convex program in the probabilities of the additive random vector. Once we have the optimal distribution, we propose an algorithm to generate pseudo-random realizations from this distribution using trajectories of a chaotic oscillator. At the other end of the channel, we have a second chaotic oscillator, which we use to generate realizations from the same distribution. Note that if we obtain the same realizations on both sides of the channel, we can simply subtract the realization from the distorted query to recover the requested query. To generate equal realizations, we need the two chaotic oscillators to be synchronized, i.e., we need them to generate exactly the same trajectories on both sides of the channel synchronously in time. We force the two chaotic oscillators into *exponential synchronization* using a driving signal. Simulations are presented to illustrate our results.

C. Murguia (✉) · I. Shames · F. Farokhi · D. Nešić
Department of Electrical and Electronic Engineering, University of Melbourne,
Melbourne, VIC, Australia
e-mail: carlos.murguia@unimelb.edu.au

I. Shames
e-mail: iman.shames@unimelb.edu.au

F. Farokhi
e-mail: farhad.farokhi@unimelb.edu.au; farhad.farokhi@csiro.au

D. Nešić
e-mail: dnesic@unimelb.edu.au

F. Farokhi
The Commonwealth Scientific and Industrial Research Organisation (CSIRO), Data61,
Canberra, Australia

© Springer Nature Singapore Pte Ltd. 2020
F. Farokhi (ed.), *Privacy in Dynamical Systems*,
https://doi.org/10.1007/978-981-15-0493-8_6

103

6.1   Introduction

In a hyperconnected world, scientific and technological advances have led to an overwhelming amount of user data being collected and processed by hundreds of companies over public networks. Companies mine this data to provide targeted advertising and personalized services. However, these new technologies have also led to an alarming widespread loss of privacy in society. Depending on adversaries' resources, opponents may infer private user information from public data available on the internet and unsecured/public servers. A motivating example of privacy loss is the potential use of data from smart electrical meters by criminals, advertising agencies, and governments, for monitoring the presence and activities of occupants [42, 51]. Other examples are privacy loss caused by information sharing in distributed control systems and cloud computing [23]; the use of travel data for traffic estimation in intelligent transportation systems [22]; and data collection and sharing by the Internet-of-Things (IoT) [54], which is, most of the time, done without the user's informed consent. These privacy concerns show that there is an acute need for privacy preserving mechanisms capable of handling the new privacy challenges induced by an interconnected world.

In this manuscript, we consider the problem of hiding private information X of users (modeled as discrete random vectors) within datasets when publicly sharing requested queries $Y(X)$ from the same source. In particular, the aim of our privacy scheme is to respond to queries with distorted queries of the form $Z = Y(X) + V$ such that, when releasing Z, the private X is "hidden". Realizations of the vector Z are transmitted over a public (unsecured) communication channel to a remote station. Then, if we do not distort $Y(X)$ before transmission, information about X is directly accessible through the public channel. The first problem that we address is the design of the probability distribution of V to maximize privacy, i.e., the distribution of V must be constructed so that $Z = Y(X) + V$ carries as little information about X as possible. Here, we follow an *information-theoretic approach* to privacy. We use the *mutual information* between private information X and distorted queries $Y(X) + V$, $I[X; Y(X) + V]$, as *privacy metric*. The design of the discrete additive vector is casted as an optimization problem where we minimize $I[X; Y(X) + V]$ using the probability mass function of V, $p_V(v)$, as optimization variables. That is, the optimal distribution, $p_V^*(v)$, is given by $p_V^*(v) := \arg\min_{p_V(v)} I[X; Y(X) + V]$, where $p_V(v)$ is taken over a class of probability mass functions. Contrary to related work [9, 18, 21, 36, 48, 53], we do not consider any sort of privacy-distortion trade-off in our formulation. We actually aim at making $I[X; Y(X) + V]$ as small as possible regardless of the distortion between $Y(X)$ and $Y(X) + V$ induced by V. Distortion is not an issue because we seek to generate exactly the same realization of V at the remote station; then, we could recover the query by simply subtracting this realization from the one of $Z = Y(X) + V$. In order to accomplish this, we propose an algorithm to generate pseudorandom realizations from $p_V^*(v)$ at both sides of the channel using trajectories of two synchronized chaotic oscillators.

There are a number of requirements that the oscillators must satisfy for our algorithm to work: (1) trajectories of the oscillators must be *bounded* and *chaotic*; (2) they must be *synchronized*, i.e., we need them to generate exactly the same trajectories on both sides of the channel synchronously in time; and (3) the synchronous solution, regarded as a random process, must be *stationary*. Before giving the algorithm, we provide general guidelines for selecting the dynamics of the oscillators so that all the aforementioned requirements are satisfied. In particular, we use a range of well-known results in the literature to provide a synthesis procedure that allows to choose suitable oscillators. For boundedness, we use the notion of *Input-to-State-Stability* (ISS); for chaos, we employ standard *largest Lyapunov exponent methods* [55] and the (0–1) *test* [19]; for synchronization, we introduce the notion of *convergent systems* [38]; and for stationarity, we use *hyperbolicity* of the chaotic trajectories [4].

To generate equal realizations, our algorithm needs trajectories of the two chaotic oscillators (one at each side of the channel) to be synchronized. We force the oscillators into *exponential synchronization* using a driving signal. Exponential synchronization implies that trajectories of the oscillators converge to each other exponentially for all admissible initial conditions and are perfectly synchronized in the limit only. Therefore, in finite time, there is always a "small" difference between their trajectories. However, because oscillators synchronize exponentially fast, and it is often possible in practice to select initial conditions from a known compact set (known to both sides of the channel), it is safe to assume that the interconnected systems have been operating for sufficiently large time such that oscillators are *practically* synchronized, i.e., the synchronization error is so small that trajectories can be assumed to be equal. This is a standard assumption that is made in most, if not all, of the existing work on chaotic encryption based on synchronization [1, 20, 27, 34, 57]. Here, we give sufficient conditions for exponential synchronization to occur, provide tools for selecting the oscillators such that these conditions are satisfied, and assume that, after transients have settled down, trajectories are perfectly synchronized to some chaotic trajectory, say $\phi(t) \in \mathbb{R}^{n_\varsigma}$, $\varsigma \in \mathbb{N}$. If $n_\varsigma > 1$, our algorithm uses any entry $\phi^s(t) \in \mathcal{S} \subset \mathbb{R}$ of $\phi(t)$ to generate realizations from $p_V^*(v)$, where $\mathcal{S}$ denotes some compact set that characterizes the support of $\phi^s(t)$. Because oscillators are selected such that $\phi(t)$, regarded as a random process, is stationary, samples from $\phi^s(t)$ follow a stationary probability density function. We obtain this density through Monte Carlo simulations [43] and divide its support $\mathcal{S}$ into a finite set of cells $C = \{c^1, \ldots, c^M\}$ such that the probability that $\phi^s(t)$ lies in these cells equals the optimal probability distribution $p_V^*(v)$. That is, we generate pseudorandom realizations from $p_V^*(v)$ by properly selecting C and evaluating if $\phi^s(t)$ lies in C at the sampling instants.

The use of additive noise to preserve privacy is common practice. There are mainly two classes of privacy metrics considered in the literature; namely, differential privacy [13, 37] and information-theoretic metrics, e.g., mutual information, conditional entropy, Kullback-Leibler divergence, and Fisher information [15–17, 45, 51]. In differential privacy, because it provides certain privacy guarantees, Laplace noise is usually used [14]. However, when maximal privacy is desired, Laplace noise is generally not the optimal solution. This raises the fundamental question: what is the noise distribution achieving maximal privacy? This question has many possible answers depending on the particular privacy metric being considered and the system configuration, see, e.g., [18, 21, 48, 53], for differential privacy based results, and [15–17, 45, 51], for information theoretic results. In general, if the data to be kept private follows continuous distributions, the problem of finding the optimal additive noise to maximize privacy is hard to solve. If a close-form solution for the distribution is desired, the problem amounts to solving a set of nonlinear partial differential equations which, in general, might not have a solution, and even if they do have a solution, it is hard to find [16]. This problem has been addressed by imposing some particular structure on the considered distributions or assuming the data to be kept private is deterministic [16, 18, 48]. The authors in [18, 48] consider deterministic input data sets and treat optimal distributions as distributions that concentrate probability around zero as much as possible while ensuring differential privacy. Under this framework, they obtain a family of piecewise constant probability density functions that achieve minimal distortion for a given level of privacy. In [16], the authors consider the problem of preserving the privacy of deterministic databases using additive continuous noise with constrained support. They use the Fisher information and the Cramer-Rao bound to construct a privacy metric between deterministic data and the one with the additive noise, and find the probability density function that minimizes it. Moreover, they prove that, in the unconstrained support case, the optimal noise distribution minimizing the Fisher information is Gaussian. This observation has been also made in [2] when using mutual information as a measure of privacy. We remark that most of the aforementioned papers consider privacy-distortion trade-offs when designing their distorting mechanisms. We do not consider this trade-off here because, at the end of the channel, we remove the distortion that we induce using our synchronization based formulation.

Existing work on chaotic encryption based on synchronization [1, 20, 27, 34, 57] directly uses the states of the chaotic oscillators to mask private information. That is, standard algorithms do not use chaotic trajectories to generate pseudorandom realization from probability distributions (as we do here); instead, they simply add the value of the sampled chaotic trajectory (or functions of it) to private messages. Although the latter succeeds in masking messages, it does not give any privacy guarantees (neither information-theoretic nor in a differential privacy sense) on the private information, and it is not optimal in any sense. Hence, the contributions of our scheme with respect to existing work on chaotic encryption [1, 20, 27, 34, 57] are the treatment of fully stochastic datasets, the information-theoretic privacy guarantees that our framework provides, and the optimal performance of the designed distorting additive vector (optimal in the sense of minimizing the mutual information $I[X; Y(X) + V]$). The work here is inspired by the experimental results presented in [25], where the authors propose a framework similar to ours for deterministic data using a electronic circuit implementation of the Mackey-Glass chaotic oscillator [35]. The contribution of our work with respect to [25] is threefold: (1) we consider fully stochastic data, which makes the privacy scheme fundamentally very different; (2) we provide a general formulation that encompasses a large class of chaotic systems, not only the electronic circuit implementation of the Mackey-Glass oscillator; and (3) we generate realizations from optimal distorting distributions, in [25], they consider uniform distributions only which is not optimal for stochastic data.

Next, we summarize the main contributions of the chapter.

Contributions:

(1) We provide a general information-theoretic privacy framework based on optimal additive distorting random vectors and synchronization of chaotic oscillators; (2) We prove that the synthesis of the probability mass function $p_V(v)$ of the distorting random vector V can be posed as a convex program in $p_V(v)$ over a class of probability mass functions; (3) We provide an algorithm to generate pseudorandom realizations from this distribution using trajectories of chaotic oscillators; (4) Using off-the-shelf results in the literature, we provide a synthesis procedure for selecting the dynamics of the oscillators so that our algorithm is guaranteed to work.

The remainder of the paper is organized as follows. In Sect. 6.2, we present some preliminaries results needed for the subsequent sections. We introduce the notion of convergent systems and the concept of mutual information. The general formulation and the specific problems to be addressed are given in Sect. 6.3. In Sect. 6.4, we pose the synthesis of the probability distribution of the optimal distorting vector. General guidelines for selecting the chaotic oscillators are given in Sect. 6.5. The algorithm for generating pseudorandom realizations from the optimal distribution is presented in Sect. 6.6. Simulation results are given in Sect. 6.7 and concluding remarks in Sect. 6.8.

6.2 Notation and Preliminaries

The symbol $\mathbb{R}$ stands for the real numbers, $\mathbb{R}_{>0}(\mathbb{R}_{\geq0})$ denotes the set of positive (non-negative) real numbers. The symbol $\mathbb{N}$ stands for the set of natural numbers. The Euclidian norm in $\mathbb{R}^n$ is denoted simply as $|\cdot|$, $|x|^2 = x^\top x$, where $\top$ defines transposition. For a given measurable function $u(t), t \in \mathbb{R}_{\geq0}$, we denote its $\mathcal{L}_\infty$ norm as $||u||_\infty := \mathrm{ess\,sup}_{t\geq0}|u(t)|$, where ess sup denotes essential supremum. Matrices composed of only ones and only zeros of dimension $n \times m$ are denoted by $\mathbf{1}_{n\times m}$ and $\mathbf{0}_{n\times m}$, respectively, or simply $\mathbf{1}$ and $\mathbf{0}$ when their dimensions are clear. For square matrices $A \in \mathbb{R}^{n\times n}$, $\rho[A]$ denotes the spectral radius of A. A continuous function $\gamma : [0, a) \to [0, \infty)$ is said to belong to class $\mathcal{K}$ if it strictly increasing and $\gamma(0) = 0$. Similarly, a continuous function $\beta : [0, a) \times [0, \infty) \to [0, \infty)$ belongs to class $\mathcal{KL}$ if, for fixed s, $\beta(r, s)$ belongs to class $\mathcal{K}$ with respect to r and, for fixed r, $\beta(r, s)$ is decreasing with respect to s and $\lim_{s\to\infty} \beta(r, s) = 0$. Consider a discrete random vector X with alphabet $\mathcal{X} = \{x_1, \ldots, x_N\}$, $x_i \in \mathbb{R}^m$, $m \in \mathbb{N}$, $i \in \{1, \ldots, N\}$, and probability mass function $p_X(x) = \Pr[X = x]$, $x \in \mathcal{X}$, where $\Pr[B]$ denotes probability of event B. Similarly, for two random vectors X and Y, taking values in the alphabets $\mathcal{X}$ and $\mathcal{Y}$, respectively, their joint probability mass function is denoted by $p_{X,Y}(x, y)$, the marginal distribution of X is given by $p_X(x) = \sum_{y\in\mathcal{Y}} p_{X,Y}(x, y)$, and the conditional distribution of X given Y as $p_{Y|X}(y|x) = p_{X,Y}(x, y)/p_X(x)$.

Analogously, for a continuous random vector Y, we denote their (multivariate) probability density function as $f_Y(y)$. The notation $X \sim f_X(x)$ $(X \sim p_X(x))$ stands for continuous (discrete) random vectors X following the probability density (mass) function $f_X(x)$ $(p_X(x))$. We denote by "Simplex" the probability simplex defined by $\sum_{x \in \mathcal{X}} p_X(x) = 1$, $p_X(x) \geq 0$ for all $x \in \mathcal{X}$. The notation $E[a]$ denotes the expected value of the random vector a. We denote independence between two random vectors, X and Y, as $X \perp\!\!\!\perp Y$.

6.2.1 Mutual Information

Definition 6.1 Consider two random vectors, X and Y, with joint probability mass function $p_{X,Y}(x, y)$ and marginal probability mass functions, $p_X(x)$ and $p_Y(y)$, respectively. Their mutual information $I[X; Y]$ is defined as the relative entropy between the joint distribution and the product distribution $p_X(x) p_Y(y)$, i.e.,

$$I[X; Y] := \sum_{x \in \mathcal{X}} \sum_{y \in \mathcal{Y}} p_{X,Y}(x, y) \log \frac{p_{X,Y}(x, y)}{p_X(x) p_Y(y)}.$$

Mutual information $I[X; Y]$ between two jointly distributed vectors, X and Y, is a measure of the dependence between X and Y.

6.2.2 Convergent Systems

Consider the dynamical system:

$$\dot{x}(t) = r(x(t), u(t)), \tag{6.1}$$

with $t \in \mathbb{R}_{\geq 0}$, state $x \in \mathbb{R}^n$, input $u \in \mathcal{U} \subseteq \mathbb{R}^m$, and vector field $r : \mathbb{R}^n \times \mathcal{U} \to \mathbb{R}^n$. The vector field $r(x, u)$ is continuously differentiable in x, and $u(t)$ is piecewise continuous in t and takes values in some compact set $\mathcal{U} \subseteq \mathbb{R}^m$.

Definition 6.2 [12] System (6.1) is said to be globally asymptotically convergent if and only if for any bounded input $u(t)$, $t \in \mathbb{R}$, there is a unique bounded globally asymptotically stable solution $\bar{x}_u(t)$, $t \in \mathbb{R}$, such that $\lim_{t \to \infty} |x(t) - \bar{x}_u(t)| = 0$ for all initial conditions.

For a *convergent system*, the limit solution is solely determined by the external excitation $u(t)$ and not by the initial conditions. A sufficient condition for convergence obtained by Demidovich [12] and later extended in [38] is presented in the following proposition.

Proposition 6.1 [12, 38] *If there exists a positive definite matrix $P \in \mathbb{R}^{n \times n}$ such that all the eigenvalues $\lambda_i(Q)$ of the symmetric matrix*

$$Q(x, u) = \frac{1}{2} \left(P \left(\frac{\partial r}{\partial x}(x, u) \right) + \left(\frac{\partial r}{\partial x}(x, u) \right)^T P \right), \tag{6.2}$$

are negative and separated from zero, i.e., there exists a constant $c \in \mathbb{R}_{>0}$ such that $\lambda_i(Q) \leq -c < 0$, for all $i \in \{1, \ldots, n\}$, $u \in \mathcal{U}$, and $x \in \mathbb{R}^n$, then system (6.1) is globally exponentially convergent; and, for any pair of solutions $x_1(t), x_2(t) \in \mathbb{R}^n$ of (6.1), the following is satisfied:

$$\frac{d}{dt} \left(\left(x_1(t) - x_2(t) \right)^T P \left(x_1(t) - x_2(t) \right) \right) \leq -\alpha \left| x_1(t) - x_2(t) \right|^2, \ t \in \mathbb{R}_{\geq 0},$$

with constant $\alpha := (c/\lambda_{\max}(P))$ and $\lambda_{\max}(P)$ being the largest eigenvalue of the symmetric matrix P.

Remark 6.1 There are other methods to verify that trajectories of system (6.1) converge to a limit solution that is independent of the initial conditions and solely determined by the external excitation $u(t)$. For instance, contraction theory [33], Lyapunov function approach to incremental stability [3], the quadratic (QUAD) inequality approach (a Lipschitz-like condition) [32], and differential dissipativity [46], which are all concepts that are closely related to notion of convergent systems [38] that we use here.

6.3 Problem Setup

Let X be a discrete random vector that must be kept private. The alphabet and probability mass function of X are denoted as $\mathcal{X} = \{x_1, \ldots, x_N\}$, $x_i \in \mathbb{R}^{n_x}$, $n_x \in \mathbb{N}$, $i \in \{1, \ldots, N\}$ and $p_X(x) = \Pr[X = x]$, $x \in \mathcal{X}$, respectively. The n_x entries of X represent, for instance, private entries of n_x users within a dataset that is stored by a trusted server. The server admits queries of the form $Y = q(X)$, $Y \in \mathbb{R}^{n_y}$, for some (stochastic or deterministic) mapping $q : \mathbb{R}^{n_x} \to \mathbb{R}^{n_y}$ characterized by the transition probabilities $p_{Y|X}(y|x)$, $x \in \mathcal{X}$, $y \in \mathcal{Y}$, where $\mathcal{Y} = \{y_1, \ldots, y_M\}$, $y_i \in \mathbb{R}^{n_y}$, $n_y \in \mathbb{N}$. The aim of our privacy scheme is to respond to queries of the form $q(X)$ with distorted queries $Z = q(X) + V$, for some discrete random vector V (with $V \perp\!\!\!\perp Y$), such that, when releasing Z, the individual entries of X are "hidden". Realizations of the vector Z are transmitted over a public (unsecured) communication channel to a remote station, see Fig. 6.1. Then, if we do not add V to $q(X)$ before transmission, information about X is directly accessible through the public channel. As a preliminary problem that we need to solve for the subsequent results, we address the design of the probability distribution of V to maximize privacy, i.e., the distribution of V must be constructed so that the sum, $Z = q(X) + V$, carries as little information

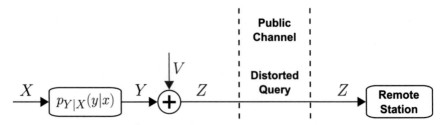

Fig. 6.1 Configuration for Problem 6.1

about X as possible. In this manuscript, we use the *mutual information* between X and $Z = Y + V$, $I[X; Z]$, as *privacy metric*. We aim at finding the probability mass function of V, $p_V(v)$, that minimizes $I[X; Z]$ over a class of probability mass functions. That is, we cast the design of $p_V(v)$ as an optimization problem with cost function $I[X; Z]$, optimization variables $p_V(v)$, and subject to $V \perp\!\!\!\perp Y$ and the usual probability simplex constraints. Note that, contrary to related work [9, 36, 45, 51, 53], we do not consider any sort of privacy-distortion trade-off in our formulation. We minimize $I[X; Y + V]$ regardless of the distortion between Y and $Y + V$ induced by V. Distortion is not an issue because, we seek to generate exactly the same realization of V at the remote station and then recover the query by subtracting this realization from the one of $Z = Y + V$. This is addressed in Problem 6.2 and Problem 6.3 below.

We let V be a discrete random vector with alphabet $\mathcal{Y}$ and probability mass function $p_V(v) = \Pr[V = v]$, $v \in \mathcal{Y}$, i.e., the alphabet of V and the one of the query $Y = q(X)$ are equal. Having equal alphabets imposes a tractable convex structure on the cost $I[X; Z]$ and reduces the optimization variables to the probabilities of each element of the alphabet. The case with arbitrary alphabet leads to a combinatorial optimization problem where the objective changes its structure for different combinations. We do not address this case in this manuscript; it is left as a future work. In what follows, we formally present the optimization problem we seek to address.

Problem 6.1 (*Optimal Distribution of the Additive Distorting Signal*) For given $p_X(x) = \Pr[X = x]$ and $p_{Y|X}(y|x) = \Pr[Y = y|X = x]$, $x \in \mathcal{X}$, $y \in \mathcal{Y}$, find the probability mass function $p_V(v) = \Pr[V = v]$, $v \in \mathcal{Y}$ solution of the problem:

$$\begin{cases} p_V^*(v) := \arg\min_{p_V(v)} I[X; V + Y], \\ s.t. \ V \perp\!\!\!\perp Y \ and \ p_V(v) \in Simplex. \end{cases} \tag{6.3}$$

Here, $p_V^*(v)$ denotes the optimal distribution solution to (6.3). To hide X, once we have obtained $p_V^*(v)$, we aim at generating realizations $v \in \mathcal{Y}$ from this distribution, add them to the required query ($Y = q(X)$), and send realizations of the sum $Z = Y + V$ to the remote station through the public channel. At the other end of the

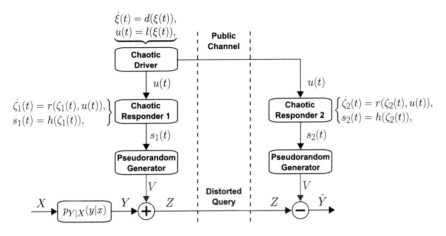

Fig. 6.2 Complete system configuration

channel, we seek to *generate the exact same realizations from* $p_V^*(v)$ so that we can recover the query by simply subtracting V from Z, see Fig. 6.2. Note that, in Fig. 6.2, we have a recovered $\hat{Y}$ at the remote station rather that the actual Y. This is because we want to remark that, due to practical errors in our algorithm–e.g., due to communication delays and transients–realizations of V that we generate at both ends of the channel might be slightly different in practice. To generate these realizations, we use trajectories, $\phi_{u,1}^\zeta(t, \zeta_1(0), u(t))$, $t \in \mathbb{R}_{\geq 0}$, $\zeta_1(0) \in \mathbb{R}^{n_\zeta}$, $u(t) \in \mathbb{R}^{n_u}$, of a chaotic dynamical system of the form:

$$\begin{cases} \dot{\zeta}_1(t) = r(\zeta_1(t), u(t)), \\ s_1(t) = h(\zeta_1(t)), \end{cases} \tag{6.4}$$

with state $\zeta_1(t) \in \mathbb{R}^{n_\zeta}$, output $s_1(t) \in \mathbb{R}$, continuous in t input $u(t) \in \mathcal{U} \subset \mathbb{R}^{n_u}$ taking values in some compact set $\mathcal{U}$, continuous function $h : \mathbb{R}^{n_\zeta} \to \mathbb{R}$, and vector field $r : \mathbb{R}^{n_\zeta} \times \mathcal{U} \to \mathbb{R}^{n_\zeta}$ continuously differentiable in its first argument, uniformly in its second argument. Hereafter, system (6.4) is referred to as *responder* 1. Responder 1 is placed at the side of the trusted server, see Fig. 6.2. The input signal $u(t)$ is generated by a chaotic autonomous exosystem:

$$\begin{cases} \dot{\xi}(t) = d(\xi(t)), \\ u(t) = l(\xi(t)), \end{cases} \tag{6.5}$$

with state $\xi(t) \in \mathbb{R}^{n_\xi}$, output $u(t) \in \mathcal{U} \subset \mathbb{R}^{n_u}$, and vector fields $d : \mathbb{R}^{n_\xi} \to \mathbb{R}^{n_\xi}$ and $l : \mathbb{R}^{n_\xi} \to \mathbb{R}^{n_u}$. The vector field $d(\xi)$ is locally Lipschitz in ξ and $l(\xi)$ is continuous. We refer to (6.5) as the *driver system*. We let $u(t)$ be connected to the remote station

via the public channel, see Fig. 6.2. At the other end of the channel, driven by the
same input signal $u(t)$, we have a third chaotic oscillator with the same dynamics
as (6.4) but with potentially different initial conditions, i.e., the second oscillator is
given by

$$
\begin{cases}
\dot{\zeta}_2(t) = r(\zeta_2(t), u(t)), \\
s_2(t) = h(\zeta_2(t)),
\end{cases}
\tag{6.6}
$$

with state $\zeta_2(t) \in \mathbb{R}^{n_\zeta}$ and output $s_2(t) \in \mathbb{R}$. We denote trajectories of (6.6) as
$\phi^\zeta_{u,2}(t, \zeta_2(0), u(t))$ with $t \in \mathbb{R}_{\geq 0}$, $\zeta_2(0) \in \mathbb{R}^{n_\zeta}$, and $u(t) \in \mathcal{U} \subset \mathbb{R}^{n_\zeta}$. System (6.6)
is referred to as *responder* 2. Note that if $\zeta_1(t) = \zeta_2(t)$, $t \in \mathbb{R}_{\geq 0}$, i.e., if systems
(6.4) and (6.6) are synchronized, and we use the synchronous chaotic solution, say
$\phi^\zeta_u(t, u(t))$, to generate realizations from $p^*_V(v)$, we could have the same realization
of V at both sides of the channel.

Problem 6.2 (*Boundedness, Chaos, and Synchronization*) State sufficient condi-
tions on the vector fields $r(\cdot)$, $h(\cdot)$, $d(\cdot)$, and $l(\cdot)$ of the coupled system (6.4)–(6.6)
such that: *(1)* trajectories of (6.4)–(6.6) exist and are bounded and chaotic; and *(2)*
systems (6.4) and (6.6) exponentially synchronize, i.e., $\lim_{t\to\infty} |\zeta_1(t) - \zeta_2(t)| = 0$,
exponentially fast.

Remark 6.2 Problem 6.2 seeks to enforce exponential synchronization by selecting
the dynamics of the oscillators. Exponential synchronization implies that trajectories
of the responders converge to each other exponentially for all initial conditions and are
perfectly synchronized in the limit only. It follows that, in finite time, there is always
a "small" difference between their trajectories. Nevertheless, because oscillators
synchronize exponentially fast, and it is often possible in practice to select initial
conditions from a known compact set (known to both the trusted server and the remote
station), it is safe to assume that the interconnected systems have been operating for
sufficiently large time such that oscillators are practically synchronized, i.e., the
synchronization error is so small that trajectories can be assumed to be equal. This is
a standard assumption that is made in most, if not all, of the existing work on chaotic
encryption based on synchronization [1, 20, 27, 34, 57].

Finally, once we have found functions solution to Problem 6.2, which guarantees
exponential synchronization of the responders, and assuming that responders are
synchronized (see Remark 6.2), we aim at designing a procedure to generate pseudorandom realizations from $p^*_V(v)$ using the synchronous chaotic solution $\phi^\zeta_u(t, u(t))$.
Note that $\zeta_1(t) = \zeta_2(t) \Rightarrow s_1(t) = h(\zeta_1(t)) = s_2(t) = h(\zeta_2(t))$, for all $t \geq 0$. Moreover, because $\zeta_1(t) = \zeta_2(t) = \phi^\zeta_u(t, u(t))$; then, $s_1(t) = s_2(t) = h(\phi^\zeta_u(t, u(t))) =:$
$\phi^s_u(t, u(t)) \in \mathcal{S} \subset \mathbb{R}$ for some compact set $\mathcal{S}$. To reduce the complexity of the algorithm, we use the lower dimensional synchronous solution $\phi^s_u(t, u(t))$ to generate
the realizations from $p^*_V(v)$.

Problem 6.3 (*Generation of Optimal Pseudorandom Numbers*) Using the lower
dimensional synchronous solution, $\phi^s_u(t, u(t))$, design an algorithm to generate pseudorandom realizations from the optimal distribution $p^*_V(v)$, $v \in \mathcal{Y}$.

6.4   Optimal Distribution of the Additive Distorting Signal

In this section, we prove that Problem 6.1 can be posed as a convex program in the probabilities $p_V(v)$, $v \in \mathcal{Y}$. We derive an explicit expression for the cost function $I[X; Z]$, $Z = Y + V$, in terms of the given $p_X(x)$ and $p_{Y|X}(y|x)$ and the variables $p_V(v)$, restricted to satisfy the independence constraint $V \perp\!\!\!\perp Y$.

Lemma 6.1 $I[X; Z]$ *with* $Z = Y + V$, $V \perp\!\!\!\perp Y$, *is a convex function of* $p_V(v)$, $v \in \mathcal{Y}$, *for given* $p_X(x)$ *and* $p_{Y|X}(y|x)$, $x \in \mathcal{X}$, $y \in \mathcal{Y}$; *and can be written compactly in terms of* $p_X(x)$, $p_{Y|X}(y|x)$, *and* $p_V(v)$, *as follows:*

$$
\begin{cases}
I[X; Z] = \displaystyle\sum_{x \in \mathcal{X}} \sum_{z \in \mathcal{Z}} p_X(x) p_{Z|X}(z|x) \log \frac{p_{Z|X}(z|x)}{p_Z(z)}, & \text{(6.7a)} \\[2ex]
p_{Z|X}(z|x) = \displaystyle\sum_{y \in \mathcal{Y}} p_{Y|X}(y|x) p_V(z - y), & \text{(6.7b)} \\[2ex]
p_Z(z) = \displaystyle\sum_{y \in \mathcal{Y}} p_Y(y) p_V(z - y). & \text{(6.7c)}
\end{cases}
$$

Proof   The expression on the right-hand side of (6.7a) follows by inspection of Definition 6.1 and the fact that $p_{Z,X}(z, x) = p_X(x) p_{Z|X}(z|x)$. By [11, Theorem 2.7.4], cost (6.7a) is convex in $p_{Z|X}(z|x)$ for given $p_X(x)$. However, our optimization variables are $p_V(v)$ and not $p_{Z|X}(z|x)$. Note that X, Y, and Z form a Markov chain in that order [44]; therefore, $p_{X,Y,Z}(x, y, z) = p_X(x) p_{Y|X}(y|x) p_{Z|Y}(z|y)$. Marginalizing $p_{X,Y,Z}(x, y, z)$ with respect to $Y \in \mathcal{Y}$ and then conditioning with respect to X yields $p_{X,Z}(x, z) = \sum_{y \in \mathcal{Y}} p_X(x) p_{Y|X}(y|x) p_{Z|Y}(z|y)$ and $p_{Z|X}(z|x) = \sum_{y \in \mathcal{Y}} p_{Y|X}(y|x) p_{Z|Y}(z|y)$, respectively. Note that $p_{Z|X}(z|x)$ is just a linear transformation of $p_{Z|Y}(z|y)$. Hence, convexity with respect to $p_{Z|X}(z|x)$ implies convexity with respect to $p_{Z|Y}(z|y)$ because convexity is preserved under affine transformations [8]. Next, consider $p_{Z|Y}(z|y) = p_{Z,Y}(z, y)/p_Y(y)$. By definition, $p_{Z,Y}(z, y) = \Pr[Z = z, Y = y]$, $z \in \mathcal{Z}$, $y \in \mathcal{Y}$. Note that

$$
\Pr[Z = z, Y = y] = \Pr[Y + V = z, Y = y] = \Pr[V = z - y, Y = y]
$$
$$
\overset{(a)}{=} \Pr[V = z - y] \Pr[Y = y] = p_V(z - y) p_Y(y),
$$

where (a) follows from independence between V and Y. Thus,

$$
\begin{aligned}
p_{Z|Y}(z|y) &= \frac{p_{Z,Y}(z, y)}{p_Y(y)} \\
&= \frac{p_V(z - y) p_Y(y)}{p_Y(y)} = p_V(z - y).
\end{aligned}
$$

We have concluded convexity of $I[X; Z]$ with respect to $p_{Z|Y}(z|y)$ above. Hence, because $p_{Z|Y}(z|y) = p_V(z - y)$ and $p_V(z - y)$ is a linear transformation of $p_V(v)$ $(p_V(z - y) = p_V(v)$ for $z - y = v$ and zero otherwise), the cost $I[X; Z]$ is convex in $p_V(v)$. Moreover, since $p_{Z|X}(z|x) = \sum_{y \in \mathcal{Y}} p_{Y|X}(y|x)p_{Z|Y}(z|y)$ and $p_{Z|Y}(z|y) = p_V(z - y)$, equality (6.7b) holds true. It remains to prove that $p_Z(z)$ can be written as (6.7c). Because $Z = Y + V$, $p_Z(z) = \Pr[Z = z]$, for a given $z \in \mathcal{Z}$, can be written as the sum of the probabilities of all $Y = y$ and $V = v$ that result in z, i.e.,

$$
\begin{aligned}
p_Z(z) = \Pr[Z = z] &= \Pr[Y + V = z] \\
&= \sum_{y \in \mathcal{Y}} \Pr[V = z - y, Y = y] \\
&\overset{(b)}{=} \sum_{y \in \mathcal{Y}} \Pr[Y = y]\Pr[V = z - y] = \sum_{y \in \mathcal{Y}} p_Y(y)p_V(z - y),
\end{aligned}
$$

where (b) follows from independence between V and Y. ∎

By Lemma 6.1, the cost $I[X; Z]$, for $V \perp\!\!\!\perp Y$, is convex in $p(v)$ and parametrized by $p_X(x)$ and $p_{Y|X}(y|x)$. In what follows, we cast the nonlinear program for solving Problem 6.1.

Theorem 6.1 *Given $p_X(x)$ and $p_{Y|X}(y|x)$, $x \in \mathcal{X}$, $y \in \mathcal{Y}$, the mapping $p_V(v)$, $v \in \mathcal{Y}$, that minimizes $I[X; Z]$, $Z = V + Y$, subject to $V \perp\!\!\!\perp Y$ can be found by solving the following convex program:*

$$
\left\{
\begin{aligned}
p_V^*(v) = \arg\min_{p_V(v)} &\sum_{x \in \mathcal{X}} \sum_{y \in \mathcal{Y}} \sum_{z \in \mathcal{Z}} p_X(x)p_{Y|X}(y|x)p_V(z - y) \\
&\times \log \frac{\sum_{y \in \mathcal{Y}} p_{Y|X}(y|x)p_V(z - y)}{\sum_{y \in \mathcal{Y}} p_Y(y)p_V(z - y)}, \\
\text{s.t. } p_V(v) \in \text{ Simplex.}
\end{aligned}
\right.
$$

$$(6.8)$$

Proof: Theorem 1 follows from Lemma 6.1.

6.5 Boundedness, Chaos, and Synchronization

6.5.1 Existence, Uniqueness, and Boundedness of Solutions

We start addressing existence, uniqueness, and boundedness of the solutions of the coupled systems (6.4)–(6.6). To be able to use synchronous solutions to generate realizations from $p_V^*(v)$, we first need these solutions to exist and be bounded. In the system description given above, we have assumed that $r(\zeta, u(t))$ is continuously differentiable in ζ uniformly in $u(t)$, $u(t)$ is continuous in t, and $d(\xi)$

is locally Lipschitz. These alone imply uniqueness and existence of solutions of (6.4)–(6.6) over some finite time interval $t \in [0, \tau]$, $\tau \in \mathbb{R}_{>0}$, [26, Theorem 2.2]. To conclude the latter for arbitrarily large τ, besides the locally Lipschitz assumption on the functions, we need boundedness of the solutions of (6.4)–(6.6) [26, Theorem 2.4]. Note that the coupled systems (6.4)–(6.6) have a cascade structure. The driver dynamics is independent of the responders states, and its output, $u(t)$, is the input of the responders. Then, an approach to conclude boundedness of the overall system is to conclude boundedness of the driver first, and then boundedness of the responders when driven by bounded inputs. In what follows, we formally introduce the notion of boundedness that we use here.

Definition 6.3 [26] The solutions of (6.5) are bounded for a bounded set of initial conditions if there exists a positive constant c, independent of the initial time instant, and for every $a \in (0, c)$, there is $b = b(a) > 0$, independent of the initial time instant, such that $|\xi(0)| \le a \Rightarrow |\xi(t)| \le b$, $\forall\, t \ge 0$. If the latter holds for arbitrarily large a; then, the solutions of (6.5) are globally bounded.

Remark 6.3 Because $l(\xi)$ is continuous, by the extreme value theorem, boundedness of $\xi(t)$ implies boundedness of $u(t) = l(\xi(t))$.

Remark 6.4 We do not give conditions for boundedness of the solutions of (6.5). It is assumed that the vector field $d(\xi)$ is such that the solutions of the driver are globally bounded. We refer the reader to, for instance, [26, Theorem 4.18], where sufficient conditions for boundedness are given in terms of Lyapunov-like results.

Next, for bounded solutions of the driver, we need the solutions of the responders to be bounded when driven by $u(t)$. To address this, we use the notion if Input-to-State-Stability (ISS) [47].

Definition 6.4 [47] System (6.4) (and thus system (6.5) as well) is said to be Input-to-State-Stable if there exist a class $\mathcal{KL}$ function $\beta(\cdot)$ and a class $\mathcal{K}$ function $\gamma(\cdot)$ such that for any initial condition $\zeta_1(0)$ and any bounded input $u(t)$, the solution $\zeta_1(t)$ exists for all $t \in \mathbb{R}_{\ge 0}$ and satisfies: $|\zeta_1(t)| \le \beta(\zeta_1(0), t) + \gamma(||u||_\infty)$.

Remark 6.5 ISS of the responders with respect to $u(t)$ guarantees that, for any bounded $u(t)$, the states $\zeta_1(t)$ and $\zeta_2(t)$ are bounded. Moreover, as t increases, $|\zeta_1(t)|$ and $|\zeta_2(t)|$ are ultimately bounded [26] by $\gamma(||u||_\infty)$, see [47] for further details.

Remark 6.6 Sufficient conditions for the responders to be ISS with input $u(t)$ are not provided here. We assume that the vector field $r(\zeta, u(t))$ is such that systems (6.4) and (6.5) are ISS with respect to $u(t)$. We refer the reader to, for instance, [26, Theorem 4.19], where sufficient conditions for ISS are given in terms of ISS-Lyapunov functions.

Remark 6.7 The weaker property of integral Input-to-State-Stable (iISS) [5] could be used to conclude boundedness of the responder's trajectories when driven by "sufficiently small" inputs. We refer the reader to [10], where sufficient conditions for iISS and related boundedness results are given.

6.5.2 Synchronization

Next, we give sufficient conditions on $r(\zeta, u(t))$ such that $\lim_{t \to \infty} |\zeta_1(t) - \zeta_2(t)| = 0$, i.e., the responders exponentially synchronize. We assume that solutions of the coupled systems (6.4)–(6.6) exist and are bounded, i.e., vector fields $r(\cdot)$, $d(\cdot)$, and $l(\cdot)$ satisfy the conditions stated in the previous subsection. Then, for bounded $u(t)$, a sufficient condition for the responders to exponentially synchronize is that systems (6.4) and (6.6) are *convergent systems* in the sense of Definition 6.2. The latter implies that, because both responders are driven by the input $u(t)$ and their dynamics are described by the same set of differential equations, trajectories of (6.4) and (6.6) converge to the same the limit solution, $\phi_u^\zeta(t, u(t))$, and this solution is solely determined by $u(t)$ and not by the initial conditions. In the following corollary of Proposition 6.1, we give a sufficient condition for the responders to be exponentially convergent (and thus to exponentially synchronize).

Corollary 6.1 *Consider the responders* (6.4) *and* (6.6). *If there exists a positive definite matrix* $P \in \mathbb{R}^{n_\zeta \times n_\zeta}$ *such that, for all* $u \in \mathbb{R}^{n_u}$ *and* $\zeta \in \mathbb{R}^{n_\zeta}$, *all the eigenvalues of the symmetric matrix:*

$$\frac{1}{2} \left(P \left(\frac{\partial r}{\partial \zeta}(\zeta, u) \right) + \left(\frac{\partial r}{\partial \zeta}(\zeta, u) \right)^T P \right), \tag{6.9}$$

are negative and separated from zero; then, responders (6.4) *and* (6.6) *are globally exponentially convergent, and thus* $\lim_{t \to \infty} |\zeta_1(t) - \zeta_2(t)| = 0$, *exponentially fast.*

Remark 6.8 If the driver's output $u(t)$ is to be sent over a network and quantization (or some sort of coding) is required, we would need to drive responders by the same quantized $u(t)$, say $u_Q(t)$, to achieve exponential synchronization. That is, if we quantize $u(t)$ to obtain $u_Q(t)$, and we drive both responders by $u_Q(t)$ (with, e.g., a Zero-Order-Hold (ZOH)), they would also exponentially synchronize. They would synchronize to a different trajectory than when driven by $u(t)$, but they would synchronize exponentially fast.

Besides the notion of convergent systems, there are other methods available in the literature that can be used to verify that trajectories of responders asymptotically synchronize to a limit solution that is independent of the initial conditions. See Remark 6.1 for details.

6.5.3 Chaotic Dynamics

There are mainly two branches of methods to identify chaotic dynamics; namely, standard largest Lyapunov exponent methods [55], and the more recent (0–1) test [19]. Both methods use trajectories (numerical or experimental) of the systems under study to decide whether they are chaotic or not. In general, there are no sufficient conditions directly on the differential equations (the vector fields $r(\cdot)$ and $d(\cdot)$) such that chaotic trajectories are guaranteed to occur. There are, however, many well known systems in the literature known to exhibit chaotic trajectories. For instance, the Lorenz system [50], Duffing [28] and van der Pol [41] oscillators, the Rössler [39] and Chua [56] systems, and neural oscillators [49] (e.g., the Hodgkin-Huxley, Morris-Lecar, Hindmarsh-Rose, and FitzHugh-Nagumo oscillators). We can use any of these chaotic systems (if they satisfy all the required extra conditions, see Sect. 6.5.4) as driver and then select a pair of responders with convergent dynamics. Indeed, we need to verify that the responders that we choose produce chaotic trajectories when driven by the chaotic driver. Moreover, to generate the pseudorandom realizations from $p_V^*(v)$ (this is addressed in the next section), we need the chaotic trajectories of the responders, regarded as a random process, to be *stationary*, i.e., after transients have settled down, trajectories must follow a stationary probability distribution [44] which is independent of the initial conditions. The latter is a strong condition that is not satisfied for all chaotic systems. The existence of stationary distributions for chaotic trajectories has been proven for hyperbolic and quasi-hyperbolic (also called singular-hyperbolic) chaotic systems [4]. The definition of (quasi) hyperbolic dynamical systems [4, 29] is technical and not needed for the subsequent results. It requires concepts from differential topology that we prefer to omit here for readability of the manuscript. It suffices to know that the chaotic system that we use for the driver must lead to stationary distributions of the responders. This can be tested numerically by Monte Carlo simulations [43]. Moreover, there are many well-known chaotic systems with (quasi) hyperbolic dynamics in the literature, e.g., the Lorenz and Chua systems [24], neural oscillators [6], the many predator-pray like systems given in [31, 52], and some mechanical nonlinear oscillators [30]. In the next subsection, we provide a synthesis procedure to choose the functions of the coupled systems (6.4)–(6.6) such that all the required conditions mentioned above are satisfied.

6.5.4 General Guidelines

Synthesis Procedure:
(1) Select a driver dynamics (6.5) (i.e., the vector field $d(\xi)$) known to be chaotic and (quasi) hyperbolic (e.g., systems in [6, 24, 30, 31, 52]).
(2) Verify that the corresponding $d(\xi)$ is locally Lipschitz and the trajectories of the driver are globally bounded, in the sense of Definition 6.3, using, e.g., [26, Theorem 4.18].
(3) In (6.5), let $\xi = (\xi^1, \ldots, \xi^{n_\xi})^\top \in \mathbb{R}^{n_\xi}$, $\xi^i \in \mathbb{R}$, and $u(t) = l(\xi(t)) = \xi^j(t)$, $i, j \in \{1, \ldots, n_\xi\}$, i.e., fix the output of the driver to be any state of (6.5). In doing this, we ensure that $u(t)$ is continuous, bounded, chaotic, and (quasi) hyperbolic.
(4) For the responders (6.4) and (6.6), select any continuously differentiable vector

field $r(\zeta, u)$ (with respect to ζ) leading to ISS dynamics, see Remark 6.6, and satisfying the conditions for convergence in Corollary 6.1, e.g., $r(\zeta, u) = A\zeta + \psi(u)$, for any matrix $A \in \mathbb{R}^{n_\zeta \times n_\zeta}$ with spectral radius $\rho[A] < 1$ and differentiable vector field $\psi : \mathbb{R}^{n_u} \to \mathbb{R}^{n_\zeta}$. Then, we ensure that the responders have bounded trajectories and exponentially synchronize.

(5) Verify that the trajectories of the responders, when driven by the chaotic driver, are chaotic (using Lyapunov exponents or the (0–1) test) and, after transients have settled down, lead to a stationary probability distribution independent of the initial conditions. See Sect. 6.5.3 for details.

(6) In (6.4) (and respectively in (6.6)), let $\zeta_1 = (\zeta_1^1, \ldots, \zeta_1^{n_\zeta})^\top \in \mathbb{R}^{n_\zeta}$, $\zeta_1^i \in \mathbb{R}$, and $s_1(t) = l(\zeta_1(t)) = \zeta_1^j(t)$, $i, j \in \{1, \ldots, n_\zeta\}$, i.e., fix the output of the responders to be any state of (6.4) and (6.6), respectively. Indeed, we need the same j for both responders, i.e., $s_1(t) = \zeta_1^j(t)$ and $s_2(t) = \zeta_2^j(t)$. In doing this, we ensure that $s_1(t)$ and $s_2(t)$ are continuous, bounded, chaotic, and lead to stationary probability distributions.

6.6 Generation of Optimal Pseudorandom Numbers

In this section, we assume that the driver and the responders dynamics have been designed following the general guidelines in Sect. 6.5.4. Then, for sufficiently large t, the chaotic trajectories of the responders are practically synchronized, i.e., for any finite $t^* \in \mathbb{R}_{>0}$, there is $\epsilon_{t^*} \in \mathbb{R}_{>0}$, such that $|s_1(t) - \phi_u^s(t, u(t))| \leq \epsilon_{t^*}$ and $|s_2(t) - \phi_u^s(t, u(t))| \leq \epsilon_{t^*}$, for all $t \geq t^*$, where $\phi_u^s(t, u(t)) \in \mathcal{S} \subset \mathbb{R}$ denotes the asymptotic synchronous solution for some compact set $\mathcal{S}$; and samples from $\phi_u^s(t, u(t))$ follow a stationary probability distribution. Here, we assume that the responders have been operating for sufficiently large time such that the synchronization error, $|s_1(t) - s_2(t)|$, is so small that trajectories of the responders can be assumed to be equal to $\phi_u^s(t, u(t))$ (see Remark 6.2), i.e., t^* is sufficiently large so that ϵ_{t^*} is practically zero. In Sect. 6.6.1, we quantify the worst-case distortion induced by assuming $s_1(t) = s_2(t) = \phi_u^s(t, u(t))$ in finite time. In particular, we give an upper bound on the mean squared error $E[|Y - \hat{Y}|^2]$, where $\hat{Y}$ denotes the estimate of realizations of Y using $s_1(t)$, $s_2(t)$, and the algorithm provided below. In the remainder of this section, we assume $s_1(t) = s_2(t) = \phi_u^s(t, u(t))$. Note that the sample space of $\phi_u^s(t, u(t))$, regarded as a random process, is some compact set $\mathcal{S} \subset \mathbb{R}$, i.e., the sample space is a subset of the real line and thus samples from $\phi_u^s(t, u(t))$ follow some stationary probability density function (pdf), say $f_S(s)$, for some virtual continuous random variable S. That is, for $s(t) := \phi_u^s(t, u(t))$, define the sampled sequence $s_k := s(t_k)$ for sampling time-instants $t_k \in \mathbb{R}_{>0}$, $t_k := \Delta k$, $k \in \mathbb{N}$, and sampling period $\Delta \in \mathbb{R}_{>0}$; then, $s_k \sim f(s)$ for all k. Because we know the dynamics (6.4)–(6.6), we can obtain $f_S(s)$ by Monte Carlo simulations [43]. If we know $f_S(s)$, we can always find a set of cells $C := \{c^1, \ldots, c^M\}$, $M \in \mathbb{N}$, $j \in \{1, \ldots, M\}$, such that $\bigcup_j c^j = \mathbb{R}$, $\bigcap_j c^j = \emptyset$, and $\Pr[s_k \in c] = \Pr[V = v] = p_V^*(v)$ for $v \in \mathcal{Y}$ and $c \in C$. In other words, using the pdf $f_S(s)$, we can select the cells C so that the probability that s_k lies in the cells equals the optimal probability distribution $p_V^*(v)$. It follows that we can generate pseudorandom realizations from $p_V^*(v)$ by properly selecting C. Note that, because realizations are being generated by a deterministic process, there would be high correlation between consecutive realizations for small sampling period Δ. However, because the s_k is a stationary process (see Sect. 6.5.3), the larger the Δ, the smaller the correlation between s_k and s_{k+1} for all $k \in \mathbb{N}$. Indeed, large Δ would introduce large time-delays for generating realizations. There is a trade-off between correlation and time-delay that should be taken into account in practice. One way to deal with this trade-off is to compute the normalized autocorrelation function [4, 27] of s_k. Then, we select the smallest time-delay $\tau \in \mathbb{N}$ that leads to a desired correlation between s_k and $s_{k+\tau}$, $k \in \mathbb{N}$, and use the delayed sequence $s^\tau(\cdot) := \{s_k, s_{k+\tau}, s_{k+2\tau}, \ldots\}$ to generate realizations from $p_V^*(v)$. In the following algorithm, we summarize the ideas introduced above.

Algorithm 1: Pseudorandom Number Generation:

1) Consider the probability mass function $p_V^*(v) = \Pr[V = v]$, $v \in \mathcal{Y} = \{y_1, \ldots, y_M\}$, solution to Problem 6.1; and the synchronous solution $s(t) = \phi_u^s(t, u(t))$ of the responders.

2) Fix the sampling period $\Delta \in \mathbb{R}_{>0}$ and obtain, by Monte Carlo simulations [43], the probability density function $f_S(s_k)$ of the sampled sequence $s_k = s(t_k)$, $t_k = \Delta k$, $k \in \mathbb{N}$.

3) Select a finite set of cells $C = \{c^1, \ldots, c^M\}$, $M \in \mathbb{N}$, $j \in \{1, \ldots, M\}$, such that $\bigcup_j c^j = \mathbb{R}$, $\bigcap_j c^j = \emptyset$, and $\Pr[s_k \in c^j] = \Pr[V = y_j]$ for all $y_j \in \mathcal{Y}$.

4) Generate realization from $p_V^*(v)$ using the piecewise function:

$$v_k = \psi(s_k) := \begin{cases} y_1 & \text{if } s_k \in c^1, \\ \quad\vdots \\ y_M & \text{if } s_k \in c^M. \end{cases} \tag{6.10}$$

6.6.1 Distortion Induced by Synchronization Errors

Algorithm 1 in Sect. 6.6 is constructed under the assumption that responders are perfectly synchronized. However, because we only have exponential synchronization, in finite time, there is always a "small" difference between $s_1(t)$ and $s_2(t)$ due to potentially different initial conditions. It follows that there is also a difference between realizations generated using $s_1(t_k)$, denoted as $v_k^1 \in \mathcal{Y}$, and realizations $v_k^2 \in \mathcal{Y}$ generated through $s_2(t_k)$, where $\mathcal{Y} = \{y_1, \ldots, y_M\}$. Exponential synchronization implies that for any finite $t^* \in \mathbb{R}_{>0}$, there is $\delta(t^*, |s_1(0) - s_2(0)|) \in \mathbb{R}_{>0}$ (denoted as δ_{t^*} for simplicity), parametrized by t^* and the initial synchronization error $|s_1(0) - s_2(0)|$, such that $|s_1(t_k) - s_2(t_k)| \leq \delta_{t^*}$ for all $t_k \geq t_k^*$, and $\lim_{k \to \infty} |s_1(t_k) - s_2(t_k)| = 0$.

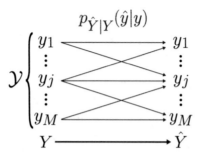

$$p_{\hat{Y}|Y}(\hat{y}|y)$$

Fig. 6.3 Transition probabilities $p_{\hat{Y}|Y}(\hat{y}|y)$

Consider the cell c^j, $c^j \in C$, with end points c_1^j and c_2^j, $c_1^j < c_2^j$, the length of c^j is defined as $l(c^j) := c_2^j - c_1^j$. If $c_1^j = \pm\infty$ (or $c_2^j = \pm\infty$), $l(c^j) = \infty$. Without loss of generality, let $l(c^2) \le l(c^3) \le \ldots \le l(c^{M-1})$, $l(c^1) = \infty$, and $l(c^M) = \infty$. Note that, if $\delta_{t^*} \le l(c^2)$, v_k^1 and v_k^2 are at most one level apart from each other, e.g., if $v_k^1 = y_1$, then either $v_k^2 = y_1$ or $v_k^2 = y_2$; and if $v_k^1 = y_3$, then $v_k^2 = y_2$, $v_k^2 = y_3$, or $v_k^2 = y_4$. It follows that $p_{\hat{Y}|Y}(\hat{y}|y)$, $y, \hat{y} \in \mathcal{Y}$, is of the form depicted in Fig. 6.3, where $\hat{Y}$ denotes the estimate of realizations of Y using $s_1(t_k)$, $s_2(t_k)$, and Algorithm 1. Similarly, if $l(c^2) < \delta_{t^*} \le l(c^3)$, v_k^1 and v_k^2 are at most two levels apart from each other and thus lead to a different structure of the transition probabilities. Here, we only consider the case where $\delta_{t^*} \le l(c^2)$. Distortion induced by larger synchronization errors can be estimated following the same methods. Note that, because responders synchronize exponentially, as $\delta_{t^*} \to 0$ ($t^* \to \infty$), $p_{\hat{Y}|Y}(\hat{y}|y) \to 1$ for $\hat{y} = y$, and $p_{\hat{Y}|Y}(\hat{y}|y) \to 0$, for $\hat{y} \ne y$, for all $y, \hat{y} \in \mathcal{Y}$. That is, distortion due to synchronization errors disappears exponentially fast. The actual value of the transition probabilities depend on the responders and driver dynamics, the initial conditions, and the cells C. However, we do not need these probabilities, only the structure of $p_{\hat{Y}|Y}(\hat{y}|y)$ depicted in Fig. 6.3 is used to derive an upper bound on the expected distortion. Let $\mathcal{V}_\delta \subseteq \mathcal{Y} \times \mathcal{Y}$ denote the set of pairs (y_j, y_i) for which there is a nonzero transition probability $p_{\hat{Y}|Y}(y_j|y_i)$ between $Y = y_j$ and $\hat{Y} = y_i$, $y_j, y_i \in \mathcal{Y}$, as depicted in Fig. 6.3. The set $\mathcal{V}_\delta$ is parametrized by the upper bound on the synchronization error $|s_1(t_k) - s_2(t_k)| \le \delta_{t^*} \le l(c^2)$. Define the distortion function $d(Y, \hat{Y}) := |Y - \hat{Y}|^2$. The function $d(Y, \hat{Y})$ is a deterministic function of two jointly distributed random vectors, Y and $\hat{Y}$, with joint distribution $p_{Y,\hat{Y}}(y, \hat{y}) = p_Y(y)p_{\hat{Y}|Y}(\hat{y}|y)$. Hence, see [44] for details, we can write the expected distortion as follows

$$
\begin{aligned}
E[d(Y, \hat{Y})] &= \sum_{y,\hat{y}\in\mathcal{Y}} p_{Y,\hat{Y}}(y, \hat{y})d(y, \hat{y}) = \sum_{y,\hat{y}\in\mathcal{Y}} p_Y(y)p_{\hat{Y}|Y}(\hat{y}|y)|y - \hat{y}|^2 \\
&= \sum_{(y,\hat{y})\in\mathcal{V}_\delta} p_Y(y)p_{\hat{Y}|Y}(\hat{y}|y)|y - \hat{y}|^2 \le \sum_{(y,\hat{y})\in\mathcal{V}_\delta} p_Y(y)|y - \hat{y}|^2 =: \bar{d}_\delta,
\end{aligned}
$$

$$(6.11)$$

where the left-hand side of (6.11) follows from the definition of $\mathcal{V}_\delta$ above, and the last inequality from the fact that $p_{\hat{Y}|Y}(\hat{y}|y) \le 1$ for all $y, \hat{y} \in \mathcal{Y}$. The constant $\bar{d}_\delta \in \mathbb{R}_{>0}$ provides an upper bound on the worst-case distortion induced by a δ_{t^*} synchronization error. Moreover, as $\delta_{t^*} \to 0$, $\mathcal{V}_\delta \to \{(y_1, y_1), (y_2, y_2), \ldots, (y_M, y_M)\}$; therefore, $\lim_{\delta_{t^*} \to 0} \bar{d}_\delta = 0$. That is, distortion due to synchronization errors is bounded by $\bar{d}_\delta$ and vanishes exponentially fast.

6.7 Simulation Results

We next present an evaluation of our algorithms on real data. We use the *adult-dataset*, available from the UCI Machine Learning Repository [7], which contains census data. Each attribute within the dataset has 3.9×10^4 entries. We use three of these attributes: race, sex, and income, which take values on finite discrete sets. We let *race* and *sex* be the private information, X, and use *income* as the information requested by the query, Y. The probability mass functions of X and Y, and part of the one of (X, Y) are given in Table 6.1. In Fig. 6.4, we depict $p_X(x)$, $p_Y(y)$, and $p_{X,Y}(x, y)$ with mass points indexed in the order given in Table 6.1. We first compute the optimal distribution $p_V^*(v)$ of the distorting additive noise V. We solve the convex program (6.8) in Theorem 6.1. The optimal distribution is depicted in Fig. 6.5 and the corresponding numerical values are given in Table 6.2. This $p_V^*(v)$ leads to $I[X; Y + V] = 0.0024$ while the mutual information without distortion is $I[X; Y] = 0.0251$, i.e., according to our metric, by optimally distorting the query, we leak about ten times less information. To generate realization from this distribution at both sides of the channel, we use trajectories of two chaotic responders as introduced in Sect. 6.2. We use the synthesis procedure in Sect. 6.5.4 to select suitable driver and responders. As driver (6.5), we use the Lorenz system:

Table 6.1 Probability mass functions of X and Y, and part of the one of (X, Y)

X	$\begin{bmatrix}0\\0\end{bmatrix}$	$\begin{bmatrix}0\\1\end{bmatrix}$	$\begin{bmatrix}0\\2\end{bmatrix}$	$\begin{bmatrix}0\\3\end{bmatrix}$	$\begin{bmatrix}0\\4\end{bmatrix}$	$\begin{bmatrix}1\\0\end{bmatrix}$	$\begin{bmatrix}1\\1\end{bmatrix}$	$\begin{bmatrix}1\\2\end{bmatrix}$	$\begin{bmatrix}1\\3\end{bmatrix}$	$\begin{bmatrix}1\\4\end{bmatrix}$
$p_X(x)$	0.5888	0.0200	0.0056	0.0560	0.0038	0.2616	0.0110	0.0042	0.0468	0.0022

Y	1	2	3	4	5	6	7	8	9
$p_Y(y)$	0.6870	0.0766	0.0364	0.0292	0.0658	0.0386	0.0002	0.0001	0.0662

(X, Y)	$\begin{bmatrix}0\\0\\1\end{bmatrix}$	$\begin{bmatrix}0\\1\\1\end{bmatrix}$	$\begin{bmatrix}0\\2\\1\end{bmatrix}$	$\begin{bmatrix}0\\3\\1\end{bmatrix}$	$\begin{bmatrix}0\\4\\1\end{bmatrix}$	$\cdots$	$\begin{bmatrix}1\\0\\9\end{bmatrix}$	$\begin{bmatrix}1\\1\\9\end{bmatrix}$	$\begin{bmatrix}1\\2\\9\end{bmatrix}$	$\begin{bmatrix}1\\3\\9\end{bmatrix}$	$\begin{bmatrix}1\\4\\9\end{bmatrix}$
$p_{X,Y}(x, y)$	0.3974	0.0130	0.0044	0.0388	0.0032	$\cdots$	0.0222	0.0014	0.0008	0.0046	0.0004

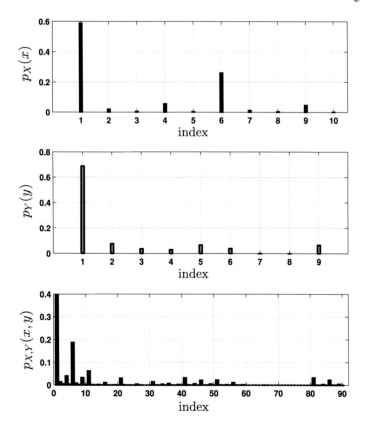

Fig. 6.4 Probability mass functions of X, Y, and (X, Y)

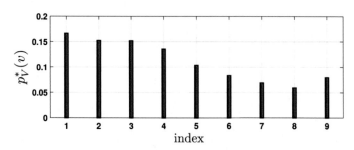

Fig. 6.5 Optimal distribution $p_V^*(v)$ solution to (6.8) in Theorem 6.1

Table 6.2 Optimal distribution $p_V^*(v)$ of the distorting additive random variable V

V	1	2	3	4	5	6	7	8	9
$p_V^*(v)$	0.1664	0.1522	0.1518	0.1355	0.1033	0.0832	0.0690	0.0591	0.0795

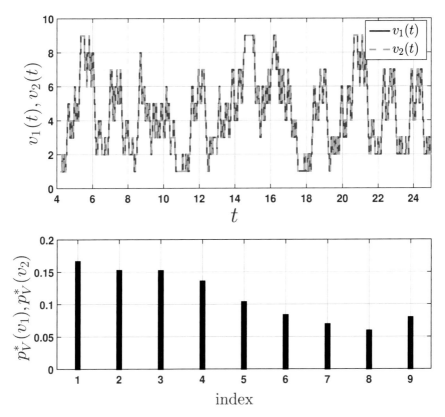

Fig. 6.6 Traces of the chaotic driver and responders trajectories. Top: trajectories of the responders converging to each other. Bottom: traces of chaotic solutions of the driver and responders

$$
\begin{cases}
\dot{\xi}_1(t) = 10(\xi_2(t) - \xi_1(t)), \\
\dot{\xi}_2(t) = 28\xi_1(t) - \xi_2(t) - \xi_1(t)\xi_3(t), \\
\dot{\xi}_3(t) = -\frac{8}{3}\xi_3(t) + \xi_1(t)\xi_2(t), \\
u(t) = \xi_1(t),
\end{cases}
\tag{6.12}
$$

with states $\xi_1, \xi_2, \xi_3 \in \mathbb{R}$ and driving signal $u \in \mathbb{R}$. The Lorenz system produces bounded trajectories [40], and is known to be chaotic and quasi-hyperbolic [24]. For the responders (6.4) and (6.6), we let $r(\zeta, u) = A\zeta + \psi(u)$, with $A = \text{diag}[-1, -2.5]$ and $\psi(u) = (-5u^2, 50 \sin(u))^\top$. Because A is diagonal and has negative eigenvalues, responders satisfy the conditions of Corollary 6.1 with $P = I_2$; hence, they are convergent systems and thus exponentially synchronize when driven by the same input $u(t)$. Moreover, since responders are linear in ζ and A is Hurwitz, systems can be proved to be ISS with input $\psi(u)$ [47]. Because u is bounded and $\psi(u)$ is continuous, by the extreme value theorem, $\psi(u)$ is bounded, which, together with ISS, imply boundedness of the responders' trajectories [47]. We let the outputs of

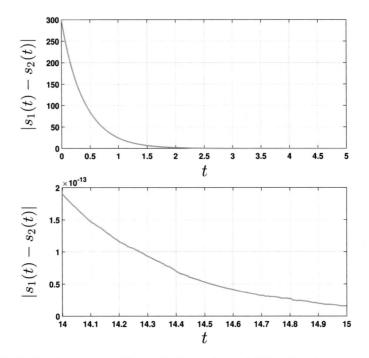

Fig. 6.7 Synchronization error $|s_1(t) - s_2(t)|$. Responders are initialized in antiphase, i.e., $s_1(0) = -s_2(0)$

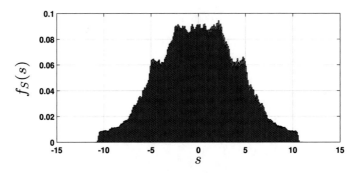

Fig. 6.8 Empirical probability densities of samples, $s(t_k)$, from the synchronous solution $s_1(t) = s_1(t) = s(t)$, for twenty different, randomly selected, initial conditions

the responders be $s_1(t) = \zeta_1^2$ and $s_2(t) = \zeta_2^2$ (their second state). In Fig. 6.6, we show traces of the chaotic driver and responders trajectories obtained by computer simulations (using Matlab from Mathworks), and in Fig. 6.7, we plot the synchronization error between the outputs of the responders. We initialized the responders in antiphase $\zeta_1(0) = -\zeta_2(0) = (150, 150)^{\top}$, and far from the limit trajectory. Note, in Fig. 6.7, that responders synchronize exponentially and are practically synchronized for $t \geq 5$. Moreover, after $t \geq 14$, the synchronization error is within Matlab's precision (10^{-12}). Because the Lorenz system is quasi-hyperbolic, samples from the driving signal $u(t)$ follow a stationary distribution that is independent of the initial conditions of the driver, see Sect. 6.5.3. Then, according to the synthesis procedure in Sect. 6.5.4, we next verify, using Monte Carlo simulations, that samples $s_k = s(t_k)$ (see Sect. 6.6), from the synchronous trajectory, $s_1(t) = s_1(t) = s(t)$, are also stationary. To do so, we compute the probability density function $f_S(s)$, $s_k \sim f_S(s)$, for different initial conditions and verify that all of them lead to the same density. In Fig. 6.8, we depict probability densities of s_k for twenty different initial conditions, sampling instants $t_k = \Delta k$, $\Delta = 0.001$, and $t \in [0, 4000]$. Note that they all lead to the same density $f_S(s)$. The support (obtained numerically) of $f_S(s)$ is given $S = [-10.8585, 10.8683]$. Finally, we use the piecewise function (6.10) to generate realizations from $p_V^*(v)$ using samples, s_k, from the synchronous trajectory. Following the algorithm given in Sect. 6.6, we have to divide the support S of $f_S(s)$ into a set of partitions $C = \{c^1, \ldots, c^M\}$, such that the probability that s_k lies in the cells equals the optimal probability distribution $p_V^*(v)$. This can be done using the empirical Cumulative Distribution Function (CDF), $F_S(s)$, corresponding to $f_S(s)$. We depict this CDF in Fig. 6.9. Then, we simply select the cells C such that $p_V^*(y_i) = \Pr[V = y_i] = \Pr[c^i \leq S \leq c^{i+1}] = F_S(c^{i+1}) - F_S(c^i)$ for all $i \in \{1, \ldots, M - 1\}$, $M = 9$ (the cardinality of the alphabet of Y). For this CDF and $p_V^*(v)$ in Table 6.2, we obtain the following cells:

$$C = \Big\{ [-\infty, -4.1739), [-4.1739, -2.0965), [-2.0965, -0.3658), \quad (6.13)$$
$$[-0.3658, 1.1408), [1.1408, 2.3321), [2.3321, 3.4341),$$
$$[3.4341, 4.5985), [4.5985, 5.7743), [5.7743, \infty] \Big\}.$$

In Fig. 6.10, we show realizations generated by the piecewise function (6.10) at both sides of the channel, and the corresponding probability mass functions. To generate this realizations, at the trusted server, we use samples from $s_1(t)$ and, at

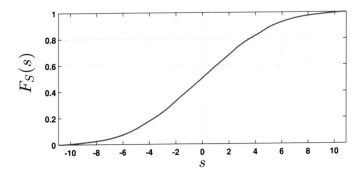

Fig. 6.9 Empirical CDF corresponding to $f_S(s)$

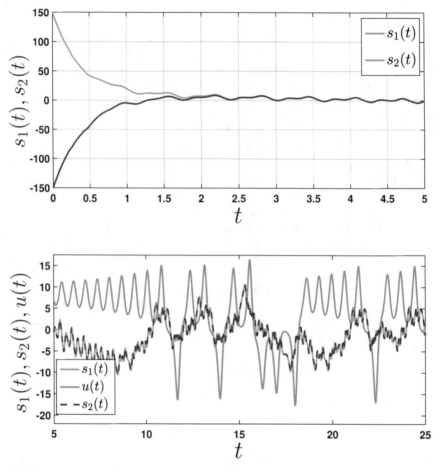

Fig. 6.10 Top: realizations of $p_V^*(v)$ generated by the piecewise function (6.10) at both sides of the channel, $v_1(t)$ at the trusted server and $v_2(t)$ at the remote station. Bottom: corresponding probability mass functions

the remote station, we sample $s_2(t)$. Note that, as expected, all samples are perfectly synchronized and their probability mass functions are equal to $p_V^*(v)$ in Fig. 6.5.

6.8 Conclusions

Using an information-theoretic privacy metric (mutual information), we have provided a general privacy framework based on additive distorting random vectors and exponential synchronization of chaotic systems. The synthesis of the optimal probability distribution, $p_V^*(v)$, of the additive distorting vector V has been posed as a convex program in $p_V(v)$. We have provided an algorithm for generating pseudorandom realizations from this distribution using trajectories of chaotic oscillators. To generate equal realizations at both sides of the channel, we have induced exponential synchronization on two chaotic oscillators (one at each side of the channel), and use their trajectories and the proposed algorithm to generate realizations. However, exponential synchronization implies that, in finite time, there is always a small error between trajectories (and thus also between realizations). We have derived an upper bound on the worst-case distortion induced by finite-time synchronization errors and showed that this distortion disappears exponentially fast. Using off-the-shelf results in the literature, we have provided general guidelines for selecting the dynamics of the responders and driver so that our algorithm for generating synchronized realizations from $p_V^*(v)$ is guaranteed to work. We have presented simulation results to illustrate our results.

References

1. Alvarez G, Li S, Montoya F, Pastor G, Romera M (2005) Breaking projective chaos synchronization secure communication using filtering and generalized synchronization. Chaos, Solitons Fractals 24:775–783
2. Akyol E, Langbort C, Basar T (2015) Privacy constrained information processing. In: 2015 54th IEEE conference on decision and control (CDC), pp 4511–4516
3. Angeli D (2000) A Lyapunov approach to incremental stability properties. IEEE Trans Automat Contr 47:410–421
4. Anishchenko VS, Astakhov V, Neiman A, Vadivasova T, Schimansky-Geier L (2007) Nonlinear dynamics of chaotic and stochastic systems: tutorial and modern developments (Springer Series in Synergetics). Springer, Berlin, Heidelberg
5. Arcak M, Angeli D, Sontag E (2002) A unifying integral iss framework for stability of nonlinear cascades. SIAM J Control Optim 40:1888–1904
6. Belykh VN, Belykh I, Mosekilde E (2005) Hyperbolic plykin attractor can exist in neuron models. I. J Bifurc Chaos 15:3567–3578
7. Blake C, Merz C (1998) UCI machine learning repository databases. http://archive.ics.uci.edu/ml
8. Boyd S, Vandenberghe L (2004) convex optimization. Cambridge University Press, New York, NY, USA
9. Calmon F, Fawaz N (2012) Privacy against statistical inference. In: 2012 50th annual Allerton conference on communication, control, and computing (Allerton), pp 1401–1408
10. Chaillet A, Angeli D, Ito H (2014) Combining iiss and iss with respect to small inputs: the strong iiss property. IEEE Trans Autom Control 59:2518–2524
11. Cover TM, Thomas JA (1991) Elements of information theory. Wiley-Interscience, New York, NY, USA
12. Demidovich B (1967) Lectures on stability theory. Russian, Moscow
13. Dwork C (2008) Differential privacy: A survey of results. In: Theory and applications of models of computation, pp 1–19. Springer, Berlin, Heidelberg
14. Dwork C, Roth A (2014) The algorithmic foundations of differential privacy. Found Trends Theor Comput Sci 9:211–407
15. Farokhi F, Nair G (2016) Privacy-constrained communication. IFAC-PapersOnLine 49:43–48
16. Farokhi F, Sandberg H (2017) Optimal privacy-preserving policy using constrained additive noise to minimize the fisher information. In: 2017 IEEE 56th annual conference on decision and control (CDC) (2017)

17. Farokhi F, Sandberg H, Shames I, Cantoni M (2015) Quadratic Gaussian privacy games. In: 2015 54th IEEE conference on decision and control (CDC), pp 4505–4510 (2015)
18. Geng Q, Viswanath P (2014) The optimal mechanism in differential privacy. In: 2014 IEEE international symposium on information theory, pp 2371–2375
19. Gottwald GA, Melbourne I (2004) A new test for chaos in deterministic systems. In: Proceedings of the royal society of London. Series A: Mathematical, Physical and Engineering Sciences, vol 460, pp 603–611
20. Grzybowski J, Rafikov M, Balthazar J (2009) Synchronization of the unified chaotic system and application in secure communication. Commun Nonlinear Sci Numer Simul 14:2793–2806
21. Han S, Topcu U, Pappas GJ (2014) Differentially private convex optimization with piecewise affine objectives. In: 53rd IEEE conference on decision and control (2014)
22. Hoh B, Xiong H, Gruteser M, Alrabady A (2006) Enhancing security and privacy in traffic-monitoring systems. IEEE Pervasive Computing 5:38–46
23. Huang Z, Wang Y, Mitra S, Dullerud GE (2014) On the cost of differential privacy in distributed control systems. In: Proceedings of the 3rd international conference on high confidence networked systems, pp 105–114
24. Kapitaniak T, Wojewoda J, Brindley J (2000) Synchronization and desynchronization in quasi-hyperbolic chaotic systems. Phys Lett A 210:283–289
25. Keuninckx L, Soriano M, Fischer I, Mirasso C, Nguimdo R, van der Sande G (2017) Encryption key distribution via chaos synchronization. Sci Reports 7:1–15
26. Khalil HK (2002) Nonlinear systems, 3rd edn. Prentice-Hall, Englewood Cliffs, NJ
27. Kocarev L, Halle K, Eckert K, Chua L, Parlitz U (1992) Experimental demonstration of secure communications via chaotic synchronization. Chua's Circuit: A Paradigm for Chaos 371–378
28. Kovacic I, Brennan M (2011) The Duffing equation: nonlinear oscillators and their behaviour. Wiley
29. Kuznetsov S (2012) Hyperbolic chaos: a physicist View. Springer, Berlin Heidelberg
30. Kuznetsov SP, Kruglov VP (2017) On some simple examples of mechanical systems with hyperbolic chaos. In: Proceedings of the Steklov Institute of Mathematics, 297
31. Kuznetsov SP, Pikovsky A (2007) Autonomous coupled oscillators with hyperbolic strange attractors. Physica D 232:87–102
32. Liu X, Chen T (2008) Boundedness and synchronization of y-coupled lorenz systems with or without controllers. Physica D 237:630–639
33. Lohmiller W, Slotine J (1998) On contraction analysis for nonlinear systems. Automatica 34:683–695
34. Lu J, Wu X, Lu J (2002) Synchronization of a unified chaotic system and the application in secure communication. Phys Lett A 305:365–370
35. Mackey M, Glass L (1977) Oscillation and chaos in physiological control systems. Science 197:287–289
36. Murguia C, Shames I, Farokhi F, Nešić D (2018) On privacy of quantized sensor measurements through additive noise. In: proceedings of the 57th IEEE conference on decision and control (CDC) (2018)
37. Ny JL, Pappas GJ (2014) Differentially private filtering. IEEE Trans Autom Control 59:341–354
38. Pavlov A, Pogromsky A, van de Wouw N, Nijmeijer H (2004) Convergent dynamics, a tribute to Boris Pavlovich Demidovich. Syst Control Lett 52:257
39. Pecora LM, Carroll TL (1990) Synchronization in chaotic systems. Phys Rev Lett 64:821–824
40. Pogromsky A (1998) Passivity based design of synchronizing systems. Int J Bifurcation Chaos 8:295–319
41. Pol Van der B, der Mark V (1927) Frequency demultiplication. Nature 120:363–364
42. Rajagopalan SR, Sankar L, Mohajer S, Poor HV (2011) Smart meter privacy: a utility-privacy framework. In: 2011 IEEE international conference on smart grid communications (SmartGridComm), pp 190–195 (2011)
43. Robert CP, Casella G (2005) Monte carlo statistical methods (Springer Texts in Statistics). Springer, Berlin, Heidelberg

44. Ross M (2006) Introduction to probability models, 9th edn. Academic Press Inc, Orlando, FL, USA
45. Salamatian S, Zhang A, du Pin Calmon F, Bhamidipati S, Fawaz N, Kveton B, Oliveira P, Taft N (2015) Managing your private and public data: bringing down inference attacks against your privacy. IEEE J Sel Topics Signal Process 9:1240–1255
46. Scardovi L, Sepulchre R (2010) Synchronization in networks of identical linear systems. IEEE Trans Automat Contr 57:2132–2143
47. Sontag E, Wang Y (1995) On characterizations of the input-to-state stability property. Syst Control Lett 24:351–359
48. Soria-Comas J, Domingo-Ferrer J (2013) Optimal data-independent noise for differential privacy. Inf Sci 250:200–214
49. Steur E, Tyukin I, Nijmeijer H (2009) Semi-passivity and synchronization of diffusively coupled neuronal oscillators. Physica D 238:2119–2128
50. Strogatz SH (2000) Nonlinear dynamics and chaos: with applications to physics. Biology Chemist Eng
51. Tan O, Gunduz D, Poor HV (2013) Increasing smart meter privacy through energy harvesting and storage devices. IEEE J Sel Areas Commun 31:1331–1341
52. Turukina L, Pikovsky A (2011) Hyperbolic chaos in a system of resonantly coupled weakly nonlinear oscillators. Phys Lett A 11:1407–1411
53. Wang Y, Huang Z, Mitra S, Dullerud GE (2014) Entropy-minimizing mechanism for differential privacy of discrete-time linear feedback systems. In: 53rd IEEE conference on decision and control, pp 2130–2135 (2014)
54. Weber RH (2010) Internet of things as new security and privacy challenges. Comput Law Secur Rev 26:23–30
55. Wiggins S (2003) Introduction to applied nonlinear dynamical systems and chaos. Texts in Applied Mathematics. Springer, New York
56. Wu C, Chua L (1995) Synchronization in an array of linearly coupled dynamical systems. IEEE Trans Circuit Sys I 42:430–447
57. Yang T, Wu C, Chua L (1997) Cryptography based on chaotic systems. IEEE Trans Circuit Syst I: Fund Theory Appl 44:469–472

Chapter 7
Differentially Private Analysis of Transportation Data

Mathilde Pelletier, Nicolas Saunier and Jerome Le Ny

Abstract To optimize the planning and operations of transportation systems, engineers analyze large amounts of data related to individual travelers, obtained through an increasing number and variety of sensors and data sources. For example, location traces collected from personal smartphones or smart cards in public transit systems can now cost-effectively complement or replace traditional data collection mechanisms such as phone surveys or vehicle detectors on highways, allowing to significantly increase the sensor coverage as well as the spatial and temporal resolution of the collected data. This trend allows for more accurate statistical estimates of the state and evolution of a transportation system, and improved responsiveness. At the same time, it raises privacy concerns, due to the possibility of making inferences on the history of visited locations and activities of individual citizens. This chapter presents some of the issues related to the privacy-preserving analysis of transportation data. We first illustrate the well-known difficulty of publishing location microdata (i.e., individual location traces) with privacy guarantees, though a case study based on the "MTL Trajet" dataset, a smartphone-based travel survey carried out in recent years in the city of Montréal. In contrast, the publication of aggregate statistics can be protected formally using state-of-the-art tools such as differential privacy, a formal notion of privacy that prevents certain types of inferences by adversaries with arbitrary side information. To illustrate the application of differential privacy to transportation data, the chapter presents a methodology for estimating the dynamic macroscopic traffic state (density, velocity) along a highway segment in real-time from single-loop detector and floating car data, while providing privacy guarantees for the individual driver trajectories. Enforcing privacy constraints impacts estimation performance (depending on the desired privacy level), but the effect is mitigated

M. Pelletier · N. Saunier
Department of Civil Engineering, Polytechnique Montreal, Montreal, QC H3T 1J4, Canada
e-mail: mathilde.pelletier@polymtl.ca

N. Saunier
e-mail: nicolas.saunier@polymtl.ca

J. Le Ny (✉)
Department of Electrical Engineering, Polytechnique Montreal, Montreal, QC H3T 1J4, Canada
e-mail: jerome.le-ny@polymtl.ca

© Springer Nature Singapore Pte Ltd. 2020
F. Farokhi (ed.), *Privacy in Dynamical Systems*,
https://doi.org/10.1007/978-981-15-0493-8_7

here by using a nonlinear model of the traffic dynamics, fused with the sensor measurements using data assimilation methods such as nonlinear Kalman filters.

7.1 Introduction

The monitoring, optimization and control of transportation systems has benefited tremendously from advances in sensing and computing technology in the past few years. Cameras with computer vision algorithms and radars can be easily deployed to monitor traffic flows, detect pedestrians, or manage intersections, complementing the role played by classical sensors such as induction loops detecting vehicles on highways. It is now also becoming easy for private companies and public organizations to collect data directly from individuals via "crowd-sensing", e.g., in the form of smart card data collected from the users of public transit systems [28], or location traces on a road network obtained from probe vehicles and individual smartphones, using cellular network location services, on-board global navigation satellite system (GNSS) devices [21] or RFID tags [35]. Data from these various sources is collected and combined by companies to provide real-time traffic estimates or by cities to plan the evolution of their transportation infrastructure. Moreover, many of these cities have open data initiatives, publishing some of their datasets with the goal of encouraging innovation in the transportation sector.

Increasing the density and variety of sensors can be very beneficial to understand and monitor the state of a transportation system, e.g., to improve the accuracy of traffic estimates and extend their coverage beyond major roads. However, this trend also raises important privacy concerns, due to the sensitive nature of location and transportation data [18]. This data can be used to track the location and activities of specific individuals, often in ways that are not immediately obvious to the engineers deploying the sensors. Indeed, multiple studies [9, 10, 17, 37] have shown that individual (microlevel) location data is highly unique and hence very hard to truly anonymize, and in particular removing obvious means of identification such as names from location traces offers little protection. As with other types of data, privacy attacks are made possible by the unavoidable existence of side-information [27, 31]. For example, one might have some knowledge about an individual's most frequently visited locations, which can then be linked with new published data to make unanticipated inferences, e.g., re-identify an anonymous location trace. Cloaking mechanisms such as k-anonymity have also been shown to offer little privacy protection when one aims to publish location microdata [30]. In fact, even publishing aggregate estimates of traffic density and velocity on a road network leaks some of the information contained in the measured driver trajectories, which could be used for sophisticated privacy attacks [5]. Previous research [27, 31] illustrates the risks of ruling out such attacks based purely on intuition. For example, Xu et al. [36] and Pyrgelis et al. [29] have recently shown that a large percentage of individual trajectories can be recovered from datasets that provide only aggregate spatio-temporal counts (number of people at given locations and given times), once general characteristics of human mobility or side-information about specific individuals are taken into account.

This chapter discusses some aspects of the broad topic of privacy-preserving analysis of location and transportation data, focusing in particular on the problem of releasing dynamic traffic estimates with differential privacy guarantees. Whereas other notions of privacy have been considered to analyze transportation data, such as k-anonymity [21, 32] and its various extensions [25], differential privacy is currently considered a state-of-the-art tool for privacy-preserving data analysis [14]. Differentially private mechanisms publish outputs that are randomized, in such a way that their distribution is not very sensitive to the data of any single individual. As a result, whether an individual provides his data or not does not change significantly the risk that any adversary, no matter how powerful, can make new inferences about him or her.

Much prior work exists on applying differential privacy to location data, see [3, 16, 20, 26] for example. Challenges remain however to produce differentially private outputs that are only moderately perturbed, e.g., traffic estimates of sufficiently high accuracy. Leveraging (publicly known) mathematical models of the dynamics of a phenomenon under consideration can help achieve better accuracy/privacy tradeoffs in estimator design, but work in this area is relatively sparse. In Sect. 7.4 of this chapter, based on the work presented in [2], we integrate traffic sensor measurements with a hydrodynamic model of traffic [33] through an Ensemble Kalman Filter (EnKF) to produce differentially private real-time traffic estimates, and introduce a spatially adaptive sampling scheme to make more efficient use of the sensitive floating car data. The EnKF [15] builds on ideas from Kalman filtering and Monte-Carlo methods, is quite popular for traffic estimation [34] and leads to a simpler estimator design than with the Extended Kalman Filter presented in [24].

This chapter is organized as follows. In Sect. 7.2 we illustrate the difficulties of publishing location microdata with privacy guarantees by analyzing the proportion of unique trajectories in a public dataset, "MTL Trajet", depending on the spatio-temporal resolution used to publish the data. The results agree with those of previous studies, showing that very few data points suffice to identify a trajectory uniquely, which constitutes a fundamental obstacle to providing meaningful privacy guarantees for such microlevel data. We then introduce the necessary technical background on differential privacy in Sect. 7.3, and illustrate a basic application of differential privacy to trip-time estimation for the MTL Trajet dataset in Sect. 7.3.3. The more complex dynamic traffic estimation problem is discussed in Sect. 7.4.

7.2 The Difficulty of Anonymizing Location Microdata: A Case Study

This section illustrates some of the difficulties that arise when attempting to publish transportation and location *microdata* (i.e., data at the level of individuals) without compromising the anonymity and privacy of the survey participants.

7.2.1 Dataset Description

Transportation planning typically relies on large household travel surveys done every couple of years. In the Greater Montréal Area, an origin-destination (OD) survey is done every five years by land-line phone. Such traditional travel surveys suffer from several issues, among which the difficulty to reach younger people who frequently only have cell phones. To address this shortcoming and complement the traditional OD survey on a more frequent basis, the City of Montréal has conducted a smartphone-based OD survey, called "MTL Trajet", every year for about a month in September and October since 2016. Other goals include measuring travel times and the impact of road construction, optimizing traffic lights and planning detours.

The MTL Trajet survey relies on a smartphone application. Participants create a profile and provide the following personal information: location of residence and place of work or study, occupation, preferred mode of transportation and socio-demographic attributes. After creating their profile, every trip taken by each participant is recorded automatically. The application tries to guess when the participant is arrived at a destination, and the participant is then prompted to provide the mode of transportation and trip purpose. The whole trip trajectory (series of GNSS positions) is saved along with this information. The survey stops when the participant has provided 25 trips during the data collection period.

The data collected in the 2016 and 2017 surveys has been released on the City of Montréal open data portal [7]. To attempt to protect the participants' privacy, all personal information constituting the participant profile is removed and trips are not linked to each participant. Other information deemed sensitive is removed: each trip's starting and ending locations are changed to the nearest intersection and trips completely outside of the Greater Montréal Area are completely removed.

Our case study relies on the 2017 MTL Trajet dataset, collected from September 18th to October 18th 2017. 4425 participants saved at least one trip in the application, and the dataset contains 477 000 trips, made of 21 million GNSS points [6]. The highest number of trips were recorded by participants during the first week of October (October 2nd to 8th): the 60436 trips are shown in Fig. 7.1. The average trip duration is approximately 30 min.

Fig. 7.1 Trips made during the week of October 2nd 2017 are displayed in blue on a map of the Greater Montréal Area (from OpenStreetMap)

7.2.2 Aggregation Study

The confidentiality of the dataset is evaluated by measuring systematically the uniqueness of each trip when discretizing the trip positions spatially and temporally. This can be motivated by a scenario where someone has some information on a survey participant, for example that he or she was in a given neighbourhood at an approximate time, and searching for that participant's trip in the dataset.

A trip is made of n positions (x_i, y_i) and instants t_i. The spatial discretization is done by assigning positions (x_i, y_i) to square cells forming a partition over the whole region, and the discretized position is denoted (x_i^d, y_i^d). The temporal discretization is done by assigning instants t_i, the time stamp at which a position is recorded, to a time interval, and the discretized instant is denoted t_i^d. A trip is defined by a subset of its positions and instants, which are discretized spatially and temporally: a trip is unique for a given spatial and temporal discretization if its subset of discretized positions and instants is unique in the whole dataset. The experimental results are obtained by testing different spatial resolutions (size of the square cells partitioning the space), temporal resolutions (length of the time intervals) and number of points drawn from each trip. For each such configuration, the observed positions are randomly drawn for each trip and are then discretized, and the proportion of unique trips is finally counted. Tests were done by repeating the results for different seeds for sampling the points and showed very little variation.

As can be seen in Fig. 7.2, the proportion of unique trips decreases as the size of the spatial and temporal quantization intervals increases (since more trips are

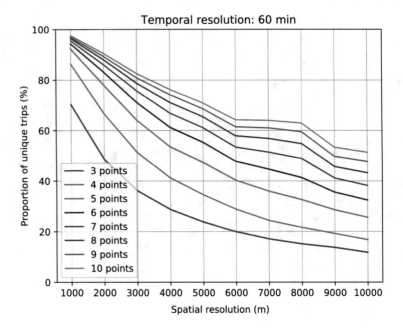

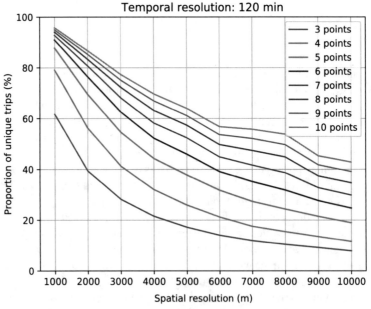

Fig. 7.2 Proportion of unique trips as a function of the spatial resolution of the cells, for two different temporal resolutions (60 and 120 min), for different numbers of points per trip

aggregated spatially and temporally) and as the number of observed positions for each trip decreases. What may be surprising however is the large share of unique trips, even at coarsest resolution with the smallest number of positions: 7.8 and 11.5% of trips are unique when representing each trip by respectively 3 and 4 positions, at spatial and temporal resolutions of respectively 10000 m (10 km) and 120 min (2 h). At the minimum spatial and temporal resolutions (1 km and 1 h), the proportion of unique trips varies from 70.2 to 97.5% when representing the trips respectively by 3 and 10 observed positions.

Other studies such as [37] have illustrated the fact that knowing a person's most visited locations, instead of sampling random points in trajectories as we do here, tends to only increase the likelihood of finding unique trajectories. In conclusion, even very coarse information about a person's trajectory is sufficient to identify it uniquely in a relatively large dataset such as MTL Trajet, and hence it appears that publishing location microdata with a resolution that remains useful for transportation studies while at the same time providing any sort of formal privacy guarantee is extremely challenging, except possibly by increasing the size of the dataset by orders of magnitude.

7.3   Differential Privacy

This section introduces the notion of differential privacy [13], which we adopt as our formal privacy definition in the following. In the standard set-up for differential privacy [4, 12, 13], a privacy-preserving data analysis mechanism has access to a database containing private information, and provides noisy answers to queries submitted by data analysts. The level of noise is carefully chosen in order to satisfy the definition of differential privacy, which is recalled below. More details about differential privacy can be found for example in [11, 14]. Naturally, one query could be to output a noisy version of the whole database, but this will typically lead to an unacceptably high level of noise, essentially making the output useless.

The standard definition of a differentially private mechanism is meant to ensure that such a mechanism publishes (aggregate) outputs that do not depend too strongly on the data of any single individual. A somewhat more general point of view that we adopt here is that a differentially private output should not be too sensitive to certain potential variations in the dataset from which the output is computed. This ensures that by publishing a result that is differentially private, one does not significantly improve the ability of a third party to discriminate between datasets that differ only by the allowed variations. In particular, if these variations consist in adding or removing the data of a single individual, publishing a differentially private output does not improve too much the ability of the third party to determine if the dataset contains the data of any specific individual. For certain tasks, achieving this goal might be too difficult, requiring an excessive amount of noise to perturb the published results. In such cases, one might be satisfied by considering weaker types of variations, for example changing the location data by a maximum amount for any single individual in a transportation dataset, rather than removing the data completely. We can then offer guarantees that one's ability to accurately track any individual contributing his or her data does not improve significantly by leveraging new information published in a differentially private way.

7.3.1 Definition of Differential Privacy

To introduce differential privacy formally, we consider a space D of possible datasets (e.g., all possible tables with a certain structure, all possible finite sets of location traces, etc.), on which we define a symmetric binary relation called adjacency and denoted Adj, which intuitively captures the variations among datasets that we want to hide in the published outputs. That is, it should be hard to determine from a differentially private output which of any two adjacent input datasets was used to produce this output. Following the terminology from the differential privacy literature, a "mechanism" refers to a randomized map M from the input space D of datasets to some output space R of published results. Mechanisms that are differentially private [12, 13], [14, Definition 2.4] necessarily randomize their outputs, in such a way that they satisfy the property of Definition 7.1. Formally, if we fix some probability space $(\Omega, \mathcal{F}, \mathbb{P})$, a mechanism M is a measurable map from D $\times \Omega$ to R. In particular, for $d \in$ D, $M(d, \cdot)$ is a measurable map from the sample space Ω to R, i.e., a random variable, and we follow the standard practice of writing this random variable simply $M(d)$.

Definition 7.1 Let D be a space equipped with a symmetric binary relation denoted Adj, and let (R, $\mathcal{R}$) be a measurable space. Let $\epsilon, \delta \geq 0$. A mechanism $M :$ D $\times \Omega$ $\to$ R is (ϵ, δ)-differentially private for Adj (and the σ-algebra $\mathcal{R}$) if for all $d, d' \in$ D such that Adj(d, d'), we have, for all sets S in $\mathcal{R}$,

$$\mathbb{P}(M(d) \in S) \leq e^\epsilon \mathbb{P}(M(d') \in S) + \delta. \tag{7.1}$$

If $\delta = 0$, the mechanism is said to be ϵ-differentially private.

Definition 7.1 bounds the allowed deviation for the output distribution of a differentially private mechanism, given any two adjacent input datasets d and d'. The bound is controlled by two parameters ϵ, δ, which set the level of privacy. Smaller values for the parameters ϵ, δ correspond to a higher level of privacy. In particular, the parameter δ should be kept as small as possible, since $\delta > 0$ allows a zero probability event for input d' to have positive probability for input d, hence its occurrence immediately allows an adversary to distinguish between d and d'. The spaces D, R depend on the application, and the adjacency relation should be defined to capture the type of deviations we want to make hard to detect from the outputs. Here we always assume R to be a metric space, for which we can then take $\mathcal{R}$ to be the σ-algebra of Borel sets.

7.3.2   Basic Differentially Private Mechanisms

Now that we have a formal privacy definition, let us describe some basic mechanisms that can enforce differential privacy in practice [12, 13]. Our focus here is on releasing numerical outputs, e.g., trip times computed from a transportation dataset. Because we also work with vector valued signals, let us define the notation $\mathsf{S}^d = (\mathbb{R}^d)^{\mathbb{N}}$, i.e., the set of sequences $\{x_k\}_{k \geq 0}$, with $x_k \in \mathbb{R}^d$. For such sequences, we define the ℓ_p norm

$$\|x\|_p := \left(\sum_{k=0}^{\infty} |x_k|_p^p \right)^{1/p},$$

provided the sum converges, with $| \cdot |_p$ the p-norm on $\mathbb{R}^d$, i.e.,

$$|v|_p = \left(\sum_{i=1}^{d} |v_i|^p \right)^{1/p}, \quad \text{for } v \in \mathbb{R}^d.$$

The following quantity plays an important role to set the level of privacy-preserving noise.

Definition 7.2 Let D be a space equipped with an binary symmetric adjacency relation Adj. Let V be a vector space equipped with a norm $\| \cdot \|$. The sensitivity of a query $q : \mathsf{U} \to V$ is defined as $\Delta q = \sup_{\text{Adj}(u,u')} \|q(u) - q(u')\|$. In particular, if V is $\mathbb{R}^d$, for some positive integer d, equipped with the p-norm, or if $V = \mathsf{S}^d$, we call Δq the p- or ℓ_p-sensitivity respectively and we use the notation $\Delta_p q$.

To introduce the next result, we define the following notation. We write $X \sim \text{Lap}^d(b)$ to mean that X is a random vector in $\mathbb{R}^d$ with components that are independent and identically distributed (iid) random variables following the centered Laplace distribution with variance $2b^2$, i.e., with probability density function (pdf) $\exp(-|x|/b)/(2b)$ for $x \in \mathbb{R}$. We write $X \sim \mathcal{N}(0, \Sigma)$ if X follows a d-dimensional normal distribution with zero mean and covariance matrix Σ. Finally, the Q-function is defined as $Q(x) := \frac{1}{\sqrt{2\pi}} \int_x^{\infty} e^{-u^2/2} du$, and we let

$$\kappa_{\delta,\epsilon} = \frac{1}{2\epsilon}(Q^{-1}(\delta) + \sqrt{(Q^{-1}(\delta))^2 + 2\epsilon}). \tag{7.2}$$

We then have the following theorem [12, 13, 23].

Theorem 7.1 (Laplace and Gaussian mechanisms) Let $\epsilon > 0$, $1 > \delta > 0$, and fix an adjacency relation Adj on some set D. Let $q : \mathsf{D} \to \mathbb{R}^d$ be a vector valued query with 1- and 2-sensitivities $\Delta_1 q$ and $\Delta_2 q$. The mechanism $M(u) = q(u) + w$, with $w \sim \text{Lap}^d(b)$ and $b \geq \Delta_1 q / \epsilon$, is ϵ-differentially private for Adj. If instead $w \sim \mathcal{N}\left(0, \sigma^2 I_d\right)$ with $\sigma \geq \kappa_{\delta,\epsilon} \Delta_2 q$ and $\kappa_{\delta,\epsilon}$ defined in (7.2), then the mechanism is (ϵ, δ)-differentially private.

Let us now assume that $q : D \to S^d$ is a signal valued query, with ℓ_1- and ℓ_2-sensitivities $\Delta_1 q$ and $\Delta_2 q$. The mechanism $M(u) = q(u) + w$, with w a white noise signal (sequence of iid zero-mean random variables) such that each sample w_k follows a $Lap^d(b)$ distribution with $b \geq \Delta_1 q / \epsilon$, is ϵ-differentially private for Adj. If the last property is replaced by $w_k \sim \mathcal{N}\left(0, \sigma^2 I_d\right)$ with $\sigma \geq \kappa_{\delta, \epsilon} \Delta_2 q$, then the mechanism is (ϵ, δ)-differentially private.

Note that the difference between the two parts of Theorem 7.1 is in the definition of the sensitivity, which includes a summation over all time periods for signal valued queries. Finally, a crucial property of differential privacy says that it is *resilient to post-processing*, i.e., transforming a differentially private output cannot weaken the privacy guarantee, as long as the input signal is not re-accessed, see [23, Theorem 1]. This property is important to ensure that an adversary performing additional processing on a published output, without having access to the original dataset, cannot weaken the provided differential privacy guarantee. Moreover, this property is also often directly useful to design mechanisms with better performance than the basic Laplace and Gaussian mechanisms of Theorem 7.1, by performing post-processing steps to smooth out the privacy-preserving noise before releasing the results of an analysis.

7.3.3 Application to MTL Trajet Data

As a first application example for the basic differentially private mechanisms introduced in Theorem 7.1, consider the same 2017 MTL Trajet dataset denoted d and a small-scale but typical query, estimating travel times for a given OD pair. We use the 1-km square cells of the aggregation study and focus on an OD pair going from an area around the Jean Talon public market to an area downtown. The dataset contains 72 such trips, with empirical average travel time of 32.78 min and standard deviation of 16.1 min. We would like for example to provide the average travel time for each business day, so that our query is a function q with output $[q_1, q_2, q_3, q_4, q_5]$ in $\mathbb{R}^5$. Let d_i be the subset of trips for the OD pair of interest happening on day i of the week (Monday to Friday). It is assumed that the number of elements $|d_i|$ in each subset d_i is not a privacy-sensitive value. The desired output i of q for day i is then

$$q_i(d) = \frac{1}{|d_i|} \sum_{j \in d_i} t_j, \ 1 \leq i \leq 5,$$

where t_j is the travel time of trip j.

Note that despite the reasonably large size of the MTL Trajet dataset overall, we have in fact relatively few observations for any single day for our OD pair (between 12 and 18), and this is what determines the sensitivity of each component q_i. We thus need to be realistic regarding what type and level of privacy guarantee we can provide. First, let us assume for the adjacency relation that we want to make it hard to distinguish between the original dataset d and any other dataset d' which is the same as d except for a *single* travel time measurement t_j, whose value would have been changed by at most ρ min (i.e., $|t_j - t'_j| \leq \rho$). An interpretation of differential privacy for this adjacency relation is that by publishing statistics we should not make it easier for a third party to infer any single trip time with an accuracy better than within a $\pm\rho$ minute interval. For example, it might be reasonable, based on physical considerations, to expect a priori for a given OD pair that the travel time is within the interval [20, 40] min, in which case taking $\rho = 20$ min would cover all possible variations. To bound the sensitivity of q for this adjacency relation, we note that only one component q_i can change as d is changed to d', and so we have here

$$\Delta_1 q = \Delta_2 q = \sum_{i=1}^{5} |q_i(d) - q_i(d')| \leq \frac{1}{\min_{1 \leq i \leq 5} |d_i|} \rho.$$

Consider now a more stringent adjacency relation, attempting to protect the data of a single individual rather than a single trip. In particular, assume that an individual can have (at most) one record for each day for the given OD pair. This is realistic (an individual might make the same trip every day) but of course requires hiding 5 data points instead of 1. If we want to protect against variations in d for each day, then the 1-sensitivity is bounded by

$$\Delta_1 = \sum_{i=1}^{5} |q_i(d) - q_i(d')| \leq \rho \sum_{i=1}^{5} \frac{1}{|d_i|},$$

because each term in the sum now contributes a variation, and for the 2-sensitivity we get

$$\Delta_2 = \sqrt{\sum_{i=1}^{5} |q_i(d) - q_i(d')|^2} \leq \rho \sqrt{\sum_{i=1}^{5} \frac{1}{|d_i|^2}}.$$

Privacy-preserving data analysis is governed by privacy-utility trade-offs. On the privacy side, the guarantee provided is captured by both the definition of the adjacency relation and by the chosen privacy level (value of ϵ, δ). Given the accuracy we want in order to provide useful answers, a given mechanism and an adjacency relation, we can then set a value for the privacy parameters, which specifies the privacy level we can guarantee with our mechanism while achieving the desired accuracy. For the basic noise additive mechanism with Laplace noise of Theorem 7.1, the standard deviation of the noise is $\sqrt{2}\Delta_1/\epsilon$, with the sensitivity Δ_1 calculated above. This standard deviation is also the desired accuracy DA. This gives for the first adjacency relation (trip level privacy) the achievable privacy parameter

$$\frac{\sqrt{2}\rho}{\epsilon \min_i\{d_i\}} = \text{DA} \Rightarrow \epsilon = \frac{\sqrt{2}\rho}{\text{DA} \min_i\{d_i\}}.$$

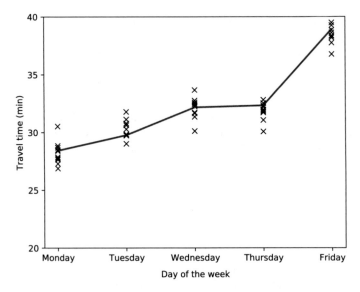

Fig. 7.3 Samples of differentially private results for the travel times on each day of the week for the chosen OD pair, with $\epsilon = 2.36$ and $\rho = 20$ min. The non-perturbed travel time per day is displayed as the red line and for illustration purposes, ten random draws per day are shown for the Laplace mechanism. In practice, *only one* randomized result would be drawn and returned to answer the query

For example, with $\rho = 20$ min and $\min_i\{d_i\} = 12$ in our dataset, the suggested value of ϵ is 2.36. If instead we consider variations for each day in the adjacency relation, then we would only be able to achieve a value of $\epsilon \approx 5 \times 2.36 \leq 12$ to get a 1 min standard deviation for each day. Note again that the privacy guarantee is determined by the adjacency relation as well as by ϵ, so that an alternative way to achieve a given accuracy target is by decreasing ρ, which also weakens the privacy guarantee. The main way to improve the privacy-utility tradeoff would be to have more data points in order to increase $\min_i\{d_i\}$, or a better mechanism that is able to achieve a given accuracy with a lower value of ϵ. In any case, the process of setting the noise level forces us to think carefully about the privacy guarantee that we are able to provide with a given mechanism, and hence reduces the risk of making invalid privacy-related claims. For illustration purposes, Fig. 7.3 presents the travel time results for the OD pair considered.

7.4 A Differentially Private Traffic Estimator

In the rest of this chapter, we focus on transportation data useful for road traffic estimation, and describe a methodology for the design of differentially private real-time traffic state estimators. Traffic data can be either cross-sectional data, obtained at fixed locations by stationary sensors such as induction loops, or trajectory data, obtained for example via computer vision algorithms tracking individual vehicles through sequences of traffic camera images, or from probe vehicles transmitting their GPS coordinates (also called floating car data). First, some necessary background on a macroscopic mathematical model of traffic used for our estimation approach is provided in Sect. 7.4.1. Section 7.4.2 then relates some typical cross-sectional and trajectory measurements to the state variables of this mathematical model, which we wish to estimate. In Sect. 7.4.3, we present a methodology to sanitize these measurements and enforce differential privacy. Finally, Sect. 7.4.4 presents an Ensemble Kalman filter (EnKF) to estimate the traffic state based on sanitized data, leveraging the dynamic traffic model and the resilience to post-processing property of differential privacy.

7.4.1   Traffic Dynamics

We introduce here some basic elements of traffic flow models and refer the reader to [33] for example for more details. Macroscopic models represent the traffic on a road as a fluid. They describe the dynamics of the variables to be estimated (traffic density and velocity), and allow us to relate traffic sensor measurements to these variables. We consider here the description of the traffic state on a single multi-lane one-way road without intersection. Let $\lambda(x)$ represent the number of lanes at position x. In general, the traffic state at time t and position x is characterized by a density $\rho^{(i)}(x, t)$ and a velocity $v^{(i)}(x, t)$ for each lane $i = 1, \ldots, \lambda(x)$. We summarize this state at each position and time by the lane-averaged density and velocity

$$\rho(x, t) = \frac{1}{\lambda(x)} \sum_{i=1}^{\lambda(x)} \rho^{(i)}(x, t), \quad v(x, t) = \frac{1}{\lambda(x)\rho(x, t)} \sum_{i=1}^{\lambda(x)} \rho^{(i)}(x, t) v^{(i)}(x, t),$$

The lane-averaged traffic flow is defined as $q(x, t) := \rho(x, t)\, v(x, t)$. First-order models postulate a static relationship between density and traffic flow, called the *fundamental diagram*, which is often assumed to be triangular [33]

$$q(\rho) = \begin{cases} v_0\, \rho, & \text{for } \rho \leq \rho_C := \frac{w}{v_0 + w} \rho_M \\ -w\, (\rho - \rho_M), & \text{for } \rho_C \leq \rho \leq \rho_M. \end{cases} \tag{7.3}$$

In (7.3), v_0 is the free traffic speed, ρ_C is the critical density where the transition occurs between free traffic (speed $v = v_0$) and congested traffic (speed $v < v_0$), ρ_M is the maximal or "jam" density, associated with a zero speed traffic, and w is the speed at which congestion waves propagate (backwards). Note that for first-order models, the density ρ becomes essentially the only fundamental state variable, with the traffic velocity given by $v(x, t) = q(\rho(x, t))/\rho(x, t)$. In particular, the triangular fundamental diagram (7.3) corresponds to the density-velocity relation

$$v(\rho) = \begin{cases} v_0, & \text{for } \rho \le \rho_C := \frac{w}{v_0+w}\rho_M, \\ w\left(\frac{\rho_M}{\rho} - 1\right), & \text{for } \rho_C \le \rho \le \rho_M. \end{cases} \tag{7.4}$$

In the following, we describe the evolution of traffic in discrete time, with time periods of length τ. The road is discretized into N cells of length Δx_j, $1 \le j \le N$, with the density $\rho_{j,t}$ in cell j at period $t \ge 0$ assumed approximately constant. Let λ_j denote the number of lanes and $f_{j,t}$ the *numerical flux* during period t (see the definition below) at the interface between cells $j-1$ and j. After discretizing a standard continuity partial differential equation [33], the dynamics of the discretized (lane-averaged) density follow the following perturbed difference equation

$$\rho_{j,t+1} = \rho_{j,t} + \left(\frac{\lambda_j}{\lambda_{j+1}}f_{j,t} - f_{j+1,t}\right)\frac{\tau}{\Delta x_j} + \nu_{j,t}, \tag{7.5}$$

where ν_j is white noise whose variance captures for example errors due to the fundamental diagram hypothesis. In the standard Cell Transmission Model (CTM) [8], which we use here, the numerical flux at the interface $(j-1) \to j$, compatible with the fundamental diagram (7.3), is given by

$$f_{j,t} := \min\{v_0\,\rho_{j-1,t},\, v_0\,\rho_C,\, w\,(\rho_M - \rho_{j,t})\},$$

and the Courant-Friedrichs-Lewy (CFL) condition $\tau \le (\Delta x_j/v_0)$ must be satisfied for all j for (7.5) to provide a valid discretization of the continuity equation. Then, (7.5) is a stochastic, piecewise linear state-space description of the traffic dynamics.

7.4.2 Traffic Sensor Measurements

As mentioned earlier, traffic data can be either cross-sectional data, most commonly captured by induction loops, or individual trajectory data obtained for example from probe vehicles. Induction loops, more precisely single-loop detectors, are common static sensors [33] placed at some fixed locations along the road, assumed here to be on some boundary between two cells in the discretized model of the previous section. A single-loop detector on the boundary between cells $j-1$ and j reports for each lane l, with a certain frequency (every 60 seconds for example):

- the vehicle counts $c_{j,t}^l$, i.e., the number of vehicles in lane l that crossed the boundary during the last sampling period, and
- the occupancies $o_{j,t}^l \in [0, 1]$, which represent the percentage of time for which a car was on top of the sensor in lane l during the last sampling period.

The set of sampling (or reporting) times for the single loop detectors is known and denoted $\mathcal{T} \subset \mathbb{N}$, and we call $\mathcal{J}$ the set of cell interfaces where a sensor is located, with $\mathcal{J} \subset \{1, \ldots, N+1\}$. Here, interface 1 is the entrance boundary of the first cell, and interface $N + 1$ is the exit boundary of cell N. We assume periodic sampling with period T_s (in seconds), with T_s a multiple of the time discretization step τ. From these basic sensor measurements, one can form measurements of the macroscopic quantities of interest. For example, a (noisy) measurement of the lane-averaged traffic flow at the interface between cells $j - 1$ and j, expressed in vehicles per second, is obtained from the vehicle counts as

$$\phi_{j,t} := \frac{1}{T_s \lambda_j} \sum_{l=1}^{\lambda_j} c_{j,t}^l, \text{ for } t \in \mathcal{T}, \ j \in \mathcal{J}. \tag{7.6}$$

One issue with flow measurements however is that they do not correspond in a one-to-one fashion to density values. Indeed, according to the fundamental diagram (7.3), a given flow value corresponds in general to two possible density values, one for the traffic being in free flow ($\rho < \rho_C$), and another when it is congested ($\rho > \rho_C$). From the occupancy measurements, one can in fact directly form a (noisy) density measurement at the interface $(j - 1) \rightarrow j$

$$\psi_{j,t} := \frac{1}{g_j \lambda_j} \sum_{l=1}^{\lambda_j} o_{j,t}^l, \text{ for } t \in \mathcal{T}, \ j \in \mathcal{J}, \tag{7.7}$$

where g_j is the so-called g-factor, i.e, the average effective vehicle length that one can observe at the sensor location during the time-interval of interest. G-factors are assumed known for the road segment under study, see [22] for a technique to estimate them. One can then assume a measurement model such as

$$\begin{aligned} \psi_{j,t} &= \rho_{j,t} + \mu_{j,t}^o, \\ \psi_{j,t} &= \rho_{j-1,t} + \mu_{j-1,t}^o, \end{aligned} \tag{7.8}$$

for $j \in \mathcal{J}$, where μ_j^o are white noise signals capturing modeling and measurement errors, and we considered $\psi_{j,t}$ as a density measurement both for the cell $j - 1$ and j. Sources of errors and bias in converting cross-sectional data to macroscopic measurements are discussed in [33, Chap. 3].

Probe vehicles are equipped with devices transmitting their velocity and position when requested. This floating car data could be obtained from private vehicles through the drivers' smartphones or an electronic toll collection system for example. For privacy reasons, we assume here a location-based sampling scheme introduced in [21] and called Virtual Trip Lines (VTL), where vehicles report their speed only when they cross specific known locations along the road. Again, we assume that all VTLs are placed at some interface between two cells, and denote the corresponding interface indices $\mathcal{J}' \subset \{1, \ldots, N + 1\}$. We compute an average velocity measurement after n cars cross a VTL, for some value n to choose, formed as the geometric mean of the individual speed measurements $v_{j,t}^{(i)}$, $1 \leq i \leq n$, i.e.,

$$\bar{v}_{j,t} := \left(\prod_{i=1}^{n} v_{j,t}^{(i)} \right)^{\frac{1}{n}}, \quad \text{for } j \in \mathcal{J}', t \in \mathcal{T}_j', \tag{7.9}$$

where $\mathcal{T}_j' \subset \mathbb{N}$ is the set of time periods at which an aggregated velocity measurement is reported for the VTL j. To link such velocity measurements to the state ρ of the dynamic model, we can invert the relation (7.4). In particular, if $\bar{v}_{j,t} \geq v_0$, we assume that the traffic is in free flow and then the velocity measurement does not provide information on ρ except for $\rho \leq \rho_C$. On the other hand, if $\bar{v}_{j,t} < v_0$, then we assume the following noisy measurement model

$$\xi_{j,t} := \ln \bar{v}_{j,t} = \ln \left(\frac{\rho_M}{\rho_{j,t}} - 1 \right) + \ln w + \mu_{j,t}^v, \quad \text{if } \xi_{j,t} < v_0,$$
$$\text{and } \xi_{j,t} := \ln \bar{v}_{j,t} = \ln \left(\frac{\rho_M}{\rho_{j-1,t}} - 1 \right) + \ln w + \mu_{j-1,t}^v, \quad \text{if } \xi_{j,t} < v_0, \tag{7.10}$$

where μ_j^v are white noise signals, and again a measurement at a VTL is interpreted as a measurement for each of the two cells adjacent to this VTL. The fact that traffic velocity measurements $\bar{v}_{j,t}$ in (7.9) are defined as geometric means and the choice of a logarithmic model in (7.10) turn out to be convenient for our privacy preserving scheme, as explained in the next section.

A standard traffic estimation problem consists in publishing in real-time an estimate of the density ρ in each cell, based on the sensor data. For this, we can rely on the dynamic model (7.5) and the measurement models introduced in this section. Moreover, since the sensor measurements capture private information about the drivers' trajectories, an additional constraint imposed here is to ensure that the published density estimate guarantees differential privacy for the input sensor data. Hence, with respect to Definition 7.1, a mechanism in this case is an algorithm that produces sequences $\{\hat{\rho}_t\}_{t \geq 0}$ of density estimates, with $\hat{\rho}_t \in \mathbb{R}^N$ for each t, computed from the sensitive traffic sensor data.

7.4.3 Traffic Data Sanitization

To instantiate the differential privacy constraint for our specific data analysis problem, we start by specifying the adjacency relation on the space of datasets (see Sect. 7.3.1), which is here the space of sequences $\{y_t\}_{t \geq 0}$ of raw traffic measurements. For each $t \in \mathcal{N}$, if $t \in \mathcal{T}$, then the vector y_t includes the values $\{c_{j,t}^l, o_{j,t}^l\}_{j \in \mathcal{J}}$, i.e., the counts and occupancy measurements collected by the single-loop detectors. If $t \in \mathcal{T}_j'$ with $j \in \mathcal{J}'$, it includes the last n vehicle speed measurements $\{v_{j,t}^{(i)}\}_{1 \leq i \leq n}$ collected at VTL j. Note that the sets $\mathcal{T}, \mathcal{T}_j'$ are not necessarily disjoint, and that y_t does not contain any measurement for the periods t outside of these sets (the sampling periods can be much larger than the discretization time-step). Recall that a differentially private mechanism should make it difficult for an adversary to decide which of any two adjacent sequences $\{\mathbf{y}_t\}_{t\geq 0}$ or $\{\tilde{\mathbf{y}}_t\}_{t\geq 0}$ was used in producing a given output, once a notion of adjacency is chosen. Here, we consider two datasets $\mathbf{y}$ and $\tilde{\mathbf{y}}$ to be adjacent if they differ by the trajectory of a single car, with some additional restrictions on the allowed impact that this single trajectory deviation can have on the occupancy and velocity measurements.

Note that a car only moves forward on the road and travels in a specific lane at each cell boundary. Changing a single vehicle trajectory can therefore change the counts $c_{j,t}^l$ at a given location $j \in \mathcal{J}$ for only two pairs (t_1, l_1) and (t_2, l_2) and by at most 1 for each pair, since the vehicle might have crossed the sensor location at a different sampling period and possibly at a different lane. If we wish to publish the sequence of flow measurements (7.6) while providing (ϵ, δ)-differential privacy, with the Gaussian mechanism, we compute the ℓ_2-sensitivity as

$$\Delta_2^\phi = \max_{\text{Adj}(c,\tilde{c})} \sqrt{\sum_{j\in\mathcal{J}}\sum_{t\in\mathcal{T}} |\phi_{j,t} - \tilde{\phi}_{j,t}|^2},$$

where the maximum is taken over count sequences that are adjacent (differ by a single vehicle trajectory) and $\phi, \tilde{\phi}$ are the corresponding flow measurement sequences. From the preceding discussion, we get

$$\Delta_2^\phi \leq \frac{1}{T_s}\sqrt{2\sum_{j\in\mathcal{J}}\frac{1}{\lambda_j^2}}.$$

Similarly, changing a single car trajectory affects the occupancy measurements $o_{j,t}^l$ at a single location j for at most two pairs (t, l), if the car crosses location j at a different time and possibly at a different lane. However, bounding the sensitivity of a mechanism publishing the sequence $\psi_{j,t}$ from (7.7) requires to make additional assumptions about the allowed trajectory deviations (in the adjacency relation), in order to obtain useful results. An occupancy measurement $o_{j,t}^l$ returned by a single-loop detector is the sum of the individual occupancies for the cars passing on top of this sensor during the last sampling period of length T_s. We protect only car trajectories that have a limited influence on occupancy measurements, which we capture through the adjacency relation. When there is a difference between some $o_{j,t}^l$ and $\tilde{o}_{j,t}^l$ due to a change of one vehicle trajectory, we assume this difference to be bounded, i.e.,

$$|o_{t,j}^l - \tilde{o}_{j,t}^l| \leq \alpha, \tag{7.11}$$

for some value of α set as discussed below. This constraint is mostly relevant when the traffic is very congested, so that vehicles travel at very low speed, and the sampling period relatively short, in which case the occupancy measurements could be very dispersed and a single car crossing the sensor or not during the sampling period could have a significant impact on the total occupancy for that period. Then, as for the flow measurements, we have for the ℓ_2-sensitivity of a mechanism publishing the density measurements

$$\Delta_2^\psi = \max_{\text{Adj}(o,\tilde{o})} \sqrt{\sum_{j\in\mathcal{J}}\sum_{t\in\mathcal{T}} |\psi_{j,t} - \tilde{\psi}_{j,t}|^2}$$

$$\Delta_2^\psi \leq \alpha \sqrt{2\sum_{j\in\mathcal{J}} \frac{1}{(g_j\lambda_j)^2}}.$$

Regarding velocity measurements at VTLs, again, since each vehicle can cross a VTL only once, we deduce that by changing a single trajectory, for any location j, there can be at most two time periods t where the value $\bar{v}_{j,t}$ changes (if the vehicle is included in a different batch of size n). As for the occupancy measurements, we need to impose further restrictions in the adjacency relation, and here we only protect against bounded relative velocity variations. Specifically, when a vehicle trajectory changes, we assume in the adjacency relation that the single possibly modified value $v_{j,t}^{(i)}$ in any $\bar{v}_{j,t}$ satisfies

$$\frac{|v_{p,t}^{(i)} - \tilde{v}_{p,t}^{(i)}|}{\min\{v_{p,t}^{(i)}, \tilde{v}_{p,t}^{(i)}\}} \leq \gamma, \tag{7.12}$$

for some value of γ to set. To publish the sequence $\{\xi_{j,t}\}_{j,t}$ (with $\xi_{j,t} = \ln \bar{v}_{j,t}$, see (7.10)) using the Gaussian mechanism, we deduce from the above discussion the following bound on ℓ_2-sensitivity

$$\Delta_2^\xi = \max_{\text{Adj}(v,\tilde{v})} \sqrt{\sum_{j\in\mathcal{J}'}\sum_{t\in\mathcal{T}'} |\ln \xi_{j,t} - \ln \tilde{\xi}_{j,t}|^2}$$

$$= \max_{\text{Adj}(v,\tilde{v})} \sqrt{\sum_{j\in\mathcal{J}'}\sum_{t\in\mathcal{T}_j'} \left|\frac{1}{n}\sum_{i=1}^{n}\left(\ln v_{j,t}^{(i)} - \ln \tilde{v}_{j,t}^{(i)}\right)\right|^2}$$

$$\leq \sqrt{\sum_{j\in\mathcal{J}'} \frac{2}{n^2}\gamma^2} = \frac{\gamma}{n}\sqrt{2|\mathcal{J}'|}, \tag{7.13}$$

where $|\mathcal{J}'|$ is the carinality of the set $\mathcal{J}'$, and we used the fact that for a, b positive numbers, we have

$$|\ln(a) - \ln(b)| \leq \frac{|a - b|}{\min\{a, b\}}.$$

The previous sensitivity calculations allow us to use the Gaussian mechanism to publish all or a subset of the macroscopic measurements. In particular, for the estimator introduced in the following, we use only the occupancy and velocity measurements. The ℓ_2 sensitivity of a mechanism publishing the two sequences ψ, ξ is bounded by

$$\Delta_2^{\psi,\xi} \leq \sqrt{(\Delta_2^{\psi})^2 + (\Delta_2^{\xi})^2}.$$

We then have the following Corollary of Theorem 7.1.

Corollary 7.1 *A mechanism publishing the sequences* $\{\psi_{j,t} + w_{j,t}\}_{j\in\mathcal{J},t\in\mathcal{T}}$ *and* $\{\xi_{j,t} + w_{j,t}\}_{j\in\mathcal{J}',t\in\mathcal{T}'_j}$, *where the sequences* w *are white Gaussian noise sequences each with standard deviation* $\kappa_{\delta,\epsilon} \Delta_2^{\psi,\xi}$, *is* (ϵ, δ)-*differentially private.*

Alternatively, it is possible to use the Gaussian mechanism twice to publish the sequences $\psi_{j,t}$ and $\ln \xi_{j,t}$ independently with $(\epsilon_\psi, \delta_\psi)$ amd $(\epsilon_\xi, \delta_\xi)$ guarantees, and then use the composition theorem for differentially private mechanisms [14] to obtain an $(\epsilon_\psi + \epsilon_\xi, \delta_\psi + \delta_\xi)$ guarantee. This approach adds more noise overall than when considering the publication of the two sequences jointly as above, but might be more flexible when the impact of the noise on one sequence is very different from the impact on the other sequence (from the point of view of overall accuracy of the final estimator taking these sequences as inputs), and hence setting different values for the privacy parameters related to each of the two sequences might be beneficial.

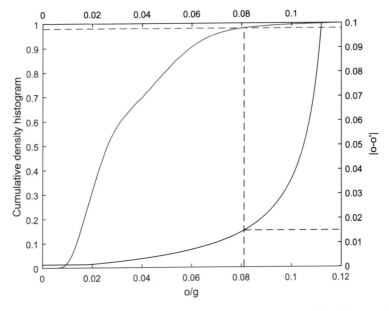

Fig. 7.4 Determination of the bound α (on the right axis). The curve on the right corresponds to a microscopic model of traffic [1] predicting for example that by setting the bound $\alpha = 0.015$, all vehicles that are captured in occupancy measurements from single-loop detectors such that $o_{j,t}^l/g_j < 0.081$ should be protected. The curve on the left gives, for illustration purposes, the proportion or measurements $o_{j,t}^l/g_j$ in the Mobile Century dataset [19] that are below a given value. ©[2017] IEEE. Reprinted, with permission, from [2]

To close this section, let us consider the question of setting appropriate values for the bounds α, γ appearing in the adjacency relation, see (7.11) and (7.12). These values should be set based on an acceptable trade-off between privacy and estimation performance. Higher values of α and γ provide better privacy (hide larger variations in individual trajectories) but require more noise to sanitize the data. In order to set α in particular, we can consider a theoretical microscopic traffic model of evenly spaced identical cars, and look at the impact of one car on occupancy measurements, see Fig. 7.4 and [1] for more details about this model. A change of trajectory for one car leads to an increasingly large variation in $o^l_{j,t}$ as $o^l_{j,t}/g_j$ (a measure of density) increases. This corresponds intuitively to the fact that as density increases (beyond ρ_C), traffic velocity decreases and each car takes more time to cross the sensor line. For example, the model tells us that by setting $\alpha = 0.015$, cars are protected as long as they appear only in single-loop detector measurements $o^l_{j,t}/g$ that are below 0.081. In the Mobile Century dataset used in Sect. 7.4.5, this represents 98% of occupancy measurements.

7.4.4   Ensemble Kalman Filter

7.4.4.1   Design Approach

When designing a differentially private traffic estimator, producing a sequence of density estimates $\{\hat{\rho}_t\}_{t \geq 0}$ from the traffic sensor measurements, a first step is to choose the set of macroscopic measurements to rely on. Whereas in absence of privacy constraint, using more measurements should typically not hurt the performance of an estimator, this is not necessarily so when privacy must be preserved, because the additional privacy-preserving noise necessary when a new data source is introduced might not be compensated by a sufficient increase in estimator performance.

In [24], we describe a traffic estimator relying only on the single-loop detector measurements. As mentioned already in Sect. 7.4.2 however, linking flow measurements (based on vehicle counts) to density values is problematic as it requires first estimating correctly the mode of traffic (free flow or congested), also in a privacy-preserving manner. In practice, we found it easier to use the velocity measurements from probe vehicles when they are available, and trying to use both sequences $\phi_{j,t}$ and $\xi_{j,t}$ led to a decrease in performance once sufficient privacy-preserving noise was added. Hence, in the following, we propose an estimator design relying only on the macroscopic measurement sequences $\psi_{j,t}$ and $\xi_{j,t}$.

Sanitizing the measurements as described in the previous section leads to new perturbed values $\bar{\psi}_{j,t}$, $\bar{\xi}_{j,t}$. These follow the same measurement models as in (7.8), (7.10), except for the additional Gaussian noise $w_{j,t}$, which is simply added to the model noises $\mu^o_{j,t}$ and $\mu^v_{j,t}$, in effect increasing their variance by the term specified in Corollary 7.1. These perturbed measurements can then combined with the dynamic model (7.5) to provide the final density estimate at each period. This amounts to a post-processing step, so the differential privacy guarantee obtained after sanitizing the measurements is preserved.

7.4.4.2 Description of the EnKF

A popular type of estimator for road traffic estimation is the Ensemble Kalman filter (EnKF) [15]. The EnKF is similar to the Kalman filter, but uses a set of particles, i.e., sampled state values, to compute the error covariance used to form the Kalman gain. The EnKF algorithm is described in Algorithm 7.1 for a generic state space system with dynamics $x_{t+1} = f(x_t, \omega_t)$ (with ω_t some noise) and linear measurements $y_t = H x_t + v_t$, where v_t has a known covariance R_t. In our case, the velocity measurement model (7.10) is nonlinear, i.e., $y_t = h(x_t) + v_t$, hence we use an extension of the EnKF discussed in more details in [15] for example, which works with augmented state vector $\hat{x} = \begin{bmatrix} x^T & h(x)^T \end{bmatrix}^T$ and the linear measurements $\hat{H} \begin{bmatrix} x \\ y \end{bmatrix} = y$.

Algorithm 7.1 EnKF algorithm

1: **for** $k = 1 \cdots n_p$ **do**
2:     $x_0^k \sim \pi_0$ ▷ Draw n_p samples from Gaussian prior π_0
3: **end for**
4: **for** $t \geq 0$ **do**
5:     **for** $k = 1 \cdots n_p$ **do**
6:         $x_t^k \leftarrow f(x_{t-1}^k, \omega_{t-1}^k)$ ▷ Prediction from the model
7:     **end for**
8:     $\bar{x}_t \leftarrow \frac{1}{n_p} \sum_{k=1}^{n_p} x_t^k$ ▷ Ensemble mean
9:     $E = \begin{bmatrix} x_t^1 - \bar{x}_t, \cdots, x_t^{n_p} - \bar{x}_t \end{bmatrix}$ ▷ Deviation from mean
10:    $P \leftarrow \frac{1}{K-1} E(E)^T$ ▷ Covariance matrix
11:    $K \leftarrow P H^T \begin{bmatrix} H P H^T + R_t \end{bmatrix}^{-1}$ ▷ Kalman gain
12:    **for** $k = 1 \cdots n_p$ **do**
13:        $\xi^k \sim N(0, R_t)$
14:        $x_t^k \leftarrow x_t^k + K \begin{bmatrix} y_t - H x_t^k + \xi^k \end{bmatrix}$ ▷ Measurement update
15:    **end for**
16:    Publish $\bar{x}_t = \frac{1}{n_p} \sum_{k=1}^{n_p} x_t^k$ ▷ Estimate
17: **end for**

Adjustment of the VTL locations. As shown by (7.13), sanitizing speed data with the Gaussian mechanism requires adding a noise with variance proportional to the number of VTLs. It is therefore important to work with a limited number of VTLs and set their locations in order to optimize the usefulness of the measurements. A scheme described in [1] adjusts the VTL locations online based on the size of the innovations $\| y_t - H \bar{x}_t \|$ in the measurement update step of the EnKF, while making sure that a vehicle does not report twice its speed for the same VTL. We adjust the positions of the VTLs, $r_1, \ldots, r_{|\mathcal{J}'|}$, using an approach based on the minimization of a potential function at each step. Given the current positions $\hat{r}_1, \ldots, \hat{r}_{|\mathcal{J}'|}$, we obtain the new positions by minimizing (locally) a function of the type

$$\sum_{s=1}^{|\mathcal{J}'|} \left[\left(\sum_{q=1}^{|\mathcal{J}'|} f_A(\hat{r}_q, r_s; h_q) \right) + \left(\sum_{\sigma \neq s} f_R(r_s, r_\sigma) \right) + f_B(r_s) \right].$$

Here $f_A(\hat{r}_q, r_s; h_q)$ is a field attracting r_s towards $\hat{r}_q$ with a strength that is increasing with h_q, the norm of the innovation for the measurement at location $\hat{r}_q$. The second term is a repulsive term maintaining sufficient spacing between the VTLs, and f_B is a term that prevents the VTL locations to escape outside of the road segment of interest.

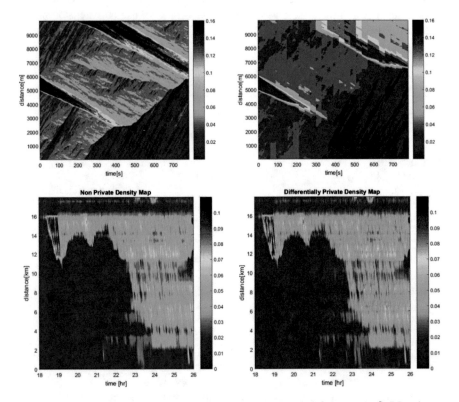

Fig. 7.5 Synthetic data (top): simulated traffic density map $\rho(x, t)$ (left) versus (ϵ, δ)-DP estimate. Data from the Mobile Century experiment (bottom): Non-private (left) versus (ϵ, δ)-DP (right) density map. The density is in vehicles/m. Some accuracy is lost with the differentially private estimate, but the traffic jams, including their front propagating backwards, remain clearly visible. Accuracy can be improved by setting the values of ϵ, δ higher. ©[2017] IEEE. Reprinted, with permission, from [2]

7.4.5   Simulation Results

Our differentially private EnKF is first validated on synthetic (simulated) data for a road with a single lane, which allows us to compare the estimator performance to the simulated ground truth, see Fig. 7.5. The simulation parameters are: $|\mathcal{J}| = 10$ single-loop detectors, $\tau = 0.5$ s, $\Delta x_j = 25$ m. We also use the values $v_0 = 90$ km/h, $w = 30$ km/h, $\rho_M = 1/7$ vehicles/m for the fundamental diagram, and $g_j = 6$ m for the g-factors. Occupation measurements are obtained periodically every 30 s. The differential privacy parameters are: $\epsilon = \ln(2 + |\mathcal{J}|)$, $\delta = 0.05$, and $\alpha = 0.015$, $\gamma = 0.4$ for the adjacency relation. It is found that an adequate number of particles for the EnKF is around 60, with more particles leading to negligible performance improvements. Figure 7.5 also shows the results of applying the differentially private EnKF to the Mobile Century dataset [19], with the same values as above for the various parameters. This dataset contains both occupancy measurements and GPS traces from floating cars for a day of traffic on Interstate 880 in California, which has either 4 or 5 lanes depending on the location.

7.5   Conclusion

To ensure individual privacy, the release of transportation microdata needs to follow general best practices established for other types of microdata, such as census data. Numerous studies, including the one carried out in this paper, have shown that it is essentially impossible to publish individual location traces with useful accuracy while providing meaningful privacy guarantees, an information that should be clearly conveyed in user agreements when recruiting survey participants. In general, the distribution of location microdata should probably remain limited to a small number of individuals or organizations for research and planning purposes, with clearly defined and restricted terms of use, such as a requirement not to attempt to re-identify individuals.

The publication of aggregate quantities, such as traffic maps or regional passenger demand, still poses privacy risks, but such risks can be managed rigorously within well-defined and formal privacy-preserving data analysis frameworks, such as ensuring a certain level of differential privacy. Ensuring differential privacy for the release of real-time data is particularly challenging. This paper has presented an Ensemble Kalman filter for differentially private traffic estimation, which benefits in particular from a macroscopic physical model of traffic to improve the estimate's accuracy for a given level of privacy. However, much work remains to be done for the various types of datasets and statistics arising in transportation research.

Acknowledgements The authors thank H. André for his work on the differentially private Ensemble Kalman filter [1, 2], which formed the basis for Sect. 7.4 of this chapter. The authors also thank François Bélisle for his help with the MTL Trajet dataset and SQL queries. This work was supported in part by FRQNT through Grant 2015-NC-181370 and by NSERC through Grant RGPAS-507950.

References

1. André H (2017) Estimation de trafic routier par filtre de Kalman d'ensemble sous contrainte de confidentialité différentielle. Master's thesis, Polytechnique Montreal
2. André H, Le Ny J (2017) A differentially private ensemble Kalman filter for road traffic estimation. In: IEEE international conference on acoustics, speech and signal processing (ICASSP), pp 6409–6413
3. Andrés ME, Bordenabe N, Chatzikokolakis K, Palamidessi C (2013) Geo-indistinguishability: differential privacy for location-based systems. In: Proceedings of the ACM SIGSAC conference on computer and communications security (CCS'13)
4. Blum A, Dwork C, McSherry F, Nissim K (2005) Practical privacy: the SuLQ framework. In: Proceedings of the twenty-fourth ACM SIGMOD-SIGACT-SIGART symposium on principles of database systems (PODS). New York, NY, USA, pp 128–138
5. Canepa ES, Claudel CG (2013) A framework for privacy and security analysis of probe-based traffic information systems. In: Proceedings of the 2nd ACM international conference on High confidence networked systems (HiCoNS), pp 25–32
6. City of Montreal: results of 2017 study | MTL trajet (2018). https://ville.montreal.qc.ca/mtltrajet/en/etude/
7. City of Montreal: déplacements MTL trajet (2019). http://donnees.ville.montreal.qc.ca/dataset/mtl-trajet
8. Daganzo CF (1994) The cell transmission model: a dynamic representation of highway traffic consistent with the hydrodynamic theory. Trans Res Part B Methodol 28(4):269–287
9. de Montjoye YA, Hidalgo CA, Verleysen M, Blondel VD (2013) Unique in the crowd: the privacy bounds of human mobility. Scientific Reports 3
10. Douriez M, Doraiswamy H, Freire J, Silva CT (2016) Anonymizing NYC taxi data: does it matter? In: 2016 IEEE international conference on data science and advanced analytics (DSAA). IEEE, pp 140–148
11. Dwork C (2006) Differential privacy. In: Proceedings of the 33rd international colloquium on automata, languages and programming (ICALP), Lecture notes in computer science, vol 4052. Venice, Italy
12. Dwork C, Kenthapadi K, McSherry F, Mironov I, Naor M (2006) Our data, ourselves: privacy via distributed noise generation. In: Proceedings of the 24th annual international conference on the theory and applications of cryptographic techniques (EUROCRYPT). St. Petersburg, Russia, pp 486–503
13. Dwork C, McSherry F, Nissim K, Smith A (2006) Calibrating noise to sensitivity in private data analysis. In: Proceedings of the third theory of cryptography conference. New York, NY, pp 265–284
14. Dwork C, Roth A (2014) The algorithmic foundations of differential privacy. Found Trends Theor Comput Sci 9(3–4):211–407
15. Evensen G (2003) The ensemble Kalman filter: theoretical formulation and practical implementation. Ocean Dyn 53(4):343–367
16. Fan L, Xiong L, Sunderam V (2013) Differentially private multi-dimensional time series release for traffic monitoring. In: 27th conference on data and applications security and privacy, Lecture notes in computer science, vol 7964. Springer, pp 33–48
17. Gambs S, Killijian MO, del Prado Cortez MN (2014) De-anonymization attack on geolocated data. J Comput Syst Sci 80(8):1597–1614. (Special issue on theory and applications in parallel and distributed computing systems)
18. Ghinita G (2013) Privacy for location-based services. Morgan & Claypool Publishers
19. Herrera JC, Work DB, Herring R, Ban X, Jacobson Q, Bayen AM (2010) Evaluation of traffic data obtained via GPS-enabled mobile phones: the Mobile Century field experiment. Trans Res Part C Emerg Technol 18(4):568–583
20. Ho SS, Ruan S (2011) Differential privacy for location pattern mining. In: Proceedings of ACM SPRINGL, pp 17–24

21. Hoh B, Iwuchukwu T, Jacobson Q, Gruteser M, Bayen A, Herrera JC, Herring R, Work D, Annavaram M, Ban J (2012) Enhancing privacy and accuracy in probe vehicle based traffic monitoring via virtual trip lines. IEEE Trans Mobile Comput 11(5)
22. Jia Z, Chen C, Coifman B, Varaiya P (2001) The PeMS algorithms for accurate, real-time estimates of g-factors and speeds from single-loop detectors. In: Proceedings of the 4th IEEE conference on intelligent transportation systems
23. Le Ny J, Pappas GJ (2014) Differentially private filtering. IEEE Trans Autom Control 59(2):341–354
24. Le Ny J, Touati A, Pappas GJ (2014) Real-time privacy-preserving model-based estimation of traffic flows. In: Proceedings of the fifth international conference on cyber-physical systems (ICCPS)
25. Li N, Li T, Venkatasubramanian S (2007) t-closeness: privacy beyond k-anonymity and l-diversity. In: Proceedings of the 23rd IEEE international conference on data engineering
26. Machanavajjhala A, Kifer D, Abowd JM, Gehrke J, Vilhuber L (2008) Privacy: theory meets practice on the map. In: Proceedings of IEEE ICDE, pp 277–286
27. Narayanan A, Shmatikov V (2008) Robust de-anonymization of large sparse datasets (how to break anonymity of the Netflix Prize dataset). In: Proceedings of the IEEE symposium on security and privacy
28. Pelletier MP, Trépanier M, Morency C (2011) Smart card data use in public transit: a literature review. Trans Res Part C Emerg Technol 19(4):557–568
29. Pyrgelis A, Troncoso C, Cristofaro ED (2017) What does the crowd say about you? evaluating aggregation-based location privacy. Proc Priv Enhanc Technol 4:156–176
30. Shokri R, Troncoso C, Diaz C, Freudiger J, Hubaux JP (2010) Unraveling an old cloak: k-anonymity for location privacy. In: Proceedings of the 9th annual ACM workshop on privacy in the electronic society. ACM, pp 115–118
31. Sweeney L (1997) Weaving technology and policy together to maintain confidentiality. J Law Med Ethics 25:98–110
32. Sweeney L (2002) k-anonymity: a model for protecting privacy. Int J Uncertain Fuzziness Knowl Based Syst 10(05):557–570
33. Treiber M, Kesting A (2013) Traffic flow dynamics. Traffic flow dynamics: data, models and simulation. Springer, Berlin
34. Work DB, Tossavainen OP, Blandin S, Bayen AM, Iwuchukwu T, Tracton K (2008) An ensemble Kalman filtering approach to highway traffic estimation using GPS enabled mobile devices. In: Proceedings of the 47th IEEE conference on decision and control, pp 5062–5068
35. Xin W, Chang J, Muthuswamy S, Talas M (2013)"Midtown in Motion": a new active traffic management methodology and its implementation in New York City. In: Transportation research board annual meeting
36. Xu F, Tu Z, Li Y, Zhang P, Fu X, Jin D (2017) Trajectory recovery from ash: user privacy is not preserved in aggregated mobility data. In: Proceedings of the 26th international conference on world wide web, pp 1241–1250
37. Zhang H, Bolot J (2011) Anonymization of location data does not work: a large-scale measurement study. In: Proceedings of the 17th annual international conference on mobile computing and networking

Chapter 8
On the Role of Cooperation in Private Multi-agent Systems

Vaibhav Katewa, Fabio Pasqualetti and Vijay Gupta

Abstract We consider a distributed quadratic optimization problem for multi-agent systems, where the agents wish to maintain privacy of their states over time. To this aim, the agents add noise to their communicated states using the differential privacy framework. We characterize the performance degradation due to the noise and show that depending on the desired level of privacy (and thus noise), the system performance is optimized by reducing the level of cooperation among the agents. The notion of cooperation level, which is formally introduced and defined in the chapter, models the trust of an agent towards the information received from neighboring agents. We characterize the optimum cooperation level and show that under certain conditions, it is always beneficial for the agents to reduce their cooperation level when the privacy level increases. We illustrate our results using the average consensus problem.

8.1 Introduction

Cooperation among the agents is a key requirement in the functioning of multi-agent systems. Typical problems in this field include consensus [3, 20], flocking [19], formation control [22], coverage control [1], and distributed optimization [18, 26]. While cooperation helps the agents to collaboratively achieve a common goal, it may lead to leakage of private information due to information sharing among the agents. For example, smart meters that transmit real-time electricity usage data to a control center can reveal the presence and daily schedules of people [16, 27]. In autonomous vehicle applications, the position and velocity data transmitted by the vehicles can

V. Katewa · F. Pasqualetti
Department of Mechanical Engineering, University of California Riverside, Riverside, CA, USA
e-mail: vkatewa@engr.ucr.edu

F. Pasqualetti
e-mail: fabiopas@engr.ucr.edu

V. Gupta (✉)
Department of Electrical Engineering, University of Notre Dame, Notre Dame, IN, USA
e-mail: vgupta2@nd.edu

© Springer Nature Singapore Pte Ltd. 2020
F. Farokhi (ed.), *Privacy in Dynamical Systems*,
https://doi.org/10.1007/978-981-15-0493-8_8

reveal travel plans of the drivers. Even if the agents are trustworthy, a possibility exists for an intruder to eavesdrop on the messages exchanged among the agents and gather their private information. Recently, noise-adding mechanisms have been proposed to implement privacy in multi-agent systems [7–12, 14], where each agent deliberately adds noise to the data communicated to other agents. This prevents the agents (or an eavesdropper) from recovering the sensitive data of individual agents by accurately processing the distorted messages.

These privacy mechanisms usually degrade the performance of dynamical systems and may even result in instability. Further, due to the cooperative and dynamical nature of the distributed multi-agent systems, the noise added by an agent adversely affects the state evolution of all the agents in the system. Thus, intuitively, there should be a tradeoff between the 'cooperation level' and performance in a distributed system when the agents are trying to keep their information private. If the noise level introduced to maintain privacy is too high, then cooperation might even impede the system functionality. On the other hand, if the agents do not cooperate and transmit information to each other, perfect privacy is achieved, at the expense of the benefits of cooperation. In this chapter, we study this fundamental trade-off between the two.

We consider a distributed quadratic optimization problem, where the goal of the agents is to cooperatively minimize the quadratic cost function of their states by sharing their state information among each other. Several problems such as consensus and formation control fall into this class. In addition, the agents wish to keep their states private during this process. We propose a noise-adding privacy mechanism for the agents to keep their states private over time, using the Differential Privacy (DP) framework [4, 14]. We introduce a method for the agents to adapt their cooperation level in response to the privacy noise. This method is motivated by the fact that in many scenarios, in addition to optimizing a common global cost, the agents also have individual goals that do not require cooperation. For example, in intelligent transport systems, the global goal would be reduce congestion in the system via cooperative routing, and individual goal would be reduce own's travel time or fuel consumption [24]. Individual objectives also exist in multi-objective optimization problems, wherein multiple conflicting goals are considered, and in optimization problems with separable cost functions. Intuitively, if the agents wish a higher level of privacy, they should reduce cooperation and focus more on their individual goals, and vice-versa. Thus, the cooperation level can be characterized based on whether the agents are willing to cooperatively minimize the global cost, or they want to selfishly minimize their individual costs. We formalize this notion by defining a new cost that is a convex combination of the global and individual costs, wherein the weighing factor represents the *cooperation level*. We then characterize the combined effect of cooperation level and privacy noise on the system performance.

Related Work Much of the development in this chapter follows [13]. Several secure multi-party computation schemes exist in literature which compute a function of agents' variables while keeping them private [15, 21]. However, in these schemes, there always exists a possibility that some agent(s) obtain auxiliary information and use it to infer other agents' private variables. Moreover, majority of agents can collude to infer the remaining agents' sensitive information. To address these issues, we use the differential privacy framework in this work. DP abstracts away from any auxiliary information that the agents might have and it is also resilient to post processing of data [5, 14].

Many recent studies have proposed privacy mechanisms for multi-agent systems. In [11], the authors present a differentially private consensus algorithm that protects the initial state of the agents, and illustrate privacy vs accuracy tradeoff. In [17], a private consensus algorithm is developed using correlated noises, that achieves perfect accuracy. We also study the consensus algorithm as an example and our DP mechanism is similar to [11]. However, we have a different goal of analyzing the cooperation vs privacy tradeoff. Some papers have addressed privacy issues in optimization problems. In [8, 12], the authors study distributed convex optimization and optimization with piecewise affine objectives respectively, and develop DP mechanisms to keep the agents' cost functions private. In [9, 10], the authors develop DP mechanisms to keep constraints of the agents private in optimization problems with convex and linear objectives, respectively. In contrast, our goal is to keep private the entire state trajectories of the agents in a quadratic optimization problem. In [7], the authors develop DP mechanisms to keep the agents' state trajectories private in a convex optimization problem with non-linear constraints. All of these works develop privacy mechanisms and analyze their effect of the on the system performance in terms of sub-optimality, accuracy, convergence etc. Thus, they are primarily concerned with the privacy vs performance tradeoff. In contrast, we study a privacy vs cooperation tradeoff by simultaneously characterizing the effect of cooperation and privacy noise on the system performance.

The cooperation level in our framework can be viewed as a weighting factor for the noisy state information received from the neighbors and used to update the agents states. Related works include [23], in which the authors analyze consensus in the presence of noise, and show that almost sure convergence can be guaranteed by using time decaying weighing factor in the updates. In [28], the authors find the optimal edge weights for the consensus problem that minimize the expected deviation among the agents. These works are specifically developed for the consensus algorithm, and may not work for other problems. In contrast, to elucidate the relation between the cooperation and privacy levels in multi-agent systems, we develop techniques that are applicable to more general quadratic optimization problems and not only limited to consensus.

Chapter Organization Section 8.2 presents the quadratic optimization setup, the privacy mechanism and its effect on the system performance. Section 8.3 presents a framework to include a cooperation parameter in the problem and quantifies the performance as a function of privacy noise and agents' cooperation level. Section 8.4 illustrates our findings by studying the consensus problem. Finally, Sect. 8.5 concludes the chapter.

Mathematical Notation $\|.\|$ denotes the euclidean norm of a vector, or the induced 2-norm of a matrix. $y[0 : \infty]$ denotes an infinite sequence/trajectory. The truncated version of y up to time $T \in \mathbb{N}$ is denoted by $y[0 : T]$. Without loss of generality, we also treat a truncated sequence as a vector of appropriate dimension. For an $N \times N$ Hermitian matrix Q, the ordered real eigenvalues are $\lambda_1(Q) \geq \lambda_2(Q) \geq \cdots \geq \lambda_N(Q)$. Further, $Q \geq 0$ (respectively $Q > 0$) denotes that Q is positive semi-definite (respectively definite). For a square matrix A, $\rho(A)$ denotes the spectral radius of A and diag(A) denotes the diagonal matrix containing the diagonal entries of A. $tr(.)$ is the trace operator. I_N denotes the $N \times N$ identity matrix, $\mathbf{1}_N = [1, 1, \ldots, 1]^T \in \mathbb{R}^N$ and, $\mathbf{0}_N = [0, 0, \ldots, 0]^T \in \mathbb{R}^N$. The Q-function is $Q(x) = \frac{1}{\sqrt{2\pi}} \int_x^\infty e^{-\frac{u^2}{2}} du$. $\mathbf{N}(0, \Sigma)$ denotes the standard normal distribution with mean 0 and covariance matrix Σ. Im($\cdot$) denotes the image of a matrix.

8.2 Problem Formulation

In this section we introduce the problem setup of multi-agent cooperative quadratic optimization and present a gradient-based distributed solution. Further, we present a noise adding mechanism that preserves the privacy of the agents' states over time.

8.2.1 A Motivating Example: Consensus of Opinion Dynamics

Opinion dynamics is the study of the evolution of the opinions of individuals in a group over time. The consensus problem requires the individuals to arrive at a common opinion by exchanging information with each other. This is a typical scenario in social networks where an individual's opinion is affected over time by the interactions and opinions of his/her friends. Consensus problems in opinion dynamics are well studied and it is known that consensus can be reached by a group through interaction among the "neighbors" [2]. Specifically, let a group with N individuals be modeled by an undirected weighted graph $\mathcal{G} = (\mathcal{N}, \mathcal{E})$, where $\mathcal{N}$ denotes the set of vertices representing the individuals, and $\mathcal{E}$ denotes the set of edges that represent interactions among them. Let $L = [l_{ij}]$ denote the weight of the edge between vertices i and j. Further, let $\mathcal{N}_i$ denote the set of "neighbors" or "friends" of individual i with whom he/she interacts directly, that is, $\mathcal{N}_i = \{j \in \mathcal{N}, j \neq i : q_{ij} \neq 0\}$. Moreover, let the opinion of individual i be denoted by a real number x_i, which, for example, can be a numerical representation of how strongly one feels about a particular social issue. Then, an algorithm to achieve consensus of opinion dynamics is as follows:

$$x_i(k+1) = x_i(k) - l_{ii}x_i(k) - \sum_{j \in \mathcal{N}_i} l_{ij}x_j(k), \qquad i = 1, 2, \ldots, n. \qquad (8.1)$$

If L is a weighted Laplacian matrix, all the individuals asymptotically converge to a common opinion, which is the average of their initial opinions. That is,

$\lim\limits_{k\to\infty} x(k) = \mu \mathbf{1}_N$, where $\mu = \frac{1}{N}\sum_{i=1}^{N} x_i(0)$. Note that the above iterations can be viewed as a distributed algorithm to minimize the following disagreement function in a distributed manner [20]

$$J(x) = \frac{1}{2}x^T Lx = \frac{1}{4}\sum_{i=1}^{N} \sum_{j:j\in\mathcal{N}_i} -l_{ij}(x_i - x_j)^2. \tag{8.2}$$

Achieving consensus requires the individuals to cooperate by sharing their personal opinions among their friends. However, in many cases, people might be uncomfortable in sharing their opinion with others and may want to keep them private. A possible solution in this scenario would be to deliberately share inaccurate opinions, which prevents individuals from knowing each other's exact opinions. A higher degree of privacy among the individuals will require a higher level of inaccuracy, and, as a result, the consensus performance will be affected to a larger extent. In such cases, it is prudent for an individual to reduce cooperation among themselves by considering the inaccurate opinions of their friends to a lesser degree. Instead, one should provide more weightage to personal opinion, which for example, can represent a personal bias. Thus, the group can perform better if individuals adapt their cooperation level according to the degree of trust among each other.

8.2.2 A Framework for Distributed Quadratic Optimization

Consider a distributed system with a set of $N \geq 2$ agents denoted by $\mathcal{N} = \{1, 2, \ldots, N\}$. The agents cooperatively aim to minimize a common quadratic objective given by

$$\mathbf{P}: \qquad \min_x \ J_{co}(x) = \frac{1}{2}x^T Qx + r^T x + s, \tag{8.3}$$

where $Q = [q_{ij}] \in \mathbb{R}^{N \times N}$, $r = [r_i] \in \mathbb{R}^N$, $s \in \mathbb{R}$, and the vector $x = [x_1, x_2, \ldots, x_N]^T \in \mathbb{R}^N$ denotes the states of all the agents. The states of the agents are coupled with each other via the quadratic term $\frac{1}{2}x^T Qx$ in the cost function. Specifically, we say that the agents i and j are uncoupled if both q_{ij} and q_{ji} are zero, and that they are coupled otherwise. The above optimization problem captures the opinion dynamics example presented in Sect. 8.2.1, as well as a number of other problems including consensus and formation control. Let $\mathcal{N}_i$ denote the *neighbor set* or the set of agents whose states are coupled to the state of agent i, and let $N_i = |\mathcal{N}_i|$. We place the following assumptions on the cost function $J_{co}(x)$:

(**A.1**) Q is symmetric and positive semi-definite. Further, if 0 is an eigenvalue of Q, then: (i) its algebraic multiplicity is 1, and (ii) $r \in \text{Im}(Q)$.

(**A.2**) Each row(or column) of $Q - \text{diag}(Q)$ has atleast one non-zero entry.

Assumption **A.2** prevents the case with any uncoupled agent(s) and Assumption **A.1** implies that the minimization problem **P** is convex and admits a finite (but not necessarily unique) solution. Let the set of all the optimum solutions of **P** be denoted by $\mathcal{X}^*$. An optimum $x^* \in \mathcal{X}^*$ can be achieved by the agents with a distributed iterative gradient descent algorithm

$$x_i(k+1) = x_i(k) - \gamma_1 \frac{\partial}{\partial x_i} J_{co}(x(k))$$

$$= x_i(k) - \gamma_1 \left(q_{ii} x_i(k) + \sum_{j \in \mathcal{N}_i} q_{ij} x_j(k) + r_i \right), \quad i = 1, \ldots, n \quad (8.4)$$

where $\gamma_1 > 0$ is the step size and $x_i(0)$ is the initial state of agent i. The algorithm requires exchange of states among the neighbors, and we assume that the agents can communicate their state to each other without any distortion. The gradient descent algorithm for all agents is collectively given by

$$\mathbf{S_1}: \quad x(k+1) = x(k) - \gamma_1(Qx(k) + r) \triangleq A_1 x(k) + b_1, \quad (8.5)$$

where $A_1 = I_N - \gamma_1 Q$, $b_1 = -\gamma_1 r$ and the initial state is given by $x(0) = [x_1(0), x_2(0), \ldots, x_N(0)]^T$. Since the gradient of the cost is linear, algorithm $\mathbf{S_1}$ can be represented as a discrete linear time-invariant system. Next, we derive the condition for the existence of the steady state solution of $\mathbf{S_1}$, and show that the steady state solution is the optimum of problem **P**.

Lemma 8.1 (Convergence of algorithm $\mathbf{S_1}$) *Let $\gamma_1 < 2\rho(Q)^{-1}$. Then, the algorithm $\mathbf{S_1}$ in (8.5) converges asymptotically to the optimum of **P**, that is $\lim_{k \to \infty} x(k) = x^*$ for an $x^* \in \mathcal{X}^*$.*

Proof First, assume $Q > 0$. For $i = 1, 2, \ldots, N$ we have $0 < \gamma_1 \lambda_i(Q) \leq \gamma_1 \lambda_1(Q) < 2$. Since $\lambda_i(A_1) = 1 - \gamma_1 \lambda_i(Q)$, the above condition is equivalent to $-1 < \lambda_i(A_1) < 1$. Thus, all eigenvalues of A_1 lie inside the unit circle and a steady state solution of (8.5) is achieved. Assume now that Q has a 0 eigenvalue. Then, by assumption **A.1**, $b_1 = 0$ and A_1 has a single eigenvalue at 1 and all other eigenvalues lie inside the unit circle. Thus, the linear system in (8.5) is marginally stable and a finite steady state solution is achieved. Further, it can be easily observed that the steady state solution of (8.5) satisfies the first order optimality condition $\nabla J_{co}(x) = Qx + r = 0_N$ of problem **P**. Thus, it minimizes $J_{co}(x)$ and is an optimum of **P**. $\square$

The steady state of (8.5), denoted by m_1 satisfies

$$m_1 = A_1 m_1 + b_1, \quad (8.6)$$

and the optimum cost achieved by the agents is given by

$$J_{co}^* \triangleq [J_{co}(x)]_{\mathbf{S}_1} = \frac{1}{2} m_1^T Q m_1 + r^T m_1 + s, \tag{8.7}$$

where the notation $[.]_{\mathbf{S}_1}$ denotes the cost achieved by the steady state of algorithm $\mathbf{S}_1$.

8.2.3 Differential Privacy Mechanism

In the cooperative algorithm $\mathbf{S}_1$, the agents update their state upon communicating the state information with their neighbors. Thus, algorithm $\mathbf{S}_1$ is not private and in fact, an agent/intruder may reconstruct the state trajectories of the agents with access to only a few messages communicated by the agents. To ensure privacy, we consider the Differential Privacy (DP) mechanism that protects the state of the agents over time, where each agent adds an artificial random noise to its state before communicating it with other agents. The noise ensures that *"any two different instances"* of the communicated state trajectories are *"statistically not very different"*, which prevents the intruder from accurately obtaining the actual state information of the agents; thus, maintaining their privacy. For the motivation and more information on DP, interested readers are referred to [4, 5].

Remark 8.1 (Privacy of state trajectory) Note that different instances of the state trajectory arise from different initial states $x(0)$. However, in addition to the initial state, agents wish to keep their states private at all times because accurate state information at any time instant can potentially reveal the complete future state trajectory. □

The noisy DP mechanism can be written as

$$\mathcal{M}: \qquad \tilde{x}_i(k) = x_i(k) + n_i(k), \tag{8.8}$$

where $\tilde{x}_i(k)$ denotes the state communicated to the neighbors of agent i and $n_i(k)$ is the random privacy noise. Let $n(k) = [n_1(k), n_2(k), \ldots, n_N(k)]^T$. We adopt the differential privacy framework developed in [14] to design the noise that ensures privacy of the state trajectories. We begin with the definition of adjacency.

Definition 8.1 (*Adjacency*) Given a finite $\beta \geq 0$, two state trajectories $x[0 : \infty]$ and $x'[0 : \infty]$ are β-adjacent (denoted by $adj(\beta)$) if

$$\left\| x[0 : \infty] - x'[0 : \infty] \right\| \leq \beta. \tag{8.9}$$

It should be noticed that in the classic definitions of DP for static databases [4] and for dynamical systems [14], adjacency is defined with respect to the change of trajectory of one agent only, while keeping the trajectories of other agents unchanged. In contrast, our definition of adjacency allows simultaneous changes in the trajectories of one or more agents. □

Remark 8.2 (Common steady state value) The adjacency definition in (8.9) implic-
itly requires that the two instances of the state trajectories (resulting from two different
initial conditions) vary only for transient periods and have a common steady state
value. This holds true if $Q > 0$, since it is easy to observe (see (8.6)) that the steady
state value does not depend on the initial condition $x(0)$. However, when $Q \geq 0$
with a single eigenvalue at 0 (see **A.1**), then the steady state value might depend on
the initial condition. Let $\mathcal{X}_0^m$ denote the set of all initial conditions that result in a
steady state value of m. Then, the privacy mechanism guarantees DP only among
those trajectories that result from initial conditions contained in the set $\mathcal{X}_0^m$.   □

Let $\tilde{x}[0 : \infty]$ and $\tilde{x}'[0 : \infty]$ denote the corresponding noisy communicated state
trajectories. Note that $\tilde{x}[0 : T] \in \mathbb{R}^{N(T+1)}$ and let $\mathcal{R}^{N(T+1)}$ denote the $\sigma - algebra$
generated by it. Next, we provide the definition of differential privacy.

Definition 8.2 (*Differential privacy*) The mechanism $\mathcal{M}$ in (8.8) is (ϵ, δ)-
differentially private if for any two β-adjacent trajectories $x[0 : \infty]$ and $x'[0 : \infty]$
and for all $S \in \mathcal{R}^{N(T+1)}$ and for all $T \geq 0$ it holds

$$\mathbb{P}[\tilde{x}[0 : T] \in S] \leq e^{\epsilon}\mathbb{P}[\tilde{x}'[0 : T] \in S] + \delta, \tag{8.10}$$

where $\epsilon > 0$ and $0 < \delta < 0.5$ are privacy parameters.   □

Definition 8.2 implies that for any two beta adjacent trajectories, the statistics of
the corresponding noisy communicated state trajectories differ only within a multiplicative factor of e^{ϵ} and an additive factor of δ. A standard way to guarantee DP is
to choose an i.i.d. Gaussian noise and scale its variance according to the adjacency
parameter β, as stated in the next lemma.

Lemma 8.2 (Ensuring differential privacy) *The mechanism $\mathcal{M}$ in (8.8) is (ϵ, δ)-
differentially private if $n(k)$ is white Gaussian noise with distribution $n(k) \sim
N(0, \sigma^2 I_N)$, where $\sigma \geq \frac{\beta}{2\epsilon}(K + \sqrt{(K^2 + 2\epsilon)})$ and $K = Q^{-1}(\delta)$.*

Proof See Theorem 3 in [14]. Since the quantity that needs to be protected and that
is communicated is same (i.e. the state trajectory), the sensitivity is trivially upper
bounded by the adjacency parameter β.   □

In Lemma 8.2, the relation between σ and the privacy parameters (ϵ, δ) implies
that the noise variance is a monotonically decreasing function of ϵ and δ. Also, note
from the definition of DP in (8.10) that a smaller value of ϵ and δ implies larger
privacy of the agents. Thus, the noise variance σ can be treated as synonymous to the
privacy level of the system, and for the ease of presentation, we present our results
directly in terms of noise level σ (instead of the privacy parameters ϵ and δ). In the
presence of privacy noise, the evolution of the algorithm $\mathbf{S_1}$ in (8.5) is modified as

$$\mathbf{S_1^{priv}} : \quad x(k + 1) = A_1 x(k) + b_1 + H_1 n(k), \tag{8.11}$$

where $H_1 \triangleq A_1 - \text{diag}(A_1)$ is obtained by replacing the diagonal elements of A_1 with zero entries, since only non-diagonal entries in A_1 represent coupling between the agents. Note that $H_1 = -\gamma_1 \tilde{Q}$ where $\tilde{Q} = Q - \text{diag}(Q)$.

While the added noise makes the states of the agents private, it also adversely affects the system performance. Due to the stochastic nature of algorithm $\mathbf{S}_1^{\text{priv}}$, we analyze the system performance by calculating the expected cost achieved by the agents in the presence of noise. The algorithm $\mathbf{S}_1^{\text{priv}}$ in (8.11) can be viewed as a linear system driven by a constant input and Gaussian privacy noise. Thus, the state of the agents at each time instant has a normal distribution, denoted by $x(k) \sim \mathbf{N}(m_1(k), P_1(k))$. The evolution of the mean and the covariance of the states of the agents is given by

$$m_1(k+1) = A_1 m_1(k) + b_1, \quad \text{and} \tag{8.12}$$
$$P_1(k+1) = A_1 P_1(k) A_1^T + \sigma^2 H_1 H_1^T, \tag{8.13}$$

with $m_1(0) = x(0)$ and $P_1(0) = 0$. If A_1 is stable (i.e. $Q > 0$ in (8.3)), then the mean and covariance reach a finite steady state value, denoted by m_1 and P_1, respectively. Note that we have overloaded the notation of m_1 with the noiseless case presented in Sect. 8.2.2 since the steady state solution of (8.5) and (8.12) are the same. Thus, the steady state mean m_1 satisfies (8.6) and the steady state covariance P_1 satisfies the following Lyapunov equation:

$$P_1 = A_1 P_1 A_1^T + \sigma^2 H_1 H_1^T. \tag{8.14}$$

If A_1 is marginally stable (that is, Q in (8.3) has a single eigenvalue at 0, see Assumption **A.1**), then m_1 exists and is finite. Yet, the covariance P_1 may become unbounded, and the system becomes unstable in the stochastic sense. We now present the performance result.

Theorem 8.1 (Performance in the presence of privacy noise) *Assume $Q > 0$. At steady state, the expected cost achieved by the agents implementing the algorithm* $\mathbf{S}_1^{\text{priv}}$ *in (8.11) is given by*

$$J_{co}^*(\sigma) \triangleq \mathbb{E}[J_{co}(x)]_{\mathbf{S}_1^{\text{priv}}} = \frac{1}{2} tr(Q P_1) + \frac{1}{2} m_1^T Q m_1 + r^T m_1 + s. \tag{8.15}$$

where expectation $\mathbb{E}[.]$ is taken w.r.t the privacy noise and P_1 depends on σ.

Proof Since $Q > 0$, its Cholesky decomposition exists, denoted by $Q = L^T L$. Further, let x denote the random steady state and let $y = Lx$. If $x \sim \mathbf{N}(m_1, P_1)$, then $y \sim \mathbf{N}(Lm_1, LP_1L^T)$. We have,

$$\mathbb{E}[J_{co}(x)]_{S_1^{priv}} = \mathbb{E}[\frac{1}{2}x^T Q x + r^T x + s] = \frac{1}{2}\mathbb{E}[y^T y] + r^T m_1 + s$$

$$= \frac{1}{2}tr(\mathbb{E}[yy^T]) + r^T m_1 + s$$

$$= \frac{1}{2}tr(L P_1 L^T + L m_1 (L m_1)^T) + r^T m_1 + s$$

$$= \frac{1}{2}(tr(Q P_1) + m_1^T Q m_1) + r^T m_1 + s,$$

where we have used the fact that $tr(.)$ is a linear and invariant under cyclic permutations.   □

The performance degradation due to privacy noise is given by (comparing (8.7) and (8.15)) $J_{co}^*(\sigma) - J_{co}^* = \frac{1}{2}tr(Q P_1)$, which increases with the noise level σ. Because the agents share noisy state information, full cooperation and use of the distorted information in the algorithm adversely affect the agents performance. In the other extreme case, if the agents forgo cooperation, then they will be completely private, since no state information will be exchanged among them, but will probably not achieve the optimum of problem **P**. Thus, a mechanism is needed for the agents to adapt their level of cooperation to maximize their performance in presence of privacy noise. In the next section, we define a notion of *cooperative level*, and present modified optimization algorithms that incorporate the cooperation level as a parameter.

8.3   Cooperation Level in Multi-agent Systems

In this section, we introduce and motivate our notion of cooperation level in private multi-agent systems. We characterize the expected cost achieved by the agents for a particular level of cooperation and privacy noise, and use it to characterize the optimum cooperation level.

8.3.1   A Notion of Cooperation Level

Agents cooperate to implement algorithm S_1. However, as discussed above, full cooperation may not be optimal if agents also want to preserve privacy. To formalize this, we introduce a cooperation parameter in the algorithm S_1. In many scenarios, in addition to minimizing the system cost J_{co}, the agents also have individual goals for which no cooperation is required. As explained in the introduction, such conflicting goals are ubiquitous in optimization and game theory. We formalize the individual agent goals by the following non-cooperative cost function

$$J_{nco}(x) = \frac{1}{2}x^T \bar{Q}x + \bar{r}^T x + \bar{s}, \qquad (8.16)$$

and assume that

(A.3) $\bar{Q}$ is diagonal and positive definite.

Assumption **A.3** implies that the states of all the agents are decoupled in $J_{nco}(x)$. As a result, no cooperation is required to minimize the decoupled cost function $J_{nco}(x)$.

We utilize these individual agent goals to introduce *cooperation level* in our framework. The costs $J_{co}(x)$ and $J_{nco}(x)$ represent two extremes on the cooperation scale. To minimize the former, full cooperation is necessary among the agents, while no cooperation is required for the latter. When the agents wish to remain private, it is prudent for them to give more weight to their individual goals as compared to the system goal. Following this reasoning and to capture the intermediate cooperation behavior, we consider a new cost function which is precisely the convex combination of $J_{co}(x)$ and $J_{nco}(x)$:

$$J_\alpha(x) = \alpha J_{co}(x) + (1 - \alpha) J_{nco}(x), \qquad (8.17)$$

where the parameter $\alpha \in [0, 1]$ is the agents' *cooperation level*. Note that the new cost $J_\alpha(x)$ is convex due to convexity of $J_{co}(x)$ and $J_{nco}(x)$. The gradient descent algorithm that minimizes J_α inherently introduces the cooperation level in our framework, and in presence of privacy noise can be written as

$$\mathbf{S}^{\mathbf{priv}} : \quad x(k+1) = x(k) - \gamma \left(\alpha(Qx(k) + r) + (1 - \alpha)(\bar{Q}x(k) + \bar{r}) \right) + H_\alpha n(k)$$
$$\triangleq A_\alpha x(k) + b_\alpha + H_\alpha n(k), \qquad (8.18)$$

where $Q_\alpha = \alpha Q + (1 - \alpha)\bar{Q}$, $A_\alpha = I_N - \gamma Q_\alpha$, $b_\alpha = -\gamma r_\alpha$, $r_\alpha = \alpha r + (1 - \alpha)\bar{r}$, $H_\alpha \triangleq A_\alpha - \text{diag}(A_\alpha) = -\gamma\alpha(Q - \text{diag}(Q))$, and $\gamma > 0$ is the step size.

Note that the decoupled cost function J_{nco}, the new cost function J_α and its minimizing algorithm $\mathbf{S}^{\mathbf{priv}}$ merely act as a means to introduce the cooperation level in our problem. The performance of the agents is measured in terms of the expected value of global cost $J_{co}(x)$ achieved when the agents follow algorithm $\mathbf{S}^{\mathbf{priv}}$. By varying the cooperation level in $\mathbf{S}^{\mathbf{priv}}$, the agents can achieve a range of private solutions of **P**. In practice, agents should select a cooperation level that maximizes the system performance, as we will show in the next subsection. Notice that agents are still required to exchange their state information for all values of the cooperation level $0 < \alpha \le 1$, and that the cooperation level determines the weight given by an agent to the information coming from its neighbors.

Remark 8.3 (Selection of the decoupled cost) There may be scenarios in which the agents do not have individual goals. In such cases, we can construct an artificial decoupled cost J_{nco} to capture the non-cooperation extreme. Several choices of the matrix $\bar{Q}$ are possible. For instance, $\bar{Q}$ can be chosen to consist of the diagonal elements of Q, or it can be an arbitrary matrix that satisfies assumption **A.3** (see example in Sect. 8.4).   □

8.3.2 Performance Analysis with Privacy and Cooperation

We analytically characterize the expected cost for a given privacy and cooperation level. Note that due to assumptions **A.1** and **A.3**, both Q_α and A_α are symmetric. Moreover, since the privacy noise has a normal distribution, the state $x(k)$ in algorithm S^{priv} is also normal. Let the steady state mean and covariance of $x(k)$ in algorithm S^{priv} be denoted by m_α and P_α, respectively. Next, we present conditions under which the steady state mean and covariance exist. Note that Lemma 8.1 presents such a condition for $\alpha = 1$.

Lemma 8.3 (Convergence of S_α^{priv} for $\alpha \neq 1$) *Let $\alpha \neq 1$. Then, the steady state mean and covariance of algorithm S^{priv} exist if*

$$\gamma < \frac{2}{\max\{\rho(Q), \rho(\bar{Q})\}}. \tag{8.19}$$

Proof The proof is similar to that of Lemma 8.1 by using the following facts: (i) $Q_\alpha > 0$ for $\alpha \neq 1$, and (ii) from Weyl's inequality [25], $\rho(Q_\alpha) \leq \max\{\rho(Q), \rho(\bar{Q})\}$.   □

Analogous to (8.6) and (8.14), the steady state mean and covariance of S^{priv} satisfy

$$m_\alpha = A_\alpha m_\alpha + b_\alpha$$
$$\Rightarrow 0 = Q_\alpha m_\alpha + r_\alpha \quad \text{and,} \tag{8.20}$$
$$P_\alpha = A_\alpha P_\alpha A_\alpha^T + \sigma^2 H_\alpha H_\alpha^T. \tag{8.21}$$

A closed form expression of P_α is given by

$$P_\alpha = \sigma^2 \gamma^2 \alpha^2 \sum_{k=0}^{\infty} A_\alpha^k (Q - diag(Q))^2 (A_\alpha^T)^k. \tag{8.22}$$

Next, we characterize the performance of the agents when they follow the algorithm S^{priv}. The performance is measured by the expected value of the global cost $J_{co}(x)$. Similarly to (8.15), the cost achieved by algorithm S^{priv} is

$$J(\alpha, \sigma) \triangleq \mathbb{E}[J_{co}(x)]_{\text{S}^{\text{priv}}} = J_{priv}(\alpha, \sigma) + J_{ico}(\alpha) \quad \text{where,} \tag{8.23}$$

$$J_{priv}(\alpha, \sigma) \triangleq \frac{1}{2} tr(Q P_\alpha) \quad \text{and,} \tag{8.24}$$

$$J_{ico}(\alpha) \triangleq \frac{1}{2} m_\alpha^T Q m_\alpha + r^T m_\alpha + s. \tag{8.25}$$

Notice that $J_{ico}(\alpha)$ represents the cost achieved by the agents for any *intermediate cooperation* level α in the absence of privacy noise. Further, the cost term $J_{priv}(\alpha, \sigma)$ quantifies the effect of the privacy noise at a given cooperation level since the covariance P_α depends on noise level σ. However, we omit that dependence in the notation for the ease of presentation. Moreover, P_α also depends on the step size γ. We do not analyze this dependence because γ dictates the number of iterations for algorithm S^{priv} to converge, which is not the primary issue addressed here. Note that the optimum for **P** is achieved only when $\alpha = 1$ and $\sigma = 0$ (that is, when the agents fully cooperate and no privacy noise is present). To clarify the notation, $J_{co}^* = J(1, 0)$, $J_{co}^*(\sigma) = J(1, \sigma)$ and $J_{ico}(\alpha) = J(\alpha, 0)$. Also note that the functions J, J_{priv} and J_{ico} are continuous functions in their respective variables.

Intuitively, in the absence of privacy noise, the performance should increase as the agents cooperate more and should equal the best performance when they cooperate fully ($\alpha = 1$). The following lemma proves this fact and justifies our definition of cooperation level.

Lemma 8.4 (Performance without privacy) *The cost $J_{ico}(\alpha)$ in (8.25) is monotonically decreasing for $\alpha \in [0, 1]$.*

Proof See the Appendix. □

Lemma 8.4 implies that, in absence of privacy noise, it is beneficial for the agents to cooperate fully. Instead, in the presence of privacy noise, agents can achieve a range of private solutions of **P** by varying the cooperation level. In practice, agents should select a cooperation level that minimizes the cost $J(\alpha, \sigma)$. The optimum cooperation level for the agents for a given level of privacy noise σ is characterized as

$$\alpha^*(\sigma) = \arg \min_\alpha J(\alpha, \sigma). \tag{8.26}$$

The optimum cooperation level $\alpha^*(\sigma)$ can be approximated numerically by discretizing the interval $[0, 1]$ for α, and evaluating the cost $J(\alpha, \sigma)$ at each point. Next, we show that under some conditions on cost functions J_{priv} and J_{ico}, we can characterize the behavior of $\alpha^*(\sigma)$.

Theorem 8.2 (Characterizing the optimum cooperation level) *For all $\sigma > 0$, let $J_{priv}(\alpha, \sigma)$ be strictly increasing for all $\alpha \in (0, 1]$. Further, let $J_{priv}(\alpha, \sigma)$ and $J_{ico}(\alpha)$ be strictly convex for $\alpha \in [0, 1)$. Then, $\alpha^*(\sigma)$ is a monotonically decreasing function of σ.*

Proof See the Appendix. □

Due to the convexity and increasing properties of the functions involved in Theorem 8.2, we are able to obtain a nice characterization of the optimum cooperation level. This result implies that under the conditions given in Theorem 8.2, it is always beneficial for the agents to reduce their cooperation level if they want to increase their privacy level. It characterizes an important and previously unidentified tradeoff between privacy and cooperation is multi-agent systems. Next, we show how to design the artificial cost function J_{nco} (in cases where individual costs are not present) which guarantees that the conditions of Theorem 8.2 hold true.

Corollary 8.1 (Design of artificial cost function) *Assume that Q satisfies the following properties:*

(i) $Q > 0$ and $\frac{\lambda_1(Q)}{\lambda_N(Q)} < 1.5$,
(ii) $diag(Q) = \mu I_N$ for some $\mu > 0$.

Then, there exists a $\bar{Q} = \delta I_N$ with $\lambda_1(Q) < \delta < 1.5\lambda_N(Q)$ and $\gamma < \delta^{-1}$ such that $J_{priv}(\alpha, \sigma)$ is strictly increasing for $\alpha \in (0, 1]$, and $J_{ico}(\alpha)$ and $J_{priv}(\alpha, \sigma)$ are strictly convex for $\alpha \in [0, 1)$.

Proof See the Appendix. □

The above corollary guarantees that $\alpha^*(\sigma)$ is a monotonically decreasing function. It should be noticed, however, that the conditions presented in Theorem 8.2 and Corollary 8.1 are not necessary, and $\alpha^*(\sigma)$ may or may not exhibit similar behavior if these conditions are not satisfied.

8.4 Private Consensus with Cooperation Level

In this section, we illustrate our results through the consensus problem that was motivated in Sect. 8.2.1. The consensus iterations (8.1) for all agents are collectively written as (including step size γ)

$$x(k + 1) = (I_N - \gamma L)x(k). \tag{8.27}$$

The Laplacian L of the underlying graph is symmetric, positive semi-definite with positive diagonal entries and non-positive non-diagonal entries. $\mathbf{1}_N$ is an eigenvector of L associated with the eigenvalue 0, that is, $L\mathbf{1}_N = 0$ and $\lambda_N(L) = 0$. Further, we assume that the graph is connected, which is equivalent to $\lambda_{N-1}(L) > 0$. Then, algorithm (8.27) asymptotically achieves average consensus [20]. Since the consensus algorithm in (8.27) can be viewed as a gradient descent algorithm to minimize the cost in (8.2), it fits into our framework (Problem (8.3) and solution (8.5)) with $Q = L, r = 0$, and $s = 0$. Further, it also satisfies assumptions **A.1** and **A.2**. To introduce the cooperation level, we select the following decoupled cost function

$$J_{nco}(x) = \frac{1}{2N}x^T x - \frac{1}{N}a^T x + \frac{1}{2N}a^T a + b^2, \tag{8.28}$$

where $a \in \mathbb{R}^N$ and $b \in \mathbb{R}$. The optimum of $J_{nco}(x)$ is achieved at $x = a$ and the optimum cost is b^2. The non-cooperative cost J_{nco} signifies the fact that each agent is stubborn/biased: i.e. it wants the opinions (states) of the other agents to converge to its own, without changing its own opinion. Consensus with stubborn (or persistent) agents is well studied and J_{nco} corresponds to the cost when all agents are fully stubborn (see [6], Eq. (1)). Since all agents are stubborn, they do not cooperate and evolve independently without using each other's state information.

Comparing the above decoupled cost function with (8.16), we obtain $\bar{Q} = \frac{1}{N}I_N, \bar{r} = -\frac{1}{N}a$, and $\bar{s} = \frac{1}{2N}a^T a + b^2$. Thus,

$$Q_\alpha = \alpha L + \frac{1-\alpha}{N}I_N, \quad A_\alpha = I_N - \gamma Q_\alpha, \quad b_\alpha = \frac{\gamma(1-\alpha)}{N}a.$$

Further, the step size can be chosen according to Lemma 8.3 to guarantee convergence of the consensus algorithm. The resulting cost with privacy mechanism becomes $J(\alpha, \sigma) = \frac{1}{2}(tr(L P_\alpha) + m_\alpha^T L m_\alpha)$. Notice that, for $\alpha = 1$, A_1 becomes marginally stable and thus, the covariance matrix P_1 becomes unbounded resulting in instability (see discussion below (8.14)). Thus, the agents have to choose a cooperation level $0 \le \alpha < 1$ for the cost to remain finite.

We now consider a specific consensus example with $N = 4$ agents, and the following Laplacian matrix

$$L = \begin{bmatrix} 3 & -1 & 0 & -2 \\ -1 & 3 & -1 & -1 \\ 0 & -1 & 2 & -1 \\ -2 & -1 & -1 & 4 \end{bmatrix}.$$

Let $a = [0, 1, 2, 3]^T$ and $x(0) = [-2, -1, 1, 2]^T$. Figure 8.1 shows the system costs J, J_{priv}, and J_{ico} in (8.23)–(8.25) as a function of α for $\sigma = 4$. We can observe that J_{ico}, which is the system cost in absence of privacy noise, is monotonically decreasing (c.f. Lemma 8.4). This validates our understanding that it is always beneficial for the agents to have full cooperation if they do not desire any privacy.

The effect of privacy noise is included in the cost J_{priv}. It is interesting to observe its behavior at different cooperation levels. Note that the noise has only a marginal effect on the cost at smaller cooperation levels. In fact, when the agents do not cooperate ($\alpha = 0$), the noise does not affect the performance at all—which is natural because the agents do not share any information. In contrast, the effect of noise is significantly higher at larger cooperation levels, because the agents use the noisy states in their updates. Thus, the cost J_{priv} is a monotonically increasing function of α. The two cost curves J_{ico} and J_{priv} highlight the trade-off between having full cooperation versus no cooperation. As evident from the resulting overall cost

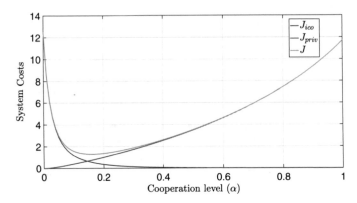

Fig. 8.1 System costs as a function of cooperation level for privacy noise level $\sigma = 4$

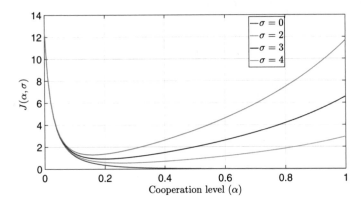

Fig. 8.2 Consensus cost as a function of cooperation and privacy

curve J, an intermediate optimum cooperation level should be chosen to achieve best performance.

Figure 8.2 shows the overall cost J achieved as a function of the cooperation level for various levels of privacy noise. Observe that for each cooperation level, the cost increases with the noise level. These curves highlight that the optimum cooperation level changes with the privacy noise level which can be seen explicitly in Fig. 8.3. Note that along with the fact that J_{priv} is an increasing function, both J_{priv} and J_{ico} are strictly convex. Thus, $\alpha^*(\sigma)$ is a monotonically decreasing function (c.f. Theorem 8.2), which implies that it is always better for the agents to reduce their cooperation level if they desire to have a higher level of privacy. Finally, note that the consensus example does not satisfy the conditions in Corollary 8.1 (see discussion Corollary 8.1).

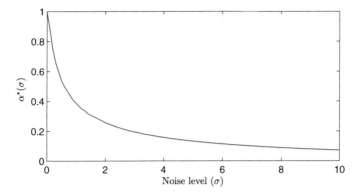

Fig. 8.3 Variation of the optimum cooperation level with noise

8.5 Conclusion

We considered a multi-agent system where the agents cooperatively optimize a quadratic cost function while ensuring privacy of their states over time. We presented a noise adding differential privacy mechanism to protect the privacy of agents' states over time. We characterized the performance degradation due to the privacy noise and argued that the degradation occurs due to cooperation between the agents. We developed a framework in which the agents can respond to the privacy noise present in the system by varying their cooperation level and studied the combined effect of privacy noise and cooperation level on the system performance. Using the performance characterization, we showed that there exists an optimum cooperation level that minimizes the system cost and obtained conditions under which the optimum cooperation level is a decreasing function of privacy noise level. The results illustrate a tradeoff between performance, privacy and cooperation, and suggest that, to optimize performance, agents should decrease their cooperation level if they want to increase the privacy level. Future research directions include extending the framework to arbitrary convex cost functions.

8.6 Appendix

Proof of Lemma 8.4 By differentiating $J_{ico}(\alpha)$ with respect to α, we obtain

$$J'_{ico}(\alpha) = (m_\alpha^T Q + r^T)m'_\alpha. \tag{8.29}$$

For $\alpha = 1$, from (8.20) we have, $Qm_1 + r = 0$. Thus, $J'_{ico}(1) = 0$.
For $\alpha \in [0, 1)$, $Q_\alpha > 0$. Thus, Q_α is invertible, and $Q_\alpha^{-1} > 0$. Differentiating (8.20), we get

$$(Q - \bar{Q})m_\alpha + Q_\alpha m'_\alpha + r - \bar{r} = 0 \overset{(a)}{\Rightarrow} m'_\alpha = -\frac{Q_\alpha^{-1}(Qm_\alpha + r)}{1 - \alpha}, \qquad (8.30)$$

where (a) follows from (8.20). Thus, we have

$$J'_{ico}(\alpha) = -\frac{1}{1 - \alpha}(Qm_\alpha + r)^T Q_\alpha^{-1}(Qm_\alpha + r).$$

and the derivative is non-positive, which completes the proof. $\qquad\square$

Proof of Theorem 8.2 Let $'$ denote the derivative or partial derivative w.r.t. α. Using (8.22), we have $J_{priv}(\alpha, \sigma) = \sigma^2 f(\alpha)$, where $f(\alpha) \triangleq \frac{1}{2}\alpha^2\gamma^2 tr[Q \sum_{k=0}^\infty A_\alpha^k(Q - diag(Q))^2(A_\alpha^T)^k]$. Since $J_{priv}(\alpha, \sigma)$ is assumed to be strictly increasing in the theorem statement, $f(\alpha)$ is also strictly increasing for $\alpha \in (0, 1]$. Also, it can be readily observed that $f'(0) = 0$ and $f''(0) > 0$. From the proof of Lemma 8.4, we have $J'_{ico}(0) \le 0$ and $J'_{ico}(1) = 0$. Thus, $J'(0, \sigma) \le 0$ and $J'(1, \sigma) > 0$. Also, $J(\alpha, \sigma)$ is strictly convex for $\alpha \in [0, 1)$. Thus, $\alpha^*(\sigma) \in [0, 1)$ is unique and $J'(\alpha^*(\sigma), \sigma) = 0$.

We prove the theorem by contradiction. Suppose $\alpha^*(\sigma)$ is not monotonically decreasing function. Then, there exist $0 < \sigma_1 < \sigma_2$ such that $0 \le \alpha^*(\sigma_1) < \alpha^*(\sigma_2) < 1$. Further,

$$J'(\alpha^*(\sigma_1), \sigma_1) = \sigma_1^2 f'(\alpha^*(\sigma_1)) + J'_{ico}(\alpha^*(\sigma_1)) = 0, \quad \text{and}$$
$$J'(\alpha^*(\sigma_2), \sigma_2) = \sigma_2^2 f'(\alpha^*(\sigma_2)) + J'_{ico}(\alpha^*(\sigma_2)) = 0.$$

Subtracting, we get

$$0 = \sigma_1^2 f'(\alpha^*(\sigma_1)) - \sigma_2^2 f'(\alpha^*(\sigma_2)) + J'_{ico}(\alpha^*(\sigma_1)) - J'_{ico}(\alpha^*(\sigma_2))$$
$$\overset{(a)}{\le} \sigma_1^2[f'(\alpha^*(\sigma_1)) - f'(\alpha^*(\sigma_2))] + J'_{ico}(\alpha^*(\sigma_1)) - J'_{ico}(\alpha^*(\sigma_2)) \overset{(b)}{<} 0,$$

where (a) follows since f is an increasing function and (b) follows from the strict convexity of f and J_{ico}. Thus, there is a contradiction and therefore, the theorem follows. $\qquad\square$

Proof of Corollary 8.1 Let $'$ denote the derivative or partial derivative w.r.t. α. First, we derive conditions under which the cost term $J_{ico}(\alpha)$ is convex w.r.t. α. By differentiating (8.29) with respect to α we obtain

$$J''_{ico}(\alpha) = (m'_\alpha)^T Q m'_\alpha + (Qm_\alpha + r)^T m''_\alpha.$$

From the proof of lemma 8.4, for $\alpha = 1$, we have $Qm_1 + r = 0$ and $m'_1 = 0$. Thus, $J''_{ico}(1) = 0$. Now let $\alpha \in [0, 1)$. By differentiating (8.30), we get

$$2(Q - \bar{Q})m'_\alpha + Q_\alpha m''_\alpha = 0 \overset{(a)}{\Rightarrow} m''_\alpha = \frac{2}{1 - \alpha}Q_\alpha^{-1}(Q - \bar{Q})Q_\alpha^{-1}(Qm_\alpha + r),$$

where (a) follows from (8.30). Substituting the derivatives m'_α and m''_α we get

$$J''_{ico}(\alpha) = \frac{1}{1-\alpha}(Qm_\alpha + r)^T Q_\alpha^{-1} \left(3Q - 2\bar{Q} + \frac{\alpha}{1-\alpha}Q\right) Q_\alpha^{-1}(Qm_\alpha + r).$$

Condition (iii) in the corollary guarantees that $3Q - 2\bar{Q} > 0$ and thus $J''_{ico}(\alpha) > 0$ for $\alpha \in (0, 1]$.

Next, we derive conditions under which the cost term $J_{priv}(\alpha, \sigma)$ is convex w.r.t. α. Recalling (8.18), let $A_\alpha = \alpha A + B$, where $A \triangleq \gamma(\bar{Q} - Q)$ and $B \triangleq I_N - \gamma\bar{Q} = (1 - \gamma\delta)I_N$. Further, let $H_\alpha = -\gamma\alpha\tilde{Q}$ where $\tilde{Q} \triangleq Q - \text{diag}(Q)$. Differentiating (8.21) and substituting the above expressions, we get

$$P'_\alpha = A_\alpha P'_\alpha A_\alpha + \underbrace{A P_\alpha B^T + B P_\alpha A^T + 2\alpha A P_\alpha A^T + 2\sigma^2\gamma^2\alpha\tilde{Q}^2}_{W}. \qquad (8.31)$$

Note that due to (iii) and (iv), $A > 0$ and $B > 0$. Further, we have the following facts (a) $QA_\alpha = A_\alpha Q$ and (b) $Q\tilde{Q} = \tilde{Q}Q$ (by (ii)). Using (a), (b) and (8.22), it can be easily observed that $A P_\alpha B^T = B P_\alpha A^T > 0$. Thus, $W > 0$ and (8.31) resembles a Lyapunov equation. Hence, we conclude that $P'_\alpha > 0$.

Taking derivative of (8.31), we get

$$P''_\alpha = A_\alpha P''_\alpha A_\alpha + \underbrace{2A P'_\alpha B^T + 2B P'_\alpha A^T + 2A(P_\alpha + 2\alpha P'_\alpha)A^T + 2\sigma^2\gamma^2\tilde{Q}^2}_{Z}.$$

Again using (a) and (b) and taking the derivative of (8.22), we get $A P'_\alpha B^T = B P'_\alpha A^T > 0$. Thus, $Z > 0$ and the above equation resembles a Lyapunov equation. Hence, we get $P''_\alpha > 0$. Thus, $J'_{priv}(\alpha, \sigma) = \frac{1}{2}tr(QP'_\alpha) > 0$ and $J''_{priv}(\alpha, \sigma) = \frac{1}{2}tr(QP''_\alpha) > 0$ and the proof is complete. $\square$

References

1. Cortes J, Martinez S, Karatas T, Bullo F (2004) Coverage control for mobile sensing networks. IEEE Trans Robot Autom 20(2):243–255
2. DeGroot M (1974) Reaching a consensus. J Am Stat Assoc 69(345):118–121
3. Du H, Wen G, Cheng Y, He Y, Jia R (2016) Distributed finitetime cooperative control of multiple high order nonholonomic mobile robots. IEEE Trans Neural Netw Learn Syst 28(12):2998–3006
4. Dwork C (2006) Differential privacy. Proc ICALP 4052:1–12
5. Dwork C (2011) A firm foundation for private data analysis. Commun ACM 54(1):86–95
6. Ghaderi J, Srikant R (2013) Opinion dynamics in social networks: a local interaction game with stubborn agents. Am Control Conf, 1982–1987
7. Hale M, Egerstedt M (2018) Cloud-enabled differentially private multiagent optimization with constraints. IEEE Trans Control Netw Syst 5(4):1693–1706
8. Han S, Topcu U, Pappas GJ (2014) Differentially private convex optimization with piecewise affine objectives. In: IEEE Conference on decision and control, pp 2160–2166

9. Han S, Topcu U, Pappas GJ (2016) Differentially private distributed constrained optimization. IEEE Trans Autom Control
10. Hsu J, Roth A, Roughgarden T, Ullman J (2014) Privately solving linear programs. Autom Lang Programm ICALP 14:612–624
11. Huang Z, Mitra S, Dullerud G (2012) Differentially private iterative synchronous consensus. In Proceedings of the ACM workshop on privacy in the electronic society, WPES, pp 81–90
12. Huang Z, Mitra S, Vaidya N (2015) Differentially private distributed optimization. In: Proceedings of ICDCN '15
13. Katewa V, Pasqualetti F, Gupta V (2018) On privacy vs. cooperation in multi-agent systems. Int J Control 91(7):1693–1707
14. Le Ny J, Pappas GJ (2014) Differentially private filtering. IEEE Trans Autom Control 59(2):341–354
15. Lindell Y, Pinkas B (2009) Secure multiparty computation for privacy-preserving data mining. J Priv Confid 1(1):59–98
16. McDaniel P, McLaughlin S (2009) Security and privacy challenges in the smart grid. IEEE Secur Priv 7(3):75–77
17. Mo Y, Murray RM (2017) Privacy preserving average consensus. IEEE Trans Autom Control 62(2):753–765
18. Nedic A, Ozdaglar A (2009) Distributed subgradient methods for multi-agent optimization. IEEE Trans Autom Control 54(1):48–61
19. Olfati-Saber R (2006) Flocking for multi-agent dynamic systems: algorithms and theory. IEEE Trans Autom Control 51(3):401–420
20. Olfati-Saber R, Fax JA, Murray RM (2007) Consensus and cooperation in networked multi-agent systems. Proc IEEE 95(2):215–233
21. Orlandi C (2011) Is multiparty computation any good in practice? In: International Conference Acoustic Speech Signal Processing, pp 5848–5851
22. Raffard RL, Tomlin CJ, Boyd SP (2004) Distributed optimization for cooperative agents: application to formation flight. In: IEEE Conference on decision and control, pp 2453–2459
23. Rajagopal R, Wainwright MJ (2011) Network-based consensus averaging with general noisy channels. IEEE Trans Signal Process 59(1):373–385
24. Rosenthal R (1973) A class of games possessing pure-strategy nash equilibria. Int J Game Theory 2(1):65–67
25. Tao T (2012) Topics in random matrix theory. Graduate Studies in Mathematics, American Mathematical Society **132**
26. Terelius H, Topcu U, Murray RM (2011) Decentralized multi-agent optimization via dual decomposition. In: Proceedings of the 18th IFAC World Congress, vol 44(1), 11245–11251
27. United States DOE (2010) Data access and privacy issues related to smart grid technologies. Tech. rep
28. Xiao L, Boyd S, Kim SJ (2007) Distributed average consensus with least-mean-square deviation. J Parallel Distrib Comput 67(1):33–46

Part II
Encryption-Based Privacy

Chapter 9
Secure Multi-party Computation for Cloud-Based Control

Andreea B. Alexandru and George J. Pappas

Abstract In this chapter, we will explore the cloud-outsourced privacy-preserving computation of a controller on encrypted measurements from a (possibly distributed) system, taking into account the challenges introduced by the dynamical nature of the data. The privacy notion used in this work is that of cryptographic multi-party privacy, i.e., the computation of a functionality should not reveal *anything* more than what can be inferred only from the inputs and outputs of the functionality. The main theoretical concept used towards this goal is Homomorphic Encryption, which allows the evaluation of sums and products on encrypted data, and, when combined with other cryptographic techniques, such as Secret Sharing, results in a powerful tool for solving a wide range of secure multi-party problems. We will rigorously define these concepts and discuss how multi-party privacy can be enforced in the implementation of a Model Predictive Controller, which encompasses computing stabilizing control actions by solving an optimization problem on encrypted data.

9.1 Introduction

Cloud computing has become a ubiquitous tool in the age of big data and geographically-spread systems, due to the capabilities of resource pooling, broad network access, rapid elasticity, measured service and on-demand self-service, as defined by NIST [44]. The computational power and storage space of a cloud service can be distributed over multiple servers. Cloud computing has been employed for machine learning applications in e.g., healthcare monitoring and social networks, smart grid control and other control engineering applications, and integration with the Internet of Things paradigm [13, 58]. However, these capabilities do not come without risks. The security and privacy concerns of cloud computing range from communication security to leaking and tampering with the stored data or interfering

A. B. Alexandru (✉) · G. J. Pappas
Department of Electrical Engineering, University of Pennsylvania, Philadelphia, PA, USA
e-mail: aandreea@seas.upenn.edu

G. J. Pappas
e-mail: pappasg@seas.upenn.edu

© Springer Nature Singapore Pte Ltd. 2020
F. Farokhi (ed.), *Privacy in Dynamical Systems*,
https://doi.org/10.1007/978-981-15-0493-8_9

with the computation, that can be maliciously or unintentionally exploited by the cloud provider or the other tenants of the service [4, 36, 63]. In this chapter, we will focus on concerns related to the privacy of the data and computation. These issues can be addressed by the cryptographic tools described below.

Secure Multi-party Computation (SMPC) encompasses a range of cryptographic techniques that facilitate joint computation over secret data distributed between multiple parties, that can be both clients and servers. The goal of SMPC is that each party is only allowed to learn its own result of the computation, and no intermediary results such as inputs or outputs of other parties or other partial information. The concept of SMPC originates from [68], where a secure solution to the millionaire's problem was proposed. Surveys on SMPC can be found in [20]. SMPC involves communication between parties and can include individual or hybrid approaches between techniques such as secret sharing [8, 53, 61], oblivious transfer [49, 55], garbled circuits [9, 33, 68], (threshold) homomorphic encryption [30, 48, 59], etc.

Homomorphic Encryption (HE), introduced in [59] as *privacy homomorphisms*, refers to a secure computation technique that allows evaluating computations on encrypted data and produces an encrypted result. HE is best suited when there is a client-server scenario with an untrusted server: the client simply has to encrypt its data and send it to the server, which performs the computations on the encrypted data and returns the encrypted result. The first HE schemes were partial, meaning that they either allowed the evaluation of additions or multiplications, but not both. Then, somewhat homomorphic schemes were developed, which allowed a limited number of both operations. One of the bottlenecks for obtaining an unlimited number of operations was the accumulation of noise introduced by one operation, which could eventually prevent the correct decryption. The first fully homomorphic encryption scheme that allowed the evaluation of both additions and multiplications on encrypted data was developed in [29], where, starting from a somewhat homomorphic encryption scheme, a bootstrapping operation was introduced. Bootstrapping allows to obliviously evaluate the scheme's decryption circuit and reduces the ciphertext noise. For a thorough history and description of HE, see the survey [42]. Privacy solutions based on HE were proposed for genome matching, national security and critical infrastructure, healthcare databases, machine learning applications and control systems, etc. [5, 6, 57]. Of particular interest to us are the the works in control applications with HE, see [2, 3, 28, 35, 40, 47, 60], to name a few.

Over the past decades, efforts in optimizing the computation and implementation of SMPC techniques, along with the improvement of network communication speed and more powerful hardware, have opened the way of SMPC deployment in real-world applications. Nevertheless, in the context of big data and the internet of things, which bring enormous amounts of data and large number of devices that want to participate in the computation, SMPC techniques lag behind plaintext computations due to the computational and communication bottleneck [17, 45, 46].

The privacy definition for SMPC stipulates that the privacy of the inputs and intermediary results is ensured, but the output, which is a function of the inputs of all parties, is revealed. For some cases [69], a different privacy definition is preferred.

Differential Privacy (DP) refers to methods of concealing the information leakage from the result of the computation, even when having access to auxiliary information [25, 26]. Intuitively, the contribution of the input of each individual on the output should be hidden from those who access the computation results. To achieve this, a carefully chosen noise is added to each entry such that the statistical properties of the database are preserved [27], which introduces a trade-off between utility and privacy. Several works combine SMPC with DP in order to achieve both computation privacy and output privacy, for instance [16, 54, 56, 62, 64].

In this chapter, we focus on proposing SMPC schemes for optimization and control problems, where we require the convergence and accuracy of the results. DP techniques can be further investigated in order to ensure output privacy.

9.1.1 Dynamical Data Challenges

Cryptographic techniques were developed mainly for static data, such as databases, or independent messages. However, dynamical systems are iterative processes that generate structured and dependent data. Moreover, output data at one iteration/time step will often be an input to the computation at the next one. Hence, special attention is needed when using cryptographic techniques in solving optimization problems and implementing control schemes. For example, values encrypted with homomorphic encryption schemes will require ciphertext refreshing or bootstrapping if the multiplicative depth of the algorithm exceeds the multiplicative depth of the scheme; when using garbled circuits, a different circuit has to be generated for different iterations/times step of the same algorithm; the controlled noise added for differential privacy at each iteration/time step will accumulate and drown the result, etc. Furthermore, privacy is guaranteed as long as the keys and randomness are never reused, but freshly generated for each time step; then, the (possibly offline) phase in which uncorrelated randomness is generated has to be repeated for a continuously running process. In this chapter, we design the solution with all these issues in mind.

9.1.2 Contribution and Roadmap

Examples of control applications that require privacy include smart metering, crucial infrastructure control, prevention of industrial espionage, swarms of robots deployed in adversarial environments, etc. In order to ensure the privacy of a control application, i.e., secure both the signals and the model, as well as the intermediate computations, most of the time, complex solutions have to be devised. We investigate a privacy-preserving cloud-based Model Predictive Control application, which, at a high level, requires privately computing additions, multiplications, comparisons and oblivious updates. Our solution encompasses several SMPC techniques: homomorphic encryption, secret sharing, oblivious transfer. The purpose of this chapter is to describe these cryptographic techniques and show how to combine them in order to build private cloud-based protocols that produce the desired control actions.

The layout of the chapter is as follows: in Sect. 9.2, we describe the model predictive control problem that we address, we show how to compute the control action when there are no privacy requirements and we outline the privacy-preserving solution. In Sect. 9.3, we formally describe the adversarial model and privacy definition we seek for our multi-party computation problem. In Sect. 9.4, we introduce the cryptographic tools that we will use in our solution and their properties. Then, in Sect. 9.5, we design the multi-party protocol for the model predictive control problem with encrypted states and model and prove its privacy. Finally, we discuss some details about the requirements of the proposed protocol in Sect. 9.6.

9.1.3 Notation

We use bold-face lower case for vectors, e.g., $\mathbf{x}$, and bold-face upper case for matrices, e.g., $\mathbf{A}$. $\mathbb{N}$ denotes the set of non-negative integers, $\mathbb{Z}$ denotes the set of integers, $\mathbb{Z}_N$ denotes the additive group of integers modulo N and $(\mathbb{Z}_N)^*$ denotes the multiplicative group of integers modulo N. For $n \in \mathbb{N}$, let $[n] := \{1, 2, \ldots, n\}$. $E(x)$ and $[[x]]$ represent encryptions of the scalar value x. We use the same notation for multidimensional objects: an encryption of a matrix $\mathbf{A}$ is denoted by $E(\mathbf{A})$ (or $[[\mathbf{A}]]$) and is performed element-wise. $\{0, 1\}^*$ defines a sequence of bits of unspecified length.

9.2 Model Predictive Control

We consider a discrete-time linear time-invariant system:

$$\mathbf{x}(t + 1) = \mathbf{A}\mathbf{x}(t) + \mathbf{B}\mathbf{u}(t),$$
$$\mathbf{x}(t) = \begin{bmatrix} \mathbf{x}^1(t)^\mathsf{T} \ldots \mathbf{x}^M(t)^\mathsf{T} \end{bmatrix}, \quad \mathbf{u}(t) = \begin{bmatrix} \mathbf{u}^1(t)^\mathsf{T} \ldots \mathbf{u}^M(t)^\mathsf{T} \end{bmatrix},$$

(9.1)

with the state $\mathbf{x} \in \mathcal{X} \subseteq \mathbb{R}^n$ and the control input $\mathbf{u} \in \mathcal{U} \subseteq \mathbb{R}^m$. The system can be either centralized or partitioned in subsytems for $i \in [M]$, $\mathbf{x}^i(t) \in \mathbb{R}^{n_i}$, $\sum_{i=1}^{M} n_i = n$ and $\mathbf{u}^i(t) \in \mathbb{R}^{m_i}$, $\sum_{i=1}^{M} m_i = m$.

The Model Predictive Control (MPC) is the optimal control receding horizon problem with constraints written as:

$$J_N^*(\mathbf{x}(t)) = \min_{\mathbf{u}_0, \ldots, \mathbf{u}_{N-1}} \frac{1}{2} \left(\mathbf{x}_N^\mathsf{T} \mathbf{P} \mathbf{x}_N + \sum_{k=0}^{N-1} \mathbf{x}_k^\mathsf{T} \mathbf{Q} \mathbf{x}_k + \mathbf{u}_k^\mathsf{T} \mathbf{R} \mathbf{u}_k \right)$$

$$s.t. \ \mathbf{x}_{k+1} = \mathbf{A}\mathbf{x}_k + \mathbf{B}\mathbf{u}_k, \ k = 0, \ldots, N - 1; \quad \mathbf{x}_0 = \mathbf{x}(t);$$
$$\mathbf{u}_k \in \mathcal{U}, \ k = 0, \ldots, N - 1,$$

(9.2)

where N is the length of the horizon and $\mathbf{P}, \mathbf{Q}, \mathbf{R} \succ 0$ are cost matrices. We consider input constrained systems with box constraints $0 \in \mathcal{U} = \{\mathbf{l}_u \preceq \mathbf{u} \preceq \mathbf{h}_u\}$, and impose stability without a terminal state constraint, but with appropriately chosen costs and horizon, such that the closed-loop system has robust performance to bounded errors due to encryption. A survey on the conditions for stability of MPC is given in [43].

Through straightforward manipulations, (9.2) can be written as the quadratic problem (9.3)—see details on obtaining the matrices $\mathbf{H}$ and $\mathbf{F}$ in [11, Chaps. 8, 11]—in the variable $\mathbf{U} := \begin{bmatrix} \mathbf{u}_0^\mathsf{T} & \mathbf{u}_1^\mathsf{T} & \dots & \mathbf{u}_{N-1}^\mathsf{T} \end{bmatrix}^\mathsf{T}$. For simplicity, we keep the same notation $\mathcal{U}$ for the augmented constraint set. After obtaining the optimal solution, the first m components of $\mathbf{U}^*(\mathbf{x}(t))$ are applied as input to the system (9.1): $\mathbf{u}^*(\mathbf{x}(t)) = \{\mathbf{U}^*(\mathbf{x}(t))\}_{1:m}$.

$$\mathbf{U}^*(\mathbf{x}(t)) = \arg\min_{\mathbf{U} \in \mathcal{U}} \frac{1}{2} \mathbf{U}^\mathsf{T} \mathbf{H} \mathbf{U} + \mathbf{U}^\mathsf{T} \mathbf{F}^\mathsf{T} \mathbf{x}(t). \tag{9.3}$$

9.2.1 Solution Without Privacy Requirements

The privacy-absent cloud-MPC problem is depicted in Fig. 9.1. The system (9.1) is composed of M subsystems, that can be thought of as different agents, which measure their states and receive control actions, and of a setup entity which holds the system's model and parameters. The control decision problem is solved at the cloud level, by a cloud controller, which receives the system's model and parameters, the measurements, as well as the constraint sets imposed by each subsystem. The control inputs are then applied by one virtual actuator. Examples include a smart building temperature control application, where the subsystems are apartments and the actuator is a machine in the basement, or the subsystems are robots in a swarm coordination application and the actuator is a ground control that sends them waypoints.

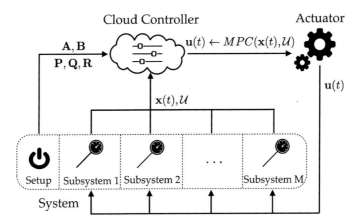

Fig. 9.1 Cloud-based MPC problem for a system composed of a number of subsystems and a setup entity. The control action is computed by a cloud controller and sent to a virtual actuator

The constraint set $\mathcal{U}$ is a hyperbox, so the projection step required for solving (9.3) has a closed form solution, denoted by $\Pi_{\mathcal{U}}(\cdot)$ and the optimization problem can be efficiently solved with the projected Fast Gradient Method (FGM) [50], given in (9.4):

For $k = 0 \ldots, K - 1$

$$\mathbf{t}_k \leftarrow \left(\mathbf{I}_{Mm} - \frac{1}{L}\mathbf{H} \right) \mathbf{z}_k - \frac{1}{L}\mathbf{F}^\mathsf{T}\mathbf{x}(t) \tag{9.4a}$$

$$\mathbf{U}_{k+1} \leftarrow \Pi_{\mathcal{U}}(\mathbf{t}_k) \tag{9.4b}$$

$$\mathbf{z}_{k+1} \leftarrow (1 + \eta)\mathbf{U}_{k+1} - \eta\mathbf{U}_k \tag{9.4c}$$

where $\mathbf{z}_0 \leftarrow \mathbf{U}_0$. The objective function is strongly convex, since $\mathbf{H} \succ 0$, therefore we can use the constant step sizes $L = \lambda_{max}(\mathbf{H})$ and $\eta = (\sqrt{\kappa(\mathbf{H})} - 1)/(\sqrt{\kappa(\mathbf{H})} + 1)$, where $\kappa(\mathbf{H})$ is the condition number of $\mathbf{H}$.

9.2.2 Privacy Objectives and Overview of Solution

The system model and MPC costs $\mathbf{A}, \mathbf{B}, \mathbf{P}, \mathbf{Q}, \mathbf{R}$ are known only to the system, but not to the cloud and actuator, hence, the matrices $\mathbf{H}, \mathbf{F}$ in (9.3) are also private. The measurements and constraints are not known to parties other than the corresponding subsystem and should remain private, such that the sensitive information of the subsystems is concealed. The control inputs should not be known by the cloud.

The goal of this work is to design a private cloud-outsourced version of the fast gradient method in (9.4) for the model predictive control problem, such that the actuator obtains the control action $\mathbf{u}^*(t)$ for system (9.1), without learning anything else in the process. At the same time, the cloud controller should not learn anything other than what was known prior to the computation about the measurements $\mathbf{x}(t)$, the control inputs $\mathbf{u}^*(t)$, the constraints $\mathcal{U}$, and the system model $\mathbf{H}, \mathbf{F}$. We formally introduce the adversarial model and multi-party privacy definition in Sect. 9.3.

As a primer to our private solution to the MPC problem, we briefly mention here the cryptographic tools used to achieve privacy. We will encrypt the data with a *labeled homomorphic encryption* scheme, which allows us to evaluate an unlimited number of additions and one multiplication over encrypted data. The labeled homomorphic encryption builds on top of an *additively homomorphic encryption* scheme, which allows only the evaluation of additions over encrypted data, a *secret sharing* scheme, which enables the splitting of a message into two random shares, that cannot be used individually to retrieve the message, and a *pseudorandom generator* that, given a key and a small seed, called *label*, outputs a larger sequence of bits that is indistinguishable from random. The right choice of labels is essential for a seamless application of labeled homomorphic encryption on dynamical data, and we choose the labels to be the time steps at which the data is generated. These tools ensure that we can evaluate polynomials on the private data. Furthermore, the computations for determining the control action also involve projections on a feasible hyperbox. To achieve this in a private way, we make use of *two-party private comparison* that involves exchanges of encrypted bits between two parties, and *oblivious transfer*, that allows us to choose a value out of many values when the index is secret. These cryptographic tools will be described in detail in Sect. 9.4, and our private cloud-based MPC solution that incorporates them will be presented in Sect. 9.5.

9.3   Adversarial Model and Privacy Definition

In cloud applications, the service provider has to deliver the contracted service that was agreed upon, otherwise the clients switch to another service provider. The clients' interest is to obtain the correct result for the service they pay for, so we may assume they do not alter the data sent to the cloud. However, the parties can locally process copies of the data they receive in any fashion they want. This adversarial model is known as semi-honest, which is defined formally as follows:

Definition 9.1 (*Semi-honest model* [32, Chap. 7]) A party is semi-honest if it does not deviate from the steps of the protocol, but may store the transcript of the messages exchanged and its internal coin tosses, as well as process the data received in order to learn more information than stipulated by the protocol.

This model also holds when considering eavesdroppers on the communication channels. Malicious and active adversaries—that diverge from the protocols or tamper with the messages—are not considered in this chapter. Privacy against malicious adversaries can be obtained by introducing zero-knowledge proofs and commitment schemes, at the cost of a computational overhead [23, 32].

We introduce some concepts necessary for the multi-party privacy definitions. An ensemble $X = \{X_\sigma\}_{\sigma \in \mathbb{N}}$ is a sequence of random variables ranging over strings of bits of length polynomial in σ, arising from distributions defined over a finite set Ω.

Definition 9.2 (*Statistical indistinguishability* [31, Chap. 3]) The ensembles $X = \{X_\sigma\}_{\sigma \in \mathbb{N}}$ and $Y = \{Y_\sigma\}_{\sigma \in \mathbb{N}}$ are **statistically indistinguishable**, denoted $\overset{s}{\equiv}$, if for every positive polynomial p, and all sufficiently large σ:

$$\text{SD}[X_\sigma, Y_\sigma] := \frac{1}{2} \sum_{\alpha \in \Omega} \left| \Pr[X_\sigma = \alpha] - \Pr[Y_\sigma = \alpha] \right| < \frac{1}{p(\sigma)}.$$

The quantity on the left is called the statistical distance between the two ensembles.

It can be proved that two ensembles are statistically indistinguishable if no algorithm can distinguish between them. Statistical indistinguishability holds against computationally unbounded adversaries. Computational indistinguishability is a weaker notion of the statistical version, as follows:

Definition 9.3 (*Computational Indistinguishability* [31, Chap. 3]) The ensembles $X = \{X_\sigma\}_{\sigma \in \mathbb{N}}$ and $Y = \{Y_\sigma\}_{\sigma \in \mathbb{N}}$ are **computationally indistinguishable**, denoted $\overset{c}{\equiv}$, if for every probabilistic polynomial-time algorithm $D : \{0, 1\}^* \to \{0, 1\}$, called the distinguisher, every positive polynomial p, and all sufficiently large σ:

$$\left| \Pr_{x \leftarrow X_\sigma} [D(x) = 1] - \Pr_{y \leftarrow Y_\sigma} [D(y) = 1] \right| < \frac{1}{p(\sigma)}.$$

Let us now look at the privacy definition that is considered in secure multi-party computation, that makes use of the real and ideal world paradigms. Consider a multi-party protocol Π that executes a functionality $f = (f_1, \ldots, f_p)$ on inputs $I = (I_1, \ldots, I_p)$ and produces an output $f(I) = (f_1(I), \ldots, f_p(I))$ in the following way: the parties have their inputs, then exchange messages between themselves in order to obtain an output. If an adversary corrupts a party (or a set of parties) in this real world, after the execution of the protocol, it will have access to its (their) input, the messages received and its (their) output. In an ideal world, there is a trusted incorruptible party that takes the inputs from all the parties, computes the functionality on them, and then sends to each party its output. In this ideal world, an adversary that corrupts a party (or a set of parties) will only have access to its (their) input and output. This concept is illustrated for three parties in Fig. 9.2. We then say that a multi-party protocol achieves computational privacy if, for each party, what a computationally bounded adversary finds out from the real world execution is equivalent to what it finds out from the ideal-world execution.

The formal definition is the following:

Definition 9.4 (*Multi-party privacy w.r.t. semi-honest behavior* [32, Chap. 7]) Let $f : (\{0, 1\}^*)^p \to (\{0, 1\}^*)^p$ be a p-ary functionality, where $f_i(x_1, \ldots, x_p)$ denotes the i-th element of $f(x_1, \ldots, x_p)$. Denote the inputs by $\bar{x} = (x_1, \ldots, x_p)$. For $I = \{i_1, \ldots, i_t\} \subset [p]$, we let $f_I(\bar{x})$ denote the subsequence $f_{i_1}(\bar{x}), \ldots, f_{i_t}(\bar{x})$, which models a coalition of a number of parties. Let Π be a p-party protocol that computes f. The **view** of the i-th party during an execution of Π on the inputs $\bar{x}$, denoted $V_i^\Pi(\bar{x})$, is $(x_i, \text{coins}, m_1, \ldots, m_t)$, where coins represents the outcome of the

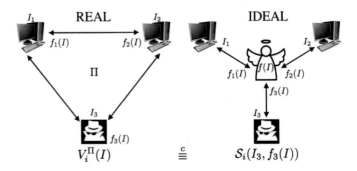

Fig. 9.2 Real and ideal paradigms for secure multi-party computation

i'th party's internal coin tosses, and m_j represents the j-th message it has received. We let the view of a coalition be denoted by $V_I^\Pi(\bar{x}) = (I, V_{i_1}^\Pi(\bar{x}), \ldots, V_{i_t}^\Pi(\bar{x}))$. For a deterministic functionality f, we say that $\boldsymbol{\Pi}$ **privately computes** f if there exist simulators S, such that, for every $I \subset [p]$, it holds that, for $\bar{x}_t = (x_{i_1}, \ldots, x_{i_t})$:

$$\left\{ S(I, \bar{x}_t, f_I(\bar{x})) \right\}_{\bar{x} \in (\{0,1\}^*)^p} \stackrel{c}{\equiv} \left\{ V_I^\Pi(\bar{x}) \right\}_{\bar{x} \in (\{0,1\}^*)^p}.$$

This definition assumes the correctness of the protocol, i.e., the probability that the output of the parties is not equal to the result of the functionality applied to the inputs is negligible [32, 41]. Auxiliary inputs, which are inputs that capture additional information available to each of the parties, (e.g., local configurations, side-information), are implicit in this definition [32, Chap. 7], [31, Chap. 4].

Definition 9.4 states that the view of any of the parties participating in the protocol, on each possible set of inputs, can be simulated based only on its own input and output. For parties that have no assigned input and output, like the cloud server in our problem, Definition 9.4 captures the strongest desired privacy model.

9.4 Cryptographic Tools

9.4.1 Secret Sharing

Secret sharing [53, 61] is a tool that distributes a private message to a number of parties, by splitting it into random shares. Then, the private message can be reconstructed only by an authorized subset of parties, which combine their shares.

One common and simple scheme is the additive 2-out-of-2 secret sharing scheme, which involves a party splitting its secret message $m \in G$, where G is a finite abelian group, into two shares, in the following way: generate uniformly at random an element $\mathfrak{b} \in G$, subtract it from the message and then distribute the shares $\mathfrak{b}$ and $m - \mathfrak{b}$. This can be also thought of as a one-time pad [10, 65] variant on G. Both shares are needed in order to recover the secret. The 2-out-of-2 secret sharing scheme achieves perfect secrecy, which means that the shares of two distinct messages are uniformly distributed on G [19].

We will also use an additive blinding scheme weaker than secret sharing (necessary for the private comparison protocol in Sect. 9.4.5): for messages of l bits, $\mathfrak{b}$ will be generated from a message space $\mathcal{M}$ with length of $\lambda + l$ bits, where λ is the security parameter, with the requirement that $\lambda + l$-bit messages can still be represented in $\mathcal{M}$, i.e. there is no wrap-around and overflow. The distribution of $m + \mathfrak{b}$ is statistically indistinguishable from a random number sampled of $l + \lambda + 1$ bits. Such a scheme is commonly employed for blinding messages, for instance in [12, 38, 66].

9.4.2   Pseudorandom Generators

Pseudorandom generators are efficient deterministic functions that expand short seeds into longer pseudorandom bit sequences, that are computationally indistinguishable from truly random sequences. More details can be found in [31, Chap. 3].

9.4.3   Homomorphic Encryption

Let $E(\cdot)$ denote a generic encryption primitive, with domain the space of private data, called **plaintexts**, and codomain the space of encrypted data, called **ciphertexts**. $E(\cdot)$ also takes as input the public key, and probabilistic encryption primitives also take a random number. The decryption primitive $D(\cdot)$ is defined on the space of ciphertexts and takes values on the space of plaintexts. $D(\cdot)$ also takes as input the private key. **Additively homomorphic** schemes satisfy the property that there exists an operator $\oplus$ defined on the space of ciphertexts such that:

$$E(a) \oplus E(b) \subset E(a+b), \qquad\qquad (9.5)$$

for any plaintexts a, b supported by the scheme. We use set inclusion instead of equality because the encryption of a message is not unique in probabilistic cryptosystems. Intuitively, Eq. (9.5) means that by performing this operation on the two encrypted messages, we obtain a ciphertext that is equivalent to the encryption of the sum of the two plaintexts. Formally, the decryption primitive $D(\cdot)$ is a homomorphism between the group of ciphertexts with the operator $\oplus$ and the group of plaintexts with addition $+$, which justifies the name of the scheme. It is immediate to see that if a scheme supports addition between encrypted messages, it will also support subtraction, by adding the additive inverse, and multiplication between an integer plaintext and an encrypted message, obtained by adding the encrypted messages for the corresponding number of times.

Furthermore, **multiplicatively homomorphic** schemes satisfy the property that there exists an operator $\otimes$ defined on the space of ciphertexts such that:

$$E(a) \otimes E(b) \subset E(a \cdot b), \qquad\qquad (9.6)$$

for any plaintexts a, b supported by the scheme. If the same scheme satisfies both (9.5) and (9.6) for an unlimited amount of operations, it is called **fully homomorphic**. If a scheme satisfies both (9.5) and (9.6) but only for a limited amount of operations, it is called **somewhat homomorphic**.

Remark 9.1 A homomorphic cryptosystem is malleable, which means that a party that does not have the private key can alter a ciphertext such that another valid ciphertext is obtained. Malleability is a desirable property in order to achieve third-party outsourced computation on encrypted data, but allows ciphertext attacks. In this work, we assume that the parties have access to authenticated channels, therefore an adversary cannot alter the messages sent by the honest parties.

9.4.3.1 Additively Homomorphic Cryptosystem

Additively Homomorphic Encryptions schemes, abbreviated as AHE, can be instantiated by various public key additively homomorphic encryption schemes such as [24, 34, 39, 52]. Let AHE $= (\text{Key}\hat{\text{G}}\text{en}, \hat{\text{E}}, \hat{\text{D}}, \hat{\text{Add}}, \text{c}\hat{\text{M}}\text{lt})$ be an instance of an asymmetric additively homomorphic encryption scheme, with $\mathcal{M}$ the message space and $\hat{\mathcal{C}}$ the ciphertext space, where we will use the following abstract notation: $\hat{\oplus}$ denotes the addition on $\hat{\mathcal{C}}$ and $\hat{\otimes}$ denotes the multiplication between a plaintext and a ciphertext. We save the notation without $(\hat{\cdot})$ for the scheme in Sect. 9.4.3.2. Asymmetric or public key cryptosystems involve a pair of keys: a public key that is disseminated publicly, and which is used for the encryption of the private messages, and a private key which is known only to its owner, used for the decryption of the encrypted messages. We will denote the encryption of a message $m \in \mathcal{M}$ by $[[m]]$ as a shorthand notation for $\hat{\text{E}}(\text{public key}, m)$.

(i) $\text{Key}\hat{\text{G}}\text{en}(1^\sigma)$: Takes the security parameter σ and outputs a public key $\hat{\text{pk}}$ and a private key $\hat{\text{sk}}$.

(ii) $\hat{\text{E}}(\hat{\text{pk}}, m)$: Takes the public key and a message $m \in \mathcal{M}$ and outputs a ciphertext $[[m]] \in \hat{\mathcal{C}}$.

(iii) $\hat{\text{D}}(\hat{\text{sk}}, c)$: Takes the private key and a ciphertext $c \in \hat{\mathcal{C}}$ and outputs the message that was encrypted $m' \in \mathcal{M}$.

(iv) $\hat{\text{Add}}(c_1, c_2)$: Takes ciphertexts $c_1, c_2 \in \hat{\mathcal{C}}$ and outputs ciphertext $c = c_1 \hat{\oplus} c_2 \in \hat{\mathcal{C}}$ such that: $\hat{\text{D}}(\hat{\text{sk}}, c) = \hat{\text{D}}(\hat{\text{sk}}, c_1) + \hat{\text{D}}(\hat{\text{sk}}, c_2)$.

(v) $\text{c}\hat{\text{M}}\text{lt}(m_1, c_2)$: Takes plaintext $m_1 \in \mathcal{M}$ and ciphertext $c_2 \in \hat{\mathcal{C}}$ and outputs ciphertext $c = m_1 \hat{\otimes} c_2 \in \hat{\mathcal{C}}$ such that: $\hat{\text{D}}(\hat{\text{sk}}, c) = m_1 \cdot \hat{\text{D}}(\hat{\text{sk}}, c_2)$.

In this chapter, we use the popular Paillier encryption [52], that has plaintext space $\mathbb{Z}_N$ and ciphertext space $(\mathbb{Z}_{N^2})^*$. Any other AHE scheme that is semantically secure [34], [32, Chap. 4] and circuit-private [14] can be employed.

9.4.3.2 Labeled Homomorphic Encryption

The model predictive control problem from Fig. 9.1, and more general control decision problems as well, can be abstracted in the following general framework in Fig. 9.3. Consider a cloud server that collects encrypted data from several clients. The data represents time series and is labeled with the corresponding time. A requester makes queries that can be written as multivariate polynomials over the data stored at the cloud server and solicits the result.

Labeled Homomorphic Encryption (LabHE) can process data from multiple users with different private keys, as long as the requesting party has a master key. This

Fig. 9.3 The clients send their private data to a cloud server. A requester sends a query to the cloud, which evaluates it on the data and sends the result to the requester

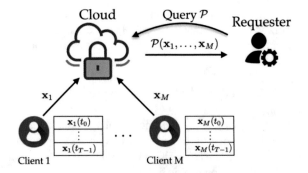

scheme makes use of the fact that the decryptor (or requester in Fig. 9.3) knows the query to be executed on the encrypted data, which we will refer to as a program. Furthermore, we want a cloud server that only has access to the encrypted data to be able to perform the program on the encrypted data and the decryptor to be able to decrypt the result. To this end, the inputs to the program need to be uniquely identified. Therefore, an encryptor (or client in Fig. 9.3) assigns a unique label to each message and sends the encrypted data along with the corresponding encrypted labels to the server. Labels can be time instances, locations, id numbers etc.

Denote by $\mathcal{M}$ the message space. An admissible function for LabHE $f : \mathcal{M}^n \to \mathcal{M}$ is a multivariate polynomial of degree 2 on n variables. A program that has labeled inputs is called a labeled program [7].

Definition 9.5 A labeled program $\mathcal{P}$ is a tuple $(f, \tau_1, \ldots, \tau_n)$ where $f : \mathcal{M}^n \to \mathcal{M}$ is an admissible function on n variables and $\tau_i \in \{0, 1\}^*$ is the label of the i-th input of f. Given t programs $\mathcal{P}_1, \ldots, \mathcal{P}_t$ and an admissible function $g : \mathcal{M}^t \to \mathcal{M}$, the composed program $\mathcal{P}^g$ is obtained by evaluating g on the outputs of $\mathcal{P}_1, \ldots, \mathcal{P}_t$, and can be denoted compactly as $\mathcal{P}^g = g(\mathcal{P}_1, \ldots, \mathcal{P}_t)$. The labeled inputs of $\mathcal{P}^g$ are all the distinct labeled inputs of $\mathcal{P}_1, \ldots, \mathcal{P}_t$.

Let $f_{id} : \mathcal{M} \to \mathcal{M}$ be the identity function and $\tau \in \{0, 1\}^*$ be a label. Denote the identity program for input label τ by $\mathcal{I}_\tau = (f_{id}, \tau)$. Any labeled program $\mathcal{P} = (f, \tau_1, \ldots, \tau_n)$, as in Definition 9.5, can be expressed as the composition of n identity programs $\mathcal{P} = f(\mathcal{I}_{\tau_1}, \ldots, \mathcal{I}_{\tau_n})$.

LabHE is constructed from an AHE scheme with the requirement that the message space must be a public ring in which one can efficiently sample elements uniformly at random. The idea is that an encryptor splits their private message as described in Sect. 9.4.1 into a random value (secret) and the difference between the message and the secret. For efficiency, instead of taking the secret to be a uniformly random value, we take it to be the output of a pseudorandom generator applied to the corresponding label. The label acts like the seed of the pseudo-random generator. The encryptor then forms the LabHE ciphertext from the encryption of the first share along with the second share, yielding $E(m) = (m - \mathfrak{b}, [[\mathfrak{b}]])$, as described in Step 1 in the following. This enables us to decrypt one multiplication of two encrypted values, using the observation (9.7). The AHE scheme allows computing $(m_1 - \mathfrak{b}_1)\hat{\otimes}[[\mathfrak{b}_2]]$, $(m_2 - \mathfrak{b}_2)\hat{\otimes}[[\mathfrak{b}_1]]$ and $[[(m_1 - \mathfrak{b}_1) \cdot (m_2 - \mathfrak{b}_2)]]$, for plaintexts $(m_i - \mathfrak{b}_i)$, $i = 1, 2$. Hence, we can obtain the AHE encryption of one multiplication $[[m_1 \cdot m_2 - \mathfrak{b}_1 \cdot \mathfrak{b}_2]]$ from $E(m_1)$ and $E(m_1)$, described in Step 4 in the following. Decryption, described in Step 5, requires that the decryptor knows the private key of the AHE scheme, and $\mathfrak{b}_i$, such that it can compute $m_1 \cdot m_2 = \hat{D}[[m_1 \cdot m_2 - \mathfrak{b}_1 \cdot \mathfrak{b}_2]] + \mathfrak{b}_1 \cdot \mathfrak{b}_2$.

$$m_1 \cdot m_2 - \mathfrak{b}_1 \cdot \mathfrak{b}_2 = (m_1 - \mathfrak{b}_1) \cdot (m_2 - \mathfrak{b}_2) + \mathfrak{b}_1 \cdot (m_2 - \mathfrak{b}_2) + \mathfrak{b}_2 \cdot (m_1 - \mathfrak{b}_1).$$
(9.7)

Let $\mathcal{M}$ be the message space of the AHE scheme, $\mathcal{L} \subset \{0, 1\}^*$ denote a finite set of labels and $F : \{0, 1\}^k \times \{0, 1\}^* \to \mathcal{M}$ be a pseudorandom function that takes as inputs a key of size k polynomial in σ the security parameter, and a label from $\mathcal{L}$. Then LabHE is defined as a tuple LabHE $= $ (Init, KeyGen, E, Eval, D):

1. Init(1^σ): Takes the security parameter σ and outputs master secret key msk and master public key mpk for AHE.
2. KeyGen(mpk): Takes the master public key mpk and outputs for each user i a user secret key usk_i and a user public key upk_i.
3. E(mpk, upk, τ, m): Takes the master public key, a user public key, a label $\tau \in \mathcal{L}$ and a message $m \in \mathcal{M}$ and outputs a ciphertext $C = (a, \beta)$. It is composed of an online and offline part:

   - Off-E(usk, τ): Computes the secret $\mathfrak{b} \leftarrow F(usk, \tau)$ and outputs $C_{\text{off}} = (\mathfrak{b}, [[\mathfrak{b}]])$.
   - On-E(C_{off}, m): Outputs $C = (m - \mathfrak{b}, [[\mathfrak{b}]]) =: (a, \beta) \in \mathcal{M} \times \hat{\mathcal{C}}$.

4. Eval(mpk, f, $C_1, \ldots, C_t$): Takes the master public key, an admissible function $f : \mathcal{M}^t \to \mathcal{M}$, t ciphertexts and returns a ciphertext C. Eval is composed of the following building blocks:

   - Mlt(C_1, C_2): Takes $C_i = (a_i, \beta_i) \in \mathcal{M} \times \hat{\mathcal{C}}$ for $i = 1, 2$ and outputs $C = [[a_1 \cdot a_2]]\hat{\oplus}(a_1\hat{\otimes}\beta_2)\hat{\oplus}(a_2\hat{\otimes}\beta_1) = [[m_1 \cdot m_2 - \mathfrak{b}_1 \cdot \mathfrak{b}_2]] =: \alpha \in \hat{\mathcal{C}}$.
   - Add(C_1, C_2): If $C_i = (a_i, \beta_i) \in \mathcal{M} \times \hat{\mathcal{C}}$ for $i = 1, 2$, then outputs $C = (a_1 + a_2, \beta_1\hat{\oplus}\beta_2) =: (a, \beta) \in \mathcal{M} \times \hat{\mathcal{C}}$. If both $C_i = \alpha_i \in \hat{\mathcal{C}}$, for $i = 1, 2$, then outputs $C = \alpha_1\hat{\oplus}\alpha_2 =: \alpha \in \hat{\mathcal{C}}$. If $C_1 = (a_1, \beta_1) \in \mathcal{M} \times \hat{\mathcal{C}}$ and $C_2 = \alpha_2 \in \hat{\mathcal{C}}$, then outputs $C = (a_1, \beta_1\hat{\oplus}\alpha_2) =: (a, \beta) \in \mathcal{M} \times \hat{\mathcal{C}}$.
   - cMlt(c, C'): Takes a plaintext $c \in \mathcal{M}$ and a ciphertext C'. If $C' = (a', \beta') \in \mathcal{M} \times \hat{\mathcal{C}}$, outputs $C = (c \cdot a', c\hat{\otimes}\beta') =: (a, \beta) \in \mathcal{M} \times \hat{\mathcal{C}}$. If $C' = \alpha' \in \hat{\mathcal{C}}$, outputs $C = c\hat{\otimes}\alpha' =: \alpha \in \hat{\mathcal{C}}$.

5. D(msk, $usk_{1,\ldots,t}$, $\mathcal{P}$, C): Takes the master secret key, a vector of t user secret keys, a labeled program $\mathcal{P}$ and a ciphertext C. It has an online and an offline part:

   - Off-D(msk, $\mathcal{P}$): Parses $\mathcal{P}$ as $(f, \tau_1, \ldots, \tau_t)$. For $i \in [t]$, it computes the secrets $\mathfrak{b}_i = F(usk_i, \tau_i)$, $\mathfrak{b} = f(\mathfrak{b}_1, \ldots, \mathfrak{b}_t)$ and outputs msk$_\mathcal{P}$(msk, $\mathfrak{b}$).
   - On-D(sk$_\mathcal{P}$, C): If $C = (a, \beta) \in \mathcal{M} \times \hat{\mathcal{C}}$: either output (i) $m = a + \mathfrak{b}$ or (ii) output $m = a + \hat{D}$(msk, β). If $C \in \hat{\mathcal{C}}$, output $m = \hat{D}$(msk, C) $+ \mathfrak{b}$.

The cost of an online encryption is the cost of an addition in $\mathcal{M}$. The cost of online decryption is independent of $\mathcal{P}$ and only depends on the complexity of $\hat{D}$.

Semantic security characterizes the security of a cryptosystem. The definition of semantic security is sometimes given as a cryptographic game [34].

Definition 9.6 (*Semantic Security* [32, Chap. 5]) An encryption scheme is **semantically secure** if for every probabilistic polynomial-time algorithm, $\mathcal{A}$, there exists a probabilistic polynomial-time algorithm $\mathcal{A}'$ such that for every two polynomially bounded functions $f, h : \{0, 1\}^* \to \{0, 1\}^*$, for any probability ensemble $\{X_\sigma\}_{\sigma \in \mathbb{N}}$ of length polynomial in σ, for any positive polynomial p and sufficiently large σ:

$$\Pr\left[\mathcal{A}(\mathrm{E}(X_\sigma), h(X_\sigma), 1^\sigma) = f(X_\sigma)\right] < \Pr\left[\mathcal{A}'(h(X_\sigma), 1^\sigma) = f(X_\sigma)\right] + \frac{1}{p(\sigma)},$$

The probability is taken over the ensemble X_σ and the internal coin tosses in $\mathcal{A}, \mathcal{A}'$.

Under the assumption of decisional composite residuosity [52], the Paillier scheme is semantically secure and has indistinguishable encryptions. Moreover, the LabHE scheme satisfies semantic security given that the underlying homomorphic encryption scheme is semantically secure and the function F is pseudorandom.

In [7] it is proved that LabHE also satisfies context-hiding (decrypting the ciphertext does not reveal anything about the inputs of the computed function, only the result of the function on those inputs).

9.4.4 Oblivious Transfer

Oblivious transfer is a technique used when one party wants to obtain a secret from a set of secrets held by another party [32, Chap. 7]. Party A has k secrets $(\sigma_0, \ldots, \sigma_{k-1})$ and party B has an index $i \in \{0, \ldots, k-1\}$. The goal of A is to transmit the i-th secret requested by the receiver without knowing the value of the index i, while B does not learn anything other than σ_i. This is called 1-out-of-k oblivious transfer. There are many constructions of oblivious transfer that achieve security as in the two-party version of the simulation definition (Definition 9.4). Many improvements in efficiency, e.g., precomputation, and security have been proposed, see e.g., [37, 51].

We will use the standard 1-out-of-2 oblivious transfer, where the inputs of party A are $[[\sigma_0]], [[\sigma_1]]$ and party B holds $i \in \{0, 1\}$ and the secret key and has to obtain σ_i. We will denote this by $\sigma_i \leftarrow \mathrm{OT}([[\sigma_0]], [[\sigma_1]], i, \hat{\mathrm{sk}})$. We will also use a variant where party A has to obliviously obtain the AHE-encrypted $[[\sigma_i]]$, and A has $[[\sigma_0]], [[\sigma_1]]$ and party B holds i, for $i \in \{0, 1\}$, and the secret key. We will denote this variant by $[[\sigma_i]] \leftarrow \mathrm{OT}'([[\sigma_0]], [[\sigma_1]], i, \hat{\mathrm{sk}})$.

The way the variant OT' works is that A chooses at random r_0, r_1 from the message space $\mathcal{M}$, and sends shares of the messages to B: $[[v_0]] := \hat{\mathrm{Add}}([[\sigma_0]], [[r_0]])$, $[[v_1]] := \hat{\mathrm{Add}}([[\sigma_1]], [[r_1]])$. B selects v_i and sends back to A the encryption of the index i: $[[i]]$ and $\hat{\text{Add}}([[v_i]], [0])$, such that A cannot obtain information about i by comparing the value it received with the values it sent. Then, A computes:

$$[[\sigma_i]] = \hat{\text{Add}}\left([[v_i]], \hat{\text{cMlt}}(r_0, \hat{\text{Add}}([[i]], [[-1]])), \hat{\text{cMlt}}(-r_1, [[i]])\right).$$

Proposition 9.1 $[[\sigma_i]] \leftarrow \text{OT}'([[\sigma_0]], [[\sigma_1]], i, \hat{\text{sk}})$ *is private w.r.t. Definition 9.4.*

We will show in the following proof how to construct the simulators in Definition 9.4 in order to prove the privacy of the oblivious transfer variant we use.

Proof Let us construct the view of A, with inputs $\hat{\text{pk}}$, $[[\sigma_0]]$, $[[\sigma_1]]$ and output $[[\sigma_i]]$:

$$V_A(\hat{\text{pk}}, [[\sigma_0]], [[\sigma_1]]) = \left(\hat{\text{pk}}, [[\sigma_0]], [[\sigma_1]], r_0, r_1, [[i]], [[v_i]], [[\sigma_i]], \text{coins}\right),$$

where coins are the random values used for encrypting r_0 and r_1 and $[[-1]]$. The view of party B, that has inputs i, $\hat{\text{pk}}$, $\hat{\text{sk}}$ and no output, is:

$$V_B(i, \hat{\text{pk}}, \hat{\text{sk}}) = \left(i, \hat{\text{pk}}, \hat{\text{sk}}, [[v_0]], [[v_1]], \text{coins}\right),$$

where coins are the random values used for encrypting v_i, i and 0.

Now let us construct a simulator S_A that generates an indistinguishable view from party A. S_A takes as inputs $\hat{\text{pk}}$, $[[\sigma_0]]$, $[[\sigma_1]]$, $[[\sigma_i]]$ and generates $\tilde{r}_0$ and $\tilde{r}_1$ as random values in $\mathcal{M}$. It then selects a random bit $\tilde{i}$ and encrypts it with $\hat{\text{pk}}$ and computes $[[\tilde{v}_i]] = \hat{\text{Add}}\left([[\sigma_i]], \hat{\text{cMlt}}(-\tilde{r}_0, \hat{\text{Add}}([[\tilde{i}]], [[-1]])), \hat{\text{cMlt}}(\tilde{r}_1, [[\tilde{i}]])\right)$. It also generates $\widetilde{\text{coins}}$ for three encryptions. S_A outputs:

$$S_A(\hat{\text{pk}}, [[\sigma_0]], [[\sigma_1]], [[\sigma_i]]) = \left(\hat{\text{pk}}, [[\sigma_0]], [[\sigma_1]], \tilde{r}_0, \tilde{r}_1, [[\tilde{i}]], [[\tilde{v}_i]], [[\sigma_i]], \widetilde{\text{coins}}\right).$$

First, $\tilde{r}_0, \tilde{r}_1$ and $\widetilde{\text{coins}}$ are statistically indistinguishable from r_0, r_1 and coins because they were generated from the same distributions. Second, $[[\tilde{i}]]$ and $[[\tilde{v}_i]]$ are indistinguishable from $[[i]]$ and $[[v_i]]$ because AHE is semantically secure and has indistinguishable encryptions, and because $[[\sigma_i]]$ is a refreshed value of $[[\sigma_0]]$ or $[[\sigma_1]]$. This means A cannot learn any information about i, hence $[[i]]$ looks like the encryption of a random bit, i.e., like $[[\tilde{i}]]$. Thus, $V_A(\hat{\text{pk}}, [[\sigma_0]], [[\sigma_1]]) \stackrel{c}{\equiv} S_A(\hat{\text{pk}}, [[\sigma_0]], [[\sigma_1]], [[\sigma_i]])$.

A simulator S_B for party B takes as inputs i, $\hat{\text{pk}}$, $\hat{\text{sk}}$ and generates two random values from $\mathcal{M}$, names them $\tilde{v}_0$ and $\tilde{v}_1$ and encrypts them. It then generates $\widetilde{\text{coins}}$ as random values for three encryptions. S_B outputs:

$$S_B(i, \hat{\text{pk}}, \hat{\text{sk}}) = \left(i, \hat{\text{pk}}, \hat{\text{sk}}, [[\tilde{v}_0]], [[\tilde{v}_1]], \widetilde{\text{coins}}\right).$$

First, $\widetilde{\text{coins}}$ are statistically indistinguishable from coins because they were generated from the same distribution. Second, $\widetilde{v}_0$ and $\widetilde{v}_1$ are also statistically indistinguishable from each other and from v_0 and v_1 due to the security of the one-time pad. Their encryptions will also be indistinguishable. Thus, $V_B(i, \hat{\text{pk}}, \hat{\text{sk}}) \overset{c}{\equiv} S_B(i, \hat{\text{pk}}, \hat{\text{sk}})$. $\qquad\square$

9.4.5 Private Comparison

Consider a two-party computation problem of two inputs encrypted under an encryption scheme that does not preserve the order from plaintexts to ciphertexts. A survey on the state of the art of private comparison protocols on private inputs owned by two parties is given in [18]. In [21, 22], Damgård, Geisler and Krøigaard describe a protocol for secure comparison and, towards that functionality, they propose the DGK additively homomorphic encryption scheme with the property that, given a ciphertext and the secret key, it is efficient to determine if the encrypted value is zero without fully decrypting it. This is useful when making decisions on bits. The authors also prove the semantic security of the DGK cryptosystem under the hardness of factoring assumption. We denote the DGK encryption of a scalar value by $[\cdot]$ and use the same operations notation $\hat{\oplus}$ and $\hat{\otimes}$.

Consider two parties A and B, each having a private value α and β. Using the binary representations of α and β, the two parties exchange l blinded and encrypted values such that each of the parties will obtain a bit $\delta_A \in \{0, 1\}$ and $\delta_B \in \{0, 1\}$ that satisfy the following relation: $\delta_A \veebar \delta_B = (\alpha \leq \beta)$, after executing Protocol 1, where $\veebar$ denotes the exclusive or operation. The protocol is described in [67, Protocol 3], where an improvement of the DGK scheme is proposed. By applying some extra steps, as in [67, Protocol 2], one can obtain a protocol for private two-party comparison where party A has two encrypted inputs with an AHE scheme $[[a]], [[b]]$, with a, b represented on l bits.

PROTOCOL 1: Private two-party comparison with plaintext inputs using DGK

Require: A: α; B: β, sk_{DGK}
Ensure: A: δ_A; B: δ_B such that $\delta_A \veebar \delta_B = (\alpha \leq \beta)$
1: B: send the encrypted bits $[\beta_i]$, $0 \leq i < l$ to A.
2: **for** each $0 \leq i < l$ **do**
3:     A: $[\alpha_i \veebar \beta_i] \leftarrow [\beta_i]$ if $\alpha_i = 0$ and $[\alpha_i \veebar \beta_i] \leftarrow [1] \hat{\oplus} (-1) \hat{\otimes} [\beta_i]$ otherwise.
4: **end for**
5: A: Choose a uniformly random bit $\delta_A \in \{0, 1\}$.
6: A: Compute the set $\mathcal{L} = \{i | 0 \leq i < l \text{ and } \alpha_i = \delta_A\}$.
7: **for** each $i \in \mathcal{L}$ **do**
8:     A: compute $[c_i] \leftarrow [\alpha_{i+1} \veebar \beta_{i+1}] \hat{\oplus} \ldots \hat{\oplus} [\alpha_l \veebar \beta_l])$.
9:     A: $[c_i] \leftarrow [1] \hat{\oplus} [c_i] \oplus (-1) \hat{\otimes} [\beta_i]$ if $\delta_A = 0$ and $[c_i] \leftarrow [1] \hat{\oplus} [c_i]$ otherwise.
10: **end for**
11: A: generate uniformly random non-zero values r_i of $2t$ bits (see [22]), $0 \leq i < l$.
12: **for** each $0 \leq i < l$ **do**
13:     A: $[c_i] \leftarrow r_i \hat{\otimes} [c_i]$ if $i \in \mathcal{L}$ and $[c_i] \leftarrow [r_i]$ otherwise.
14: **end for**
15: A: send the values $[c_i]$ in random order to B.
16: B: if at least one of the values c_i is decrypted to zero, set $\delta_B \leftarrow 1$, otherwise set $\delta_B \leftarrow 0$.

PROTOCOL 2: Private two-party comparison with encrypted inputs using DGK

Require: A: $[[a]]$, $[[b]]$; B: $\hat{sk}$, sk_{DGK}

Ensure: B: $(\delta = 1) \equiv (a \leq b)$

1: A: choose uniformly at random r of $l + 1 + \lambda$ bits, compute $[[z]] \leftarrow [[b]]\hat{\ominus}[[a]]\hat{\oplus}[[2^l + r]]$ and send it to B. Then compute $\alpha \leftarrow r \bmod 2^l$.
2: B: decrypt $[[z]]$ and compute $\beta \leftarrow z \bmod 2^l$.
3: A,B: execute Protocol 1.
4: B: send $[[z \div 2^l]]$ and $[[\delta_B]]$ to A.
5: A: $[[(\beta < \alpha)]] \leftarrow [[\delta_B]]$ if $\delta_A = 1$ and $[[(\beta < \alpha)]] \leftarrow [[1]]\hat{\ominus}[[\delta_B]]$ otherwise.
6: A: compute $[[\delta]] \leftarrow [[z \div 2^l]]\hat{\ominus}([[r \div 2^l]]\hat{\oplus}[[(\beta < \alpha)]])$ and send it to B.
7: B: decrypts δ.

Proposition 9.2 *([21, 67]) Protocol 2 is private w.r.t. Definition 9.4.*

9.5 MPC with Encrypted Model and Encrypted Signals

As remarked in the Introduction, in many situations it is important to protect not only the signals (e.g., the states, measurements), but also the system model. To this end, we propose a solution that uses Labeled Homomorphic Encryption to achieve encrypted multiplications and the private execution of MPC on encrypted data. LabHE has a useful property called unbalanced efficiency that can be observed from Sect. 9.4.3.2 and was described in [15], which states that only the evaluator is required to perform operations on ciphertexts, while the decryptor performs computations only on the plaintext messages. We will employ this property by having the cloud perform the more complex operations and the actuator the more efficient ones.

Our protocols will consist of three phases: the offline phase, in which the computations that are independent from the specific data of the users are performed, the initialization phase, in which the computations related to the constant parameters in the problem are performed, and the online phase, in which computations on the variables of the problem are performed.

Figure 9.4 represents the private version of the MPC diagram from Fig. 9.1, where the quantities will be briefly described next and more in detail in Sect. 9.5.1. The actuator holds the MPC query functionality, denoted by f_{MPC}. Offline, the actuator generates a pair of master keys, as described in Sect. 9.4.3.2 and publishes the master public key. The setup and subsystems generate their secret keys and send the corresponding public keys to the actuator. Still offline, these parties generate the labels corresponding to their data with respect to the time stamp and the size of the data. As explained in Sect. 9.4.3.2, the labels are crucial to achieving the encrypted multiplications. Moreover, when generating them, it is important to make sure that no two labels that will be encrypted with the same key are the same. When the private data are times series, as in our problem, the labels can be easily generated using the time

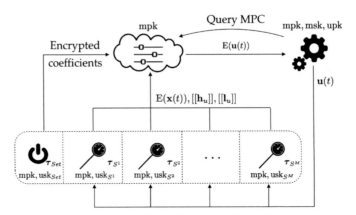

Fig. 9.4 The setup and subsystems send their encrypted data to the cloud. The cloud has to run the MPC algorithm on the private measurements and the system's private matrices and send the encrypted result to the actuator. The latter then actuates the system with the decrypted inputs

steps and sizes corresponding to each signal, with no other complex synchronization process necessary between the actors. This is shown in Protocol 3.

The setup entity sends the LabHE encryptions of the state matrices and costs to the cloud controller before the execution begins. The subsystems send the encryptions of the input constraints to the cloud controller, also before the execution begins. Online, at every time step, the subsystems encrypt their measurements and send them to the cloud. After the cloud performs the encrypted MPC query for one time step, it sends the encrypted control input at the current time step to the actuator, which decrypts it and inputs it to the system. In Protocol 4, we describe how the encrypted MPC query is performed by the parties.

We will now show how to transform the FGM (9.4) into a private version. The message space $\mathcal{M}$ we choose for the encryption schemes is $\mathbb{Z}_N$, the additive group of integers modulo a large value N. This means that, prior to encryption, the values have to be represented as integers. For now, assume that this preprocessing step has been already performed. We postpone the details to Sect. 9.5.3.

First, let us write $\mathbf{t}_k$ in (9.4a) as a function of $\mathbf{U}_k$ and $\mathbf{U}_{k-1}$:

$$
\begin{aligned}
\mathbf{t}_k &= \left(\mathbf{I}_{Mm} - \frac{1}{L}\mathbf{H}\right) z_k - \frac{1}{L}\mathbf{F}^\mathsf{T}\mathbf{x}(t) \\
&= \left(\mathbf{I}_{Mm} - \frac{1}{L}\mathbf{H}\right)\left[(1+\eta)\mathbf{U}_k - \eta\mathbf{U}_{k-1}\right] - \frac{1}{L}\mathbf{F}^\mathsf{T}\mathbf{x}(t) \\
&= \mathbf{U}_k + \eta(\mathbf{U}_k - \mathbf{U}_{k-1}) - \frac{1}{L}\mathbf{H}\mathbf{U}_k - \frac{\eta}{L}\mathbf{H}(\mathbf{U}_k - \mathbf{U}_{k-1}) - \frac{1}{L}\mathbf{F}^\mathsf{T}\mathbf{x}(t).
\end{aligned}
$$

If we consider the composite variables $\frac{1}{L}\mathbf{H}$, $\frac{\eta}{L}\mathbf{H}$, $\frac{1}{L}\mathbf{F}$ and variables $\mathbf{U}_k, \mathbf{U}_{k-1}, \mathbf{x}(t)$, then $\mathbf{t}_k$ can be written as a degree-two multivariate polynomial. This allows us to compute $[[\mathbf{t}_k]]$ using LabHE.

Then, one encrypted iteration of the FGM, where we assume that the cloud has access to $E\left(-\frac{1}{L}\mathbf{H}\right)$, $E\left(-\frac{\eta}{L}\mathbf{H}\right)$, $E\left(\frac{1}{L}\mathbf{F}^\mathsf{T}\right)$, $E(\mathbf{x}(t))$, $[[\mathbf{h}_u]]$, $[[\mathbf{l}_u]]$ can be written as follows. Denote the computation on the inputs mentioned previously as f_{iter}. We use both Add and $\oplus$, Mlt and $\otimes$ for a better visual representation.

$$
\begin{aligned}
[[\mathbf{t}_k - \rho_k]] = {}&\text{Mlt}\left(E\left(\frac{-1}{L}\mathbf{F}^\mathsf{T}\right), E(\mathbf{x}(t))\right) \\
&\oplus \text{Mlt}\left(\text{Add}\left(\mathbf{I}_{Mm}, E\left(\frac{1}{L}\mathbf{H}\right)\right), E\left(\mathbf{U}_k\right)\right) \oplus \\
&\oplus \text{Mlt}\left(\text{Add}\left(E(\eta) \otimes \mathbf{I}_{Mm}, E\left(\frac{-\eta}{L}\mathbf{H}\right)\right), \left(E(\mathbf{U}_k) \ominus E(\mathbf{U}_{k-1})\right)\right),
\end{aligned}
\tag{9.8}
$$

where ρ_k is the secret obtained by applying f_{iter} on the LabHE secrets of the inputs of f_{iter}. When the actuator applies the LabHE decryption primitive on $[[\mathbf{t}_k - \rho_k]]$, ρ_k is removed. Hence, for simplicity, we will write $[[\mathbf{t}_k]]$ instead of $[[\mathbf{t}_k - \rho_k]]$.

Second, let us address how to perform (9.4b) in a private way. We have to perform the projection of $\mathbf{t}_k$ over the feasible domain described by $\mathbf{h}_u$ and $\mathbf{l}_u$, where all the variables are encrypted, as well as the private update of $\mathbf{U}_{k+1}$ with the projected iterate. One solution to the private comparison was described in Sect. 9.4.5. The cloud has $[[\mathbf{t}_k]]$ and assume it also has the AHE encryptions of the limits $[[\mathbf{h}_u]]$ and $[[\mathbf{l}_u]]$. The cloud and the actuator will engage in two instances of the DGK protocol and oblivious transfer: first, to compare $\mathbf{t}_k$ to $\mathbf{h}_u$ and obtain an intermediate value of $\mathbf{U}_{k+1}$, and second, to compare $\mathbf{U}_{k+1}$ to $\mathbf{l}_u$ and to update the iterate $\mathbf{U}_{k+1}$.

Before calling the comparison Protocol 2, described in Sect. 9.4.5, the cloud should randomize the order of the inputs, such that, after obtaining the comparison bit, the actuator does not learn whether the iterate was feasible or not. This is done in lines 8 and 11 in Protocol 4. Upon the completion of the comparison, the cloud and actuator perform the oblivious transfer variant, described in Sect. 9.4.4, such that the cloud obtains the intermediate value of $[[\mathbf{U}_{k+1}]]$ and subsequently, update the AHE encryption of the iterate $[[\mathbf{U}_{k+1}]]$. At the last iteration, the cloud and actuator perform the standard oblivious transfer for the first m positions in the values such that the actuator obtains $\mathbf{u}(t)$. Finally, because the next iteration can proceed only if the cloud has access to the full LabHE encryption $E(\mathbf{U}_{k+1})$, instead of $[[\mathbf{U}_{k+1}]]$, the cloud and actuator have to refresh the encryption. Specifically, the cloud secret-shares $[[\mathbf{U}_{k+1}]]$ in $[[\mathbf{U}_{k+1} - \mathbf{r}_{k+1}]]$ and $\mathbf{r}_{k+1}$, and sends $[[\mathbf{U}_{k+1} - \mathbf{r}_{k+1}]]$ to the actuator. The actuator decrypts it, and, using a previously generated secret, sends back $E(\mathbf{U}_{k+1} - \mathbf{r}_{k+1}) = \left(\mathbf{U}_{k+1} - \mathbf{r}_{k+1}, [[\mathbf{b}_{k+1}^U]]\right)$. Then, the cloud recovers the LabHE encryption as $E(\mathbf{U}_{k+1}) = \text{Add}(E(\mathbf{U}_{k+1} - \mathbf{r}_{k+1}), \mathbf{r}_{k+1})$. In what follows, we will outline the private protocols obtained by integrating the above observations.

9.5.1 Private Protocol

Assume that K, N are fixed and known by all parties. Subscript S^i stands for the i-th subsystem, for $i \in [M]$, subscript Set for the Setup, subscript A for the actuator and subscript C for the cloud.

PROTOCOL 3: Initialization of encrypted MPC

Require: Actuator: f_{MPC}; Subsystems: $\mathcal{U}^i$; Setup: $\mathbf{A}, \mathbf{B}, \mathbf{P}, \mathbf{Q}, \mathbf{R}$.

Ensure: Subsytems: mpk, upk$_{S^i}$, usk$_{S^i}$, τ_{S^i}, $\mathbf{b}_{S^i}$, $[[\mathbf{b}_{S^i}]]$, $\mathfrak{R}_{S^i}$; Setup: $\mathbf{H}, \mathbf{F}, \eta, L$, mpk, upk$_{Set}$, usk$_{Set}$, τ_{Set}, $\mathbf{b}_{Set}$, $[[\mathbf{b}_{Set}]]$, $\mathfrak{R}_{Set}$; Actuator: usk, upk, τ_A, $\mathbf{b}_A$, $[[\mathbf{b}_A]]$, $\mathfrak{R}_A$; Cloud: mpk, $\mathrm{E}\left(-\frac{1}{L}\mathbf{H}\right)$, $\mathrm{E}\left(-\frac{\eta}{L}\mathbf{H}\right)$, $\mathrm{E}\left(\frac{1}{L}\mathbf{F}^{\mathsf{T}}\right)$, $\mathrm{E}(\eta)$, $[[\mathbf{h}_u]]$, $[[\mathbf{l}_u]]$, $\mathfrak{R}_C$.

Offline:
1: Actuator: Generate (mpk, msk) $\leftarrow$ Init(1^σ) and distribute mpk to the others. Also generate a key usk for itself.
2: Subsystems, Setup: Each get (usk, upk) $\leftarrow$ KeyGen(mpk) and send upk to the actuator.
3: Subsystems, Setup, Actuator: Allocate labels to the inputs of function f_{MPC}: $\tau_1, \ldots, \tau_v$, where v is the total number of inputs, as follows:
  Subsystem i: for $\mathbf{x}^i(t)$ of size n^i, where i denotes a subsystem, generate the corresponding labels $\tau_{\mathbf{x}^i(t)} = [0 \ 1 \ n^i \ \ldots \ n^i - 1]^{\mathsf{T}}$.
  Setup: for matrix $\mathbf{H} \in \mathbb{R}^{Mm \times Mm}$, set $l = 0$, generate
4: $\tau_{\mathbf{H}} = \begin{bmatrix} l & l+1 & \ldots & l+Mm-1 \\ \vdots & & & \vdots \\ l+(Mm-1)Mm & l+(Mm-1)Mm+1 & \ldots & l+M^2m^2-1 \end{bmatrix}$ and update $l = M^2m^2$, then follow the same
  steps for $\mathbf{F}$, starting from l and updating it.
  Actuator: follow the same steps as the subsystems and setup, and then generate similar labels for the iterates $\mathbf{U}_k$ starting from the last l, for $k = 0, \ldots, K-1$.
5: Subsystems, Setup, Actuator, Cloud: Generate randomness for blinding and encryptions $\mathfrak{R}$.
6: Subsystems, Setup, Actuator: Perform the offline part of the LabHE encryption primitive. The actuator also performs the offline part for the decryption. The parties thus obtain $\mathbf{b}$, $[[\mathbf{b}]]$.
7: Actuator: Generate initializations for the initial iterate $\mathbf{U}'_0$.
8: Actuator: Form the program $\mathcal{P} = (f_{\text{MPC}}, \tau_1, \ldots, \tau_v)$.

Initialization:
9: Setup: Compute $\mathbf{H}$ and $\mathbf{F}$ from $\mathbf{A}, \mathbf{B}, \mathbf{P}, \mathbf{Q}, \mathbf{R}$ and then $L = \lambda_{\max}(\mathbf{H})$ and $\eta = (\sqrt{\kappa(\mathbf{H})} - 1)/(\sqrt{\kappa(\mathbf{H})} + 1)$. Perform the online part of LabHE encryption and send to the cloud: $\mathrm{E}\left(-\frac{1}{L}\mathbf{H}\right)$, $\mathrm{E}\left(-\frac{\eta}{L}\mathbf{H}\right)$, $\mathrm{E}\left(\frac{1}{L}\mathbf{F}^{\mathsf{T}}\right)$, $\mathrm{E}(\eta)$.
10: Subsystems: Perform the online part of LabHE encryption and send to the cloud, which aggregates what it receives into: $[[\mathbf{h}_u]]$, $[[\mathbf{l}_u]]$.

PROTOCOL 4: Encrypted MPC step

Require: Actuator: f_{MPC}; Subsystems: $\mathbf{x}^i(t), \mathcal{U}^i$; Setup: $\mathbf{A}, \mathbf{B}, \mathbf{P}, \mathbf{Q}, \mathbf{R}$.

Ensure: Actuator: $\mathbf{u}(t)$

Offline + Initialization:
1: Subsystems, Setup, Cloud, Actuator: Run Protocol 3.

Online:
2: Cloud: $\left[\left[\frac{1}{L}\mathbf{F}^{\mathsf{T}}\mathbf{x}(t)\right]\right] \leftarrow \text{Mlt}\left(\mathrm{E}\left(\frac{1}{L}\mathbf{F}^{\mathsf{T}}\right), \mathrm{E}(\mathbf{x}(t))\right)$.
3: Actuator: Send the initial iterate to the cloud: $\mathrm{E}(\mathbf{U}'_0)$.
4: Cloud: Change the initial iterate: $\mathrm{E}(\mathbf{U}_0) = \text{Add}\left(\mathrm{E}\left(\mathbf{U}'_0\right), \mathbf{r}_0\right)$.
5: Cloud: $\mathrm{E}\left(\mathbf{U}_{-1}\right) \leftarrow \mathrm{E}\left(\mathbf{U}_0\right)$.
6: **for** $k = 0, \ldots, K-1$ **do**
7:   Cloud: Compute $[[\mathbf{t}_k]]$ as in Equation (9.8).

8:    Cloud: $([[\mathbf{a}_k]], [[\mathbf{b}_k]]) \leftarrow$ randomize $([[\mathbf{h}_u]], [[\mathbf{t}_k]])$.

9:    Cloud, Actuator: Execute comparison Protocol 2; Actuator obtains δ_k.

10:    Cloud, Actuator: $[[\mathbf{U}_{k+1}]] \leftarrow$ OT$'$ $([[\mathbf{a}_k]], [[\mathbf{b}_k]], \delta_k, \text{msk})$.

11:    Cloud: $([[\mathbf{a}_k]], [[\mathbf{b}_k]]) \leftarrow$ randomize $([[\mathbf{l}_u]], [[\mathbf{U}_{k+1}]])$.

12:    Cloud, Actuator: Execute comparison Protocol 2; Actuator obtains δ_k.

13:    **if** k! = K − 1 **then**

14:       Cloud, Actuator: $[[\mathbf{U}_{k+1}]] \leftarrow$ OT$'$ $([[\mathbf{a}_k]], [[\mathbf{b}_k]], \delta_k \veebar \mathbf{1}, \text{msk})$. Cloud receives $[[\mathbf{U}_{k+1}]]$.

15:       Cloud: Send to the actuator $[[\mathbf{U}'_{k+1}]] \leftarrow$ Add $([[\mathbf{U}_{k+1}]], [[-\mathbf{r}_k]])$, where $\mathbf{r}_k$ is randomly selected from $\mathcal{M}^{Mm}$.

16:       Actuator: Decrypt $[[\mathbf{U}'_{k+1}]]$ and send back E$(\mathbf{U}'_{k+1})$.

17:       Cloud: E $(\mathbf{U}_{k+1}) \leftarrow$ Add $(\text{E}(\mathbf{U}'_{k+1}), \mathbf{r}_k)$.

18:    **else**

19:       Cloud, Actuator: $\mathbf{u}(t) \leftarrow$ OT $([[\mathbf{a}_k]]_{1:m}, [[\mathbf{b}_k]]_{1:m}, \{\delta_k\}_{1:m} \veebar \mathbf{1}, \text{msk})$. Actuator receives $\mathbf{u}(t)$.

20:    **end if**

21: **end for**

22: Actuator: Input $\mathbf{u}(t)$ to the system.

Lines 3 and 4 ensure that neither the cloud nor the actuator knows the initial point of the optimization problem.

9.5.2   Privacy of Protocol 4

Assumption 2 An adversary cannot corrupt at the same time both the cloud controller and the virtual actuator or more than $M − 1$ subsystems.

Theorem 9.1 *Under Assumption 2, the encrypted MPC solution presented in Protocol 4 achieves multi-party privacy (Definition 9.4).*

Proof The components of the protocol: AHE, secret sharing, pseudorandom generator, LabHE, oblivious transfer and the comparison protocol are individually secure, meaning either their output is computationally indistinguishable from a random output or they already satisfy Definition 9.4. We are going to build the views of the allowed coalitions and their corresponding simulators and use the previous results to prove they are computationally indistinguishable.

The cloud has no inputs and no outputs, hence its view is composed solely from received messages and coins:

$$V_C(\emptyset) = \left(\text{mpk}, \text{E}\left(-\frac{1}{L}\mathbf{H}\right), \text{E}\left(-\frac{\eta}{L}\mathbf{H}\right), \text{E}\left(\frac{1}{L}\mathbf{F}^\mathsf{T}\right), \text{E}(\eta), \right.$$
$$[[\mathbf{h}_u]], [[\mathbf{l}_u]], \mathfrak{R}_C, \text{E}(\mathbf{U}'_0),$$
$$\left. \left\{ \text{E}(\mathbf{U}_k), [[\mathbf{a}_k]], [[\mathbf{b}_k]], [[\mathbf{U}_{k+1}]], \text{msg}_{\text{Pr.2}}, \text{msg}_{\text{OT}} \right\}_{k \in \{0, \ldots, K-1\}} \right). \quad (9.9)$$

The actuator's input is the function f_{MPC} and the output is $\mathbf{u}(t)$. Then, its view is:

$$V_A(f_{\mathrm{MPC}}) = \left(f_{\mathrm{MPC}}, \mathrm{mpk}, \mathrm{msk}, \mathrm{upk}, \mathfrak{R}_A, \right.$$

$$\left. \left\{ \mathbf{U}'_{k+1}, \mathrm{msg}_{\mathrm{Pr.2}}, \mathrm{msg}_{\mathrm{OT}} \right\}_{k \in \{0,\dots,K-1\}}, \mathbf{u}(t) \right), \qquad (9.10)$$

which includes the keys mpk, msk because their generation involve randomness.

The setup's inputs are the model and costs of the system and no output after the execution of Protocol 4, since it is just a helper entity. Its view is:

$$V_{Set}(\mathbf{A}, \mathbf{B}, \mathbf{P}, \mathbf{Q}, \mathbf{R}) = \left(\mathbf{A}, \mathbf{B}, \mathbf{P}, \mathbf{Q}, \mathbf{R}, \mathrm{mpk}, \mathrm{usk}_{Set}, \mathfrak{R}_{Set} \right). \qquad (9.11)$$

Finally, for a subsystem i, $i \in [M]$, the inputs are the local control action constraints and the measured states and there is no output obtained through computation after the execution of Protocol 4. Its view is:

$$V_{S^i}(\mathcal{U}^i, \mathbf{x}^i(t)) = \left(\mathcal{U}^i, \mathbf{x}^i(t), \mathrm{mpk}, \mathrm{usk}_{S^i}, \mathfrak{R}_{S^i} \right). \qquad (9.12)$$

In general, the indistinguishability between the view of the adversary corrupting the real-world parties and the simulator is proved through sequential games in which some real components of the view are replaced by components that could be generated by the simulator, which are indistinguishable from each other. In our case, we can directly make the leap between the real view and the simulator by showing that the cloud only receives encrypted messages, and the actuator receives only messages blinded by one-time pads. In [1], the proof for the privacy of a quadratic optimization problem solved in the same architecture is given with sequential games.

For the cloud, consider a simulator S_C that generates $\widetilde{\mathrm{mpk}}, \widetilde{\mathrm{msk}} \leftarrow \mathrm{Init}(1^\sigma)$, generates $\widetilde{\mathrm{usk}}_j \leftarrow \mathrm{KeyGen}(\widetilde{\mathrm{mpk}})$, for $j \in \{S^i, Set, A\}, i \in [M]$ and then $(\widetilde{\tau}, \widetilde{\mathbf{b}}, \widetilde{[[\mathbf{b}]]})_j$. We use $\widetilde{(\cdot)}$ also over the encryptions to show that the keys are different from the ones in the view. Subsequently, the simulator encrypts random values of appropriate sizes to obtain $\mathrm{E}\left(-\frac{1}{L}\mathbf{H}\right), \mathrm{E}\left(-\frac{\eta}{L}\mathbf{H}\right), \mathrm{E}\left(\frac{1}{L}\mathbf{F}^{\mathsf{T}}\right), \widetilde{\mathrm{E}(\eta)}, \widetilde{[[\mathbf{h}_u]]}, \widetilde{[[\mathbf{l}_u]]}, \widetilde{\mathrm{E}(\mathbf{U}'_0)}$. S_C generates the coins $\widetilde{\mathfrak{R}}_C$ as in line 4 in Protocol 3 and obtains $\mathrm{E}(\mathbf{U}_0)$ as in line 4 in Protocol 4. Then, for each $k = \{0, \dots, K-1\}$, it computes $[[\mathbf{t}_k]]$ as in line 7 in Protocol 4 and shuffles $\widetilde{[[\mathbf{t}_k]]}$ and $\widetilde{[[\mathbf{h}_u]]}$ into $\widetilde{[[\mathbf{a}_k]]}, \widetilde{[[\mathbf{b}_k]]}$. S_C then performs the same steps as the simulator for party A in Protocol 2 and gets $\widetilde{\mathrm{msg}}_{\mathrm{Pr.2}}$. Furthermore, S_C generates an encryption of random bits $\widetilde{\delta}_k$ and of $\mathrm{E}(\mathbf{U}_k)$ and performs the same steps as the simulator for party A as in the proof of Proposition 9.1 (or the simulator for the standard OT) and gets $\widetilde{\mathrm{msg}}_{\mathrm{OT}}$. It then outputs:

$$S_C(\emptyset) = \left(\widetilde{\mathrm{mpk}}, \mathrm{E}\widetilde{\left(-\frac{1}{L}\mathbf{H}\right)}, \mathrm{E}\widetilde{\left(-\frac{\eta}{L}\mathbf{H}\right)}, \mathrm{E}\widetilde{\left(\frac{1}{L}\mathbf{F}^{\mathsf{T}}\right)}, \widetilde{\mathrm{E}(\eta)},\right.$$

$$[\![\widetilde{\mathbf{h}_u}]\!], [\![\widetilde{\mathbf{l}_u}]\!], \widetilde{\mathfrak{R}}_C, \widetilde{\mathrm{E}(\mathbf{U}_0')},$$

$$\left.\left\{\widetilde{\mathrm{E}(\mathbf{U}_k)}, \left([\![\widetilde{\mathbf{a}_k}]\!], [\![\widetilde{\mathbf{b}_k}]\!]\right), [\![\widetilde{U_{k+1}}]\!], \widetilde{\mathrm{msg}}_{\mathrm{Pr.2}}, \widetilde{\mathrm{msg}}_{\mathrm{OT}}\right\}_{k\in\{0,\dots,K-1\}}\right). \tag{9.13}$$

All the values in the view of the cloud in (9.9)—with the exception of the random values $\mathfrak{R}_C$ and the key mpk, which are statistically indistinguishable from $\widetilde{\mathfrak{R}}_C$ and $\widetilde{\mathrm{mpk}}$ because they are drawn from the same distributions—are encrypted with semantically secure encryptions schemes (AHE and LabHE). This means they are computationally indistinguishable from the encryptions of random values in (9.13), even with different keys. This happens even when the values from different iterations are encryptions of correlated quantities. Thus, $V_C(\emptyset) \overset{c}{\equiv} S_C(\emptyset)$.

We now build a simulator S_A for the actuator that takes as input $f_{\mathrm{MPC}}, \mathbf{u}(t)$. S_A will take the same steps as in lines 1, 3–7 in Protocol 3, obtaining $\widetilde{\mathrm{mpk}}, \widetilde{\mathrm{msk}}, \widetilde{\mathrm{upk}}, \widetilde{\mathrm{usk}}, \widetilde{\tau}_A$, $\widetilde{\mathbf{b}}_A, [\![\widetilde{\mathbf{b}_A}]\!], \widetilde{\mathfrak{R}}_A, \widetilde{\mathbf{U}}_0'$ and instead of line 2, it generates $\widetilde{\mathrm{upk}}_j, \widetilde{\mathrm{usk}}_j \leftarrow \mathrm{KeyGen}(\widetilde{\mathrm{mpk}})$ itself, for $j \in \{S^i, Set\}, i \in [M]$. For $k = 0, \dots, K-1$, S_A performs the same steps as the simulator for party B in Protocol 2 and gets $\widetilde{\mathrm{msg}}_{\mathrm{Pr.2}}$. Furthermore, S_A performs the same steps as the simulator for party B as in the proof of Proposition 9.1 (or the simulator for the standard OT) and gets $\widetilde{\mathrm{msg}}_{\mathrm{OT}}$. It then outputs:

$$S_A(f_{\mathrm{MPC}}, \mathbf{u}(t)) = \left(f_{\mathrm{MPC}}, \widetilde{\mathrm{mpk}}, \widetilde{\mathrm{msk}}, \widetilde{\mathrm{upk}}, \widetilde{\mathfrak{R}}_A,\right.$$

$$\left.\left\{\widetilde{\mathbf{U}_{k+1}'}, \widetilde{\mathrm{msg}}_{\mathrm{Pr.2}}, \widetilde{\mathrm{msg}}_{\mathrm{OT}}\right\}_{k\in\{0,\dots,K-1\}}, \mathbf{u}(t)\right). \tag{9.14}$$

All the values in the view of the actuator in (9.10)—with the exception of the random values $\mathfrak{R}_A$ and the keys upk, which are statistically indistinguishable from $\widetilde{\mathfrak{R}}_A$ and $\widetilde{\mathrm{upk}}$ because they are drawn from the same distributions and $\mathbf{u}(t)$—are blinded by random numbers, different at every iteration, which means that they are statistically indistinguishable from the random values in (9.14). This again holds even when the values that are blinded at different iterations are correlated and the actuator knows the solution $\mathbf{u}(t)$, because the values of interest are drowned in large noise. Thus, $V_A(f_{\mathrm{MPC}}) \overset{c}{\equiv} S_A(f_{\mathrm{MPC}}, \mathbf{u}(t))$.

The setup and subsystems do not receive any other messages apart from the master public key (9.11), (9.12). Hence, a simulator S_{Set} for the setup and a simulator S_{S^i} for a subsystem i can simply generate $\widetilde{\mathrm{mpk}} \leftarrow \mathrm{Init}(1^\sigma)$ and then proceed with the execution of lines 2–5 in Protocol 3 and output their inputs, messages and coins. The outputs of the simulators are trivially indistinguishable from the views.

When an adversary corrupts a coalition, the view of the coalition contains the inputs of all parties, and a simulator takes the coalition's inputs and outputs. The view of the coalition between the cloud, the setup, and a number l of subsystems is:

$$V_{CSl}\big(\mathbf{A}, \mathbf{B}, \mathbf{P}, \mathbf{Q}, \mathbf{R}, \{\mathcal{U}^i, \mathbf{x}^i(t)\}_{i \in i_1, \dots, i_l}\big) = V_C(\emptyset) \cup V_{Set}(\mathbf{A}, \mathbf{B}, \mathbf{P}, \mathbf{Q}, \mathbf{R}) \cup$$
$$\cup V_{S^{i_1}}(\mathcal{U}^{i_1}, \mathbf{x}^{i_1}(t)) \cup \dots \cup V_{S^{i_l}}(\mathcal{U}^{i_l}, \mathbf{x}^{i_l}(t)).$$

A simulator S_{CSl} for this coalition takes in the inputs of the coalition and no output and performs almost the same steps as S_C, S_{Set}, S_{S^i}, without randomly generating the quantities that are known by the coalition. The same argument of having the messages drawn from the same distributions and encrypted with semantically secure encryption schemes proves the indistinguishability between $V_{CSl}(\cdot)$ and $S_{CSl}(\cdot)$.

The view of the coalition between the actuator, the setup, and a number l of subsystems is the following:

$$V_{ASl}\big(f_{\text{MPC}}, \mathbf{u}(t), \mathbf{A}, \mathbf{B}, \mathbf{P}, \mathbf{Q}, \mathbf{R}, \{\mathcal{U}^i, \mathbf{x}^i(t)\}_{i \in i_1, \dots, i_l}\big) = V_A(f_{\text{MPC}}, \mathbf{u}(t)) \cup$$
$$\cup V_{Set}(\mathbf{A}, \mathbf{B}, \mathbf{P}, \mathbf{Q}, \mathbf{R}) \cup V_{S^{i_1}}(\mathcal{U}^{i_1}, \mathbf{x}^{i_1}(t)) \cup \dots \cup V_{S^{i_l}}(\mathcal{U}^{i_l}, \mathbf{x}^{i_l}(t)).$$

A simulator S_{ASl} for this coalition takes in the inputs of the coalition and $\mathbf{u}(t)$ and performs almost the same steps as S_A, S_{Set}, S_{S^i}, without randomly generating the quantities that are now known. The same argument of having the messages drawn from the same distributions and blinded with one-time pads proves the indistinguishability between $V_{ASl}(\cdot)$ and $S_{ASl}(\cdot)$.

The proof is now complete.   □

We can also have the private MPC scheme run for multiple time steps. Protocol 3 can be modified to also generate the labels and secrets for T time steps. Protocol 4 can be run for multiple time steps, and warm starts can be included by adding two lines such that the cloud obtains $\mathrm{E}\left(\{\mathbf{U}_K^t\}_{m+1:M}\right)$ and sets $\mathrm{E}(\mathbf{U}_0^{t+1}) = \left[\mathbf{U}_K^{t\,\mathsf{T}}\ \mathbf{0}_m^{\mathsf{T}}\right]^{\mathsf{T}}$.

9.5.3 Analysis of Errors

As mentioned at the beginning of the section, the values that are encrypted, added to or multiplied with encrypted values have to be integers. We consider fixed-point representations with one sign bit, l_i integer bits and l_f fractional bits. We multiply the values by 2^{l_f} then truncate to obtain integers prior to encryption, and, after decryption, we divide the result by the appropriate quantity (e.g., we divide the result of a multiplication by 2^{2l_f}). Furthermore, the operations can increase the number of bits of the result, hence, before the comparisons in Protocol 4 an extra step that performs interactive truncation has to be performed, because Protocol 2 requires a fixed number of bits for the inputs. Also, notice that when the encryption is refreshed, i.e., a ciphertext is decrypted and re-encrypted, the accumulation of bits due to the operations is truncated back to the desired size.

Working with fixed-point representations can lead to overflow, quantization and arithmetic round-off errors. Thus, we want to compute the deviation between the fixed-point solution and optimal solution of the FGM algorithm in (9.4). In order to bound it, we need to ensure that the number of fractional bits l_f is sufficiently large such that the feasibility of the fixed-point precision solution is preserved, that the strong convexity of the fixed-point objective function still holds and that the fixed-point step size is such that the FGM converges. The errors can be written as states of a stable linear system with bounded disturbances. Bounds on the errors for the case of public model are derived in [2] and similar bounds can be obtained for the private model. The bounds on this deviation can be used in an offline step to choose an appropriate fixed-point precision for the desired performance of the system.

9.6   Discussion

Secure multi-party computation protocols require many rounds of communication in general. This can also be observed in Protocol 4. Therefore, in order to be able to use this proposed protocol, we need fast and reliable communication between the cloud and the actuator.

In the architecture we considered, the subsystems are computationally and memory constrained devices, hence, they are only required to generate the encryptions for their measurements. The setup only holds constant data, and it only has to compute the matrices in (9.4) and encrypt them in the initialization step. Furthermore, we considered the existence of the setup entity for clarity in the exposition of the protocols, but the data held by the setup could be distributed to the other participants in the computation. In this case, the cloud would have to perform some extra steps in order to aggregate the encrypted system parameters (see [3] for a related solution). Notice that the subsystems and setup do not need to generate labels for the number of iterations, only the actuator does. The actuator is supposed to be a machine with enough memory to store the labels and reasonable amount of computation power such that the encryptions and decryptions are performed in a practical amount of time (that will be dependent on the sampling time of the system), but less powerful than the cloud. The cloud controller is assumed to be a server, which has enough computational power and memory to be capable to deal with the computations on the ciphertexts, which can be large, depending on the encryption schemes employed.

If fast and reliable communication is not available or if the actuator is a highly constrained device, then a fully homomorphic encryption solution that is solely executed by the cloud might be more appropriate, although its execution can be substantially slower.

Compared to the two-server MPC with private signals but public model from [2], where only AHE is required, the MPC with private signals and private model we considered in this chapter is only negligibly more expensive. Specifically, the ciphertexts are augmented with one secret that has the number of bits substantially smaller than the number of bits in an AHE ciphertext, and each online iteration only incurs one extra round of communication, one decryption and one encryption. All the other computations regarding the secrets are done offline. This shows the efficiency and suitability of LabHE for encrypted control applications.

References

1. Alexandru AB, Gatsis K, Shoukry Y, Seshia SA, Tabuada P, Pappas GJ (2018) Cloud-based quadratic optimization with partially homomorphic encryption. arXiv preprint arXiv:1809.02267
2. Alexandru AB, Morari M, Pappas GJ (2018) Cloud-based MPC with encrypted data. In: IEEE conference on decision and control (CDC), pp 5014–5019
3. Alexandru AB, Pappas GJ (2019) Encrypted LQG using Labeled Homomorphic Encryption. In: 10th ACM/IEEE International Conference on Cyber-Physical Systems (ICCPS), pp 129–140
4. Ali M, Khan SU, Vasilakos AV (2015) Security in cloud computing: opportunities and challenges. Inf Sci 305:357–383
5. Archer D, Chen L, Cheon JH, Gilad-Bachrach R, Hallman RA, Huang Z, Jiang X, Kumaresan R, Malin BA, Sofia H, Song Y, Wang S (2017) Applications of homomorphic encryption. Technical report, Microsoft Research
6. Aslett LJ, Esperança PM, Holmes CC (2015) A review of homomorphic encryption and software tools for encrypted statistical machine learning. arXiv preprint arXiv:1508.06574
7. Barbosa M, Catalano D, Fiore D (2017) Labeled homomorphic encryption. In: European Symposium on Research in Computer Security, pp 146–166. Springer, Cham
8. Beimel A (2011) Secret-sharing schemes: a survey. In: International conference on coding and cryptology, pp 11–46. Springer, Berlin
9. Bellare M, Hoang VT, Rogaway P (2012) Foundations of garbled circuits. In: Conference on computer and communications security, pp 784–796. ACM
10. Bellovin SM (2011) Frank Miller: inventor of the one-time pad. Cryptologia 35(3):203–222
11. Borrelli F, Bemporad A, Morari M (2017) Predictive control for linear and hybrid systems. Cambridge University Press
12. Bost R, Popa RA, Tu S, Goldwasser S (2015) Machine learning classification over encrypted data. In: Network & distributed system security symposium (NDSS)
13. Botta A, De Donato W, Persico V, Pescapé A (2016) Integration of cloud computing and internet of things: a survey. Future Gener Comput Syst 56:684–700
14. Catalano D, Fiore D (2015) Boosting linearly-homomorphic encryption to evaluate degree-2 functions on encrypted data. Cryptology ePrint Archive, Report 2014/813. https://eprint.iacr.org/2014/813
15. Catalano D, Fiore D (2015) Using linearly-homomorphic encryption to evaluate degree-2 functions on encrypted data. In: 22nd ACM SIGSAC conference on computer and communications security, pp 1518–1529. ACM
16. Chase M, Gilad-Bachrach R, Laine K, Lauter K, Rindal P (2017) Private collaborative neural network learning. Technical report, Cryptology ePrint Archive, Report 2017/762. https://eprint.iacr.org/2017/762
17. Chen H, Gilad-Bachrach R, Han K, Huang Z, Jalali A, Laine K, Lauter K (2018) Logistic regression over encrypted data from fully homomorphic encryption. BMC Med Genomics 11(4):81
18. Couteau G (2016) Efficient secure comparison protocols. Cryptology ePrint Archive, Report 2016/544. http://eprint.iacr.org/2016/544

19. Cramer R, Damgård I, Nielsen JB (2012) Secure multiparty computation and secret sharing-an information theoretic approach. Book draft
20. Cramer R, Damgård IB, Nielsen JB (2015) Secure multiparty computation. Cambridge University Press
21. Damgård I, Geisler M, Krøigaard M (2007) Efficient and secure comparison for on-line auctions. In: Australasian conference on information security and privacy, pp 416–430. Springer, Berlin
22. Damgård I, Geisler M, Krøigaard M (2009) A correction to "Efficient and secure comparison for on-line auctions". Int J Appl Cryptogr 1(4):323–324
23. Damgård I, Orlandi C (2010) Multiparty computation for dishonest majority: from passive to active security at low cost. In: Annual cryptology conference, pp 558–576. Springer
24. Damgård IB, Jurik M (2001) A generalisation, a simplification and some applications of Paillier's probabilistic public-key system. In: International workshop on public key cryptography, pp 119–136. Springer, Berlin
25. Dwork C (2008) Differential privacy: a survey of results. In: International conference on theory and applications of models of computation, pp 1–19. Springer, Berlin
26. Dwork C, Kenthapadi K, McSherry F, Mironov I, Naor M Our data, ourselves: privacy via distributed noise generation. In: Annual international conference on the theory and applications of cryptographic techniques, pp 486–503. Springer (2006)
27. Dwork C, Roth A et al (2014) The algorithmic foundations of differential privacy. Found Trends® Theor Comput Sci **9**(3–4), 211–407
28. Farokhi F, Shames I, Batterham N (2017) Secure and private control using semi-homomorphic encryption. Control Eng Pract 67:13–20
29. Gentry C (2009) A fully homomorphic encryption scheme. Ph.D. thesis, Department of Computer Science, Stanford University. http://www.crypto.stanford.edu/craig
30. Gentry C, Boneh D (2009) A fully homomorphic encryption scheme, vol 20, no 09. Stanford University Stanford
31. Goldreich O (2003) Foundations of cryptography: basic tools, vol 1. Cambridge University Press, New York
32. Goldreich O (2004) Foundations of cryptography: basic applications, vol 2. Cambridge University Press, New York
33. Goldreich O, Micali S, Wigderson A (1987) How to play any mental game. In: 19th annual ACM symposium on theory of computing, pp 218–229. ACM
34. Goldwasser S, Micali S (1982) Probabilistic encryption & how to play mental poker keeping secret all partial information. In: 14th annual ACM symposium on Theory of Computing, pp 365–377. ACM
35. Gonzalez-Serrano FJ, Amor-Martın A, Casamayon-Anton J (2014) State estimation using an extended Kalman filter with privacy-protected observed inputs. In: IEEE international workshop on information forensics and security (WIFS), pp 54–59. IEEE
36. Hamlin A, Schear N, Shen E, Varia M, Yakoubov S, Yerukhimovich A (2016) Cryptography for big data security. In: Hu F (ed) Big data: storage, sharing, and security, Chap 10, pp 241–288. Taylor & Francis LLC, CRC Press
37. Ishai Y, Prabhakaran M, Sahai A (2008) Founding cryptography on oblivious transfer—efficiently. In: Annual international cryptology conference, pp 572–591. Springer, Berlin
38. Jeckmans A, Peter A, Hartel P (2013) Efficient privacy-enhanced familiarity-based recommender system. In: Proceedings of European symposium on research in computer security, pp 400–417. Springer, Berlin
39. Joye M, Libert B (2013) Efficient cryptosystems from 2^k-th power residue symbols. In: International conference on the theory and applications of cryptographic techniques, pp 76–92. Springer, Berlin
40. Kim J, Lee C, Shim H, Cheon JH, Kim A, Kim M, Song Y (2016) Encrypting controller using fully homomorphic encryption for security of cyber-physical systems. IFAC-PapersOnLine 49(22):175–180

41. Lindell Y (2017) How to simulate it–a tutorial on the simulation proof technique. In: Tutorials on the foundations of cryptography, pp 277–346. Springer International Publishing
42. Martins P, Sousa L, Mariano A (2018) A survey on fully homomorphic encryption: an engineering perspective. ACM Comput Surv (CSUR) 50(6):83
43. Mayne DQ, Rawlings JB, Rao CV, Scokaert PO (2000) Constrained model predictive control: stability and optimality. Automatica 36(6):789–814
44. Mell P, Grance T et al (2011) The NIST definition of cloud computing
45. Mirhoseini A, Sadeghi AR, Koushanfar F (2016) Cryptoml: secure outsourcing of big data machine learning applications. In: IEEE International symposium on hardware oriented security and trust (HOST), pp 149–154. IEEE
46. Mohassel P, Zhang Y (2017) SecureML: a system for scalable privacy-preserving machine learning. Cryptology ePrint Archive, Report 2017/396. http://eprint.iacr.org/2017/396
47. Murguia C, Farokhi F, Shames I (2018) Secure and private implementation of dynamic controllers using semi-homomorphic encryption. arXiv preprint arXiv:1812.04168
48. Naehrig M, Lauter K, Vaikuntanathan V (2011) Can homomorphic encryption be practical? In: 3rd ACM workshop on cloud computing security workshop, pp 113–124. ACM
49. Naor M, Pinkas B (2001) Efficient oblivious transfer protocols. In: 12th annual ACM-SIAM symposium on discrete algorithms, pp 448–457. SIAM
50. Nesterov Y (2013) Introductory lectures on convex optimization: a basic course, vol 87. Springer Science & Business Media
51. Nielsen JB, Nordholt PS, Orlandi C, Burra SS (2012) A new approach to practical active-secure two-party computation. In: Advances in cryptology–CRYPTO, pp 681–700. Springer, Berlin
52. Paillier P (1999) Public-key cryptosystems based on composite degree residuosity classes. In: Annual international conference on the theory and applications of cryptographic techniques, pp 223–238. Springer, Berlin
53. Pedersen TP (1991) Non-interactive and information-theoretic secure verifiable secret sharing. In: Annual international cryptology conference, pp 129–140. Springer, Berlin
54. Pettai M, Laud P (2015) Combining differential privacy and secure multiparty computation. In: 31st Annual computer security applications conference, pp 421–430. ACM
55. Rabin MO (2005) How to exchange secrets with oblivious transfer. Cryptology ePrint Archive, Report 2005/187. https://eprint.iacr.org/2005/187
56. Rastogi V, Nath S (2010) Differentially private aggregation of distributed time-series with transformation and encryption. In: ACM SIGMOD International Conference on Management of data, pp 735–746. ACM
57. Riazi MS, Rouhani BD, Koushanfar F (2018) Deep learning on private data. IEEE Secur Privacy Mag
58. Rittinghouse JW, Ransome JF (2016) Cloud computing: implementation, management, and security. CRC Press
59. Rivest RL, Adleman L, Dertouzos ML (1978) On data banks and privacy homomorphisms. Found Secure Comput 4(11):169–180
60. Schulze Darup M, Redder A, Shames I, Farokhi F, Quevedo D (2018) Towards encrypted MPC for linear constrained systems. IEEE Control Syst Lett 2(2):195–200
61. Shamir A (1979) How to share a secret. Commun ACM 22(11):612–613
62. Shi E, Chan HTH, Rieffel E, Chow R, Song D (2011) Privacy-preserving aggregation of time-series data. In: Network & distributed system security symposium (NDSS)
63. Singh S, Jeong YS, Park JH (2016) A survey on cloud computing security: issues, threats, and solutions. J Netw Comput Appl 75:200–222
64. Vadhan S (2018) Multiparty differential privacy. In: Differential privacy meets multi-party computation (DPMPC) workshop. https://www.bu.edu/hic/dpmpc-2018/
65. Vernam GS (1926) Cipher printing telegraph systems: for secret wire and radio telegraphic communications. J AIEE 45(2):109–115
66. Veugen T (2010) Encrypted integer division. In: International workshop on information forensics and security, pp 1–6. IEEE

67. Veugen, T.: Improving the DGK comparison protocol. In: International workshop on information forensics and security, pp 49–54. IEEE (2012)
68. Yao AC (1982) Protocols for secure computations. In: 23rd Annual symposium on foundations of computer science, pp 160–164. IEEE
69. Zhu T, Li G, Zhou W, Philip SY (2017) Differential privacy and applications. Springer, Cham

Chapter 10
Comprehensive Introduction to Fully Homomorphic Encryption for Dynamic Feedback Controller via LWE-Based Cryptosystem

Junsoo Kim, Hyungbo Shim and Kyoohyung Han

Abstract The cryptosystem based on the Learning-with-Errors (LWE) problem is considered as a post-quantum cryptosystem, because it is not based on the factoring problem with large primes which is believed to be easily solved by a quantum computer. Moreover, the LWE-based cryptosystem allows fully homomorphic arithmetics so that two encrypted variables can be added and multiplied without decrypting them. This chapter provides a comprehensive introduction to the LWE-based cryptosystem with examples. A key to the security of the LWE-based cryptosystem is the injection of random errors in the ciphertexts, which however hinders unlimited recursive operation of homomorphic arithmetics on ciphertexts due to the growth of the error. We show how this limitation can be overcome for dynamic feedback controllers that guarantee stability of the closed-loop system when the system matrix of the controller consists of integers. Finally, we illustrate through MATLAB codes how the LWE-based cryptosystem can be customized to build a secure feedback control system. This chapter is written for the control engineers who do not have background on cryptosystems.

10.1 Introduction

Applications of homomorphic cryptography to the feedback controller are relatively new. To the authors' knowledge, the first contribution was made by Kogiso and Fujita [1] in 2015, followed by Farokhi et al. [2] and Kim et al. [3] both in 2016. Interestingly, each of them uses different homomorphic encryption schemes; El-Gamal [4],

J. Kim · H. Shim (✉)
ASRI, Department of Electrical & Computer Engineering, Seoul National University, Seoul, South Korea
e-mail: hshim@snu.ac.kr

J. Kim
e-mail: kjs9044@snu.ac.kr

K. Han
Department of Mathematical Sciences, Seoul National University, Seoul, South Korea
e-mail: satanigh@snu.ac.kr

© Springer Nature Singapore Pte Ltd. 2020   209
F. Farokhi (ed.), *Privacy in Dynamical Systems*,
https://doi.org/10.1007/978-981-15-0493-8_10

Paillier [5], and LWE [6] are employed, respectively. Because other two schemes are introduced in other chapters in this book, this chapter is written for introducing the LWE-based cryptosystem and its customization for building a *dynamic* feedback controller.

Homomorphic encryption implies a cryptographic scheme in which arithmetic operations can be performed directly on the encrypted data (i.e., ciphertexts) without decrypting them. When applied to the control systems, security increases because there is no need to keep the secret key inside the controller (see Fig. 10.1), which is supposed to be a vulnerable point in the feedback loop. After the idea of homomorphic encryption appeared in 1978 by Rivest et al. [7], two *semi-homomorphic* encryption schemes were developed. One is the multiplicatively homomorphic scheme by El-Gamal [4] developed in 1985, and the other is the additively homomorphic scheme by Paillier [5] developed in 1999. Homomorphic encryption schemes that allow both addition and multiplication appeared around 2000, and they are called *somewhat-homomorphic* because, even if both arithmetics are enabled, the arithmetic operations can be performed only finite times on an encrypted variable. In 2009, Gentry [6] developed an algorithm called 'bootstrapping' which finally overcame the restriction of finite number of operations. By performing the bootstrapping regularly on the encrypted variable, the variable becomes like a newborn ciphertext and so it allows more operations on it. The encryption scheme with this algorithm is called *fully homomorphic*. However, fully homomorphic encryption sometimes simply implies a scheme that allows both addition and multiplication, and we follow this convention.

In this chapter, we introduce the LWE-based fully homomorphic encryption scheme. We illustrate that, if the scheme is used with a stable closed-loop system then, interestingly, there is no need to employ the bootstrapping for infinite number of arithmetic operations as long as the system matrix of the controller consists of integer numbers, and the actuator and the sensor sacrifice their resolutions a little bit.[1] Moreover, by utilizing the fully homomorphic arithmetics, we are able to encrypt all the parameters in the controller, as seen in Fig. 10.1.

This chapter consists of three parts. In the first part (Sect. 10.2), we present an introduction to the LWE-based encryption, discuss the homomorphic arithmetics, and illustrate the error growth in the ciphertexts. The second part (Sect. 10.3) is about customization of the LWE-based cryptosystem for the linear time-invariant dynamic feedback controllers. In the last part (Sect. 10.4), we show the error growth can be handled by the stability, so that the dynamic controller operates seamlessly with unlimited times fully homomorphic arithmetics. In Sect. 10.5, we conclude the chapter with a discussion on the need for integer system matrix of the controller, which is related to one of future research issues. For pedagogical purposes, we simplify many issues that should be considered in practice, and instead, focus on the key ingredients of homomorphic encryptions. In the same respect, the codes

---

[1]If one needs to use the bootstrapping because his/her application does not satisfies these conditions, then he/she may refer to [3] in which a method to orchestrate the bootstrapping and the *dynamic* feedback controller has been presented.

Fig. 10.1 The control system configuration considered in this chapter. Note that the secret key sk is kept only in the plant side, and there is no need to store sk beyond the network. The parameters of the controllers (such as **F**, **G**, **H**, and **J**, as well as the initial condition **x**[0]) are also encrypted. In this chapter, bold fonts imply encrypted variables

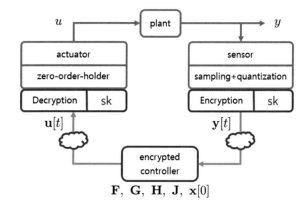

presented in this chapter consist of simple MATLAB commands that may not be used in real applications even if it works for simple examples.

10.2 Cryptosystem Based on Learning-with-Errors Problem

We first present how to encrypt and decrypt a message, in order for the reader to look and feel the ciphertexts in the LWE-based cryptosystem. Then, we briefly introduce the learning-with-errors (LWE) problem because the security of the presented cryptosystem is based on the hardness of this problem. We also explain how the homomorphic arithmetics are performed in the LWE-based cryptosystem.

Now, with $p \in \mathbb{N}$, let the set of integers bounded by $p/2$ be denoted by

$$[p] := \left\{ i \in \mathbb{Z} : -\frac{p}{2} \leq i < \frac{p}{2} \right\} \tag{10.1}$$

so that the cardinality of $[p]$ is p. In addition, we need the following set.

Set of integers modulo q

Let $\mathbb{Z}_q$ be the set of integers modulo[2] $q \in \mathbb{N}$. This means that any two integers a and b are regarded as the same elements of $\mathbb{Z}_q$ if $(a - b) \bmod q = 0$.[3] By this rule, all the integers are related with other integers that have the same remainder when divided by q, and thus, $[q]$ can represent $\mathbb{Z}_q$ if any integer $a \in \mathbb{Z}_q$ is treated as $b \in [q]$ such

---

[2] In this chapter, all the modulus are chosen as powers of 10 for convenience of understanding while they are often powers of 2 in practice.

[3] $x \bmod q$ is the remainder after division of x by q. In this chapter, we suppose the remainder is an element of $[q]$; for example, if the remainder is greater than or equal to $q/2$, make it negative by subtracting q, e.g., $17 \bmod 10 = -3 \in [10]$.

that $(a - b) \bmod q = 0$. In this sense, $\mathbb{Z}_q$ is closed under addition, subtraction, and multiplication.

The following subsections are mainly based on [9]. In this chapter, we only consider the symmetric-key encryption for simplicity. For the use of public-key encryption, see [8] or [9].

10.2.1 Basics of LWE-Based Cryptosystem

Let the set $[p]$ be where the integer to be encrypted belongs to, and let us denote it by $\mathcal{P}$ and call it by the plaintext space. An element $m \in \mathcal{P}$ is called a message or a *plaintext*. The value of p can be chosen as a power of 10 such that $|m| < p/2$ for all messages to be used. Now, let $\mathbb{Z}_q$ be a set of integers modulo q where $q = Lp$ with L being a power of 10. Finally, in order to encrypt and decrypt a message m, choose a *secret key* sk which is an integer vector of size N such that $\mathsf{sk} \in \mathbb{Z}_q^N$. These L and N are the parameters of the LWE-based cryptosystem and their selection is discussed in more detail in Sect. 10.2.1.2.

Now, let us encrypt a column vector $m \in [p]^n$ having n elements in $[p]$. Whenever a new message m is encrypted, a new random matrix $\mathsf{A} \in \mathbb{Z}_q^{n \times N}$ is sampled from the uniform distribution over $\mathbb{Z}_q^{n \times N}$, and a new vector $\mathsf{e} \in [r]^n$ is randomly sampled where $r < L$, so that each component e_i satisfies that $|\mathsf{e}_i| < L/2$ for $i = 1, \ldots, n$. With them, compute

$$\mathsf{b} \leftarrow (-\mathsf{A} \cdot \mathsf{sk} + Lm + \mathsf{e}) \quad \bmod q \tag{10.2}$$

where $\cdot$ is the standard multiplication of a matrix and a vector. Then, the *ciphertext* $\mathbf{m}$ of the plaintext vector m is obtained by the matrix $\mathbf{m} = [\mathsf{b}, \mathsf{A}] \in \mathcal{C} = \mathbb{Z}_q^{n \times (N+1)}$ where $\mathcal{C}$ is called the ciphertext space. Define the secret key vector $\mathsf{s} := [1, \mathsf{sk}^T]^T$ and let $\lceil \cdot \rfloor$ be the rounding operation for vectors.[4] Then, the ciphertext $\mathbf{m}$ is decrypted as

$$\left\lceil \frac{(\mathbf{m} \cdot \mathsf{s}) \bmod q}{L} \right\rfloor = \left\lceil \frac{Lm + \mathsf{e}}{L} \right\rfloor \rightarrow m \tag{10.3}$$

because the size of each element of e is less than $L/2$. One of the key ingredients in the LWE-based scheme is the vector e, which is intentionally injected in the ciphertext by (10.2) (and is eliminated by the decryption (10.3)). This vector is called "error," and it will be seen in Sect. 10.2.1.1 that this error makes this encryption scheme secure.

The discussions so far yield the MATLAB codes with an example parameter set

---

[4]We round away from zero for negative numbers, e.g., $\lceil -2.5 \rfloor = -3$ while $\lceil -2.4 \rfloor = -2$. When a is a vector, $\lceil a \rfloor$ implies component-wise rounding.

```
env.p = 1e4;   env.L = 1e4;   env.r = 1e1;   env.N = 4;
```

which are put in a structure variable env. We randomly select a secret key by[5, 6]

```
sk = Mod(randi(env.p*env.L,[env.N, 1]), env.p*env.L);
```

Under these parameters, the codes are as follows.

Functions Enc, Dec

```
function ciphertext = Enc(m,sk,env)
n = length(m);    q = env.L*env.p;
A = randi(q, [n, env.N]);
e = Mod(randi(env.r, [n,1]), env.r);
b = -A*sk + env.L*m + e;
ciphertext = Mod([b,A], q);
end

function plaintext = Dec(c,sk,env)
s = [1; sk];
plaintext = round( Mod(c*s, env.L*env.p)/env.L );
end
```

An example run shows encryption and decryption of a number 30:

```
sk =
-13203881
-22462885
-28840424
4713455

c = Enc(30,sk,env)
c =
```

---

[5]Since the mod function in MATLAB always returns non-negative remainders, we use our customized function (starting with the capital M)

```
function y = Mod(x,p), y = mod(x,p); y = y - (y >= p/2)*p; end
```

in order to have signed results in the set like (10.1).

[6]In order to run the codes in this chapter, please choose small numbers for the secret key in order not to cause the overflow of the double variable in MATLAB. An example is to replace env.p*env.L by 10 for example.

```
    -43264645        21696438      -15923263      46660236      20129426

    m = Dec(c,sk,env)

    m =
    30
```

It is seen from the outcome that the message $30L$ is hiding in the number $-\mathsf{A} \cdot \mathsf{sk} + \mathsf{e}$ that is the first element of $\mathbf{m}$.

10.2.1.1 Necessity of Error Injection and the Learning-with-Errors Problem

A measure for the security of a cryptosystem is how hard it is to find the secret key sk when arbitrarily many pairs $(m_i, \mathbf{m}_i)$ are given. In fact, the ciphertexts $\mathbf{m}_i$ are easily available to the adversary by eavesdropping the communication line, and there are many cases that the plaintexts m_i are also obtainable. (For example, one may guess an email begins with the word "Dear" even if it is encrypted.) When the pair m_i and $\mathbf{m}_i = [\mathsf{b}_i, \mathsf{A}_i]$ is available, the adversary can easily obtain $(-\mathsf{A}_i \cdot \mathsf{sk} + \mathsf{e})$ as well as A_i by subtracting Lm_i from b_i (see (10.2)). Hence, if there is no error injection of e, then the problem of searching sk simply becomes solving a linear equation in $\mathbb{Z}_q^N$, which is not difficult.

Interestingly, with the error e injected, it was proved that solving (or, 'learning') sk becomes extremely difficult. This problem is called 'learning-with-error (LWE)' problem, which has been introduced in [9]. For example, with $\bar{\mathsf{s}} = [\mathsf{s}_1, \mathsf{s}_2, \ldots, \mathsf{s}_4]^T = [3, -5, 1, 0]^T \in \mathbb{Z}_{100}^4$, consider a sequence of linear equations with errors:

$$32\mathsf{s}_1 + 17\mathsf{s}_2 - 5\mathsf{s}_3 + 8\mathsf{s}_4 + \mathsf{e}_1 = 6 + \mathsf{e}_1 = 7 \qquad (\mathrm{mod}\ 100)$$
$$-6\mathsf{s}_1 + 17\mathsf{s}_2 + 1\mathsf{s}_3 + 18\mathsf{s}_4 + \mathsf{e}_2 = -2 + \mathsf{e}_2 = -3 \qquad (\mathrm{mod}\ 100)$$
$$44\mathsf{s}_1 + 32\mathsf{s}_2 + 12\mathsf{s}_3 + 28\mathsf{s}_4 + \mathsf{e}_3 = -16 + \mathsf{e}_3 = -15 \qquad (\mathrm{mod}\ 100)$$

$$\vdots$$

With the error (which need not be large; just a small error is enough, e.g., the error of $1, -1, 1, \cdots$ is used in the above example), finding $\bar{\mathsf{s}}$ (and e_i as well) in the set $\mathbb{Z}_{100}^4$ is a very difficult problem. In fact, this problem is known to be as hard as the worst-case "lattice problem" so that the cryptosystem based on it becomes secure at the same level of difficulty to solve the problem. Actually, the cryptosystem based on the LWE problem is known to be as much secure as even the quantum computer takes long time to solve (and thus, it is known as a post-quantum cryptosystem [10]). This is because the difficulty is not based on the factoring problem,[7] which has been a basis for many other cryptosystems.

---

[7]Factoring problem is to find large prime numbers p and q when $N = pq$ is given. This problem is supposed to be easily solved (i.e., in polynomial time) by quantum computers.

10.2.1.2 How to Choose Parameters for Desired Level of Security?

An encryption scheme is called λ-bit secure, whose meaning is briefly introduced in this subsection. For this, let us consider a game between an adversary and a challenger. The rule is that, whenever the adversary submits two messages m_1 and m_2 to the challenger, the challenger randomly chooses one of them with equal probability and returns it back to the adversary after encrypting the chosen message. Then, the adversary guesses which one of m_1 and m_2 is encrypted by inspecting the received ciphertext. If the ratio of the adversary's correct guess is not meaningfully greater than 0.5 as the game repeats, then the encryption scheme is said to be *indistinguishable*. Let $\mathcal{A}$ denote the algorithm that the adversary uses in the game to guess. Then, the encryption scheme is called λ-*bit secure* if, for all available adversary's algorithm $\mathcal{A}$, it holds that

$$(\text{Computation complexity of } \mathcal{A}) \times \left(\frac{1}{|0.5 - \text{Success probability of } \mathcal{A}|} \right) > 2^{\lambda}.$$

Clearly, large λ implies that the adversary needs high computational complexity while the success rate is not very different from 0.5 for all possible attack algorithms.

For the case of the LWE-based cryptosystem used in this chapter, it is rather convenient to assess its level of security with a useful tool called "LWE estimator." This tool is implemented using Sage program language and an on-line version is also available.[8] When the parameters $p, L, r,$ and N of the LWE-based cryptosystem are given, the estimator computes expected number of operations to attack the encryption scheme by various attack algorithms $\mathcal{A}$, and finally returns λ. Below is an example for using the estimator with a specific parameter set.

```
load(estimator.py)
p = 1e4; L = 1e4; r = 1e2; N = 20
_ = estimate_lwe(N, (r / (L * p)), (L * p))
```

The following is the output of the estimator with the parameter set.

```
usvp: rop: = 2^29.7, ...
dec: rop: = 2^32.3, ...
dual: rop: = 2^31.3, ...
```

Those values of rop mean the number of operations for each attack called usvp, dec, and dual. Therefore, we can say that the parameter set has at least 29.7-bit security. It is known that the security level λ roughly has the following property:

$$\frac{N}{\log q - \log r} \propto \frac{\lambda}{\log \lambda}.$$

Therefore, increasing N may easily lead to higher security.

---

[8]https://bitbucket.org/malb/lwe-estimator.

10.2.2   Homomorphic Property of LWE-Based Cryptosystem

As a control engineer, a reason for particular interest on the LWE-based cryptosystem is that it allows homomorphic arithmetics. By homomorphic arithmetics, we mean, for two plaintexts m_1 and m_2, it holds that

$$\mathsf{Dec}\big(\mathsf{Enc}(m_1) *_C \mathsf{Enc}(m_2)\big) = m_1 *_P m_2$$

where $*_P$ and $*_C$ are binary operations on the plaintext space P and the ciphertext space C, respectively, and Dec and Enc symbolize the encryption and the decryption functions, respectively. The LWE-based cryptosystem provides both the homomorphic addition and the homomorphic multiplication.

The addition is defined in the following way:

$$\mathsf{Enc}(m_1) +_C \mathsf{Enc}(m_2) = \mathbf{m_1} + \mathbf{m_2}$$
$$= \big[-\mathsf{A}_1 \cdot \mathsf{sk} + Lm_1 + \mathsf{e}_1, \ \mathsf{A}_1\big] + \big[-\mathsf{A}_2 \cdot \mathsf{sk} + Lm_2 + \mathsf{e}_2, \ \mathsf{A}_2\big] \quad (10.4)$$
$$= \big[-(\mathsf{A}_1 + \mathsf{A}_2) \cdot \mathsf{sk} + L(m_1 + m_2) + (\mathsf{e}_1 + \mathsf{e}_2), \ \mathsf{A}_1 + \mathsf{A}_2\big]$$

where $+_C$ is the addition on the ciphertext space and $+$ is the standard matrix addition. To see the homomorphic property, observe that

$$\mathsf{Dec}(\mathbf{m_1} + \mathbf{m_2}) = \left\lceil \frac{(\mathbf{m_1} + \mathbf{m_2}) \cdot \mathsf{s} \mod q}{L} \right\rfloor$$
$$= \left\lceil \frac{L(m_1 + m_2) + \mathsf{e}_1 + \mathsf{e}_2}{L} \right\rfloor$$
$$= m_1 + m_2 \quad (10.5)$$

as long as $m_1 + m_2 \in [p]$ and each element of $|\mathsf{e}_1 + \mathsf{e}_2|$ is less than $L/2$.

Let us now consider the homomorphic multiplication of two *scalar* plaintexts $m_1 \in [p]$ and $m_2 \in [p]$. Without loss of generality, let m_1 be the multiplicand and m_2 be the multiplier for the product $m_2 m_1$. For the multiplicand m_1, we use the previous encryption function $\mathbf{m_1} = \mathsf{Enc}(m_1) \in \mathbb{Z}_q^{1 \times (N+1)}$ but for the multiplier m_2, we slightly change the encryption method[9, 10] as

$$\mathbf{M_2} = \mathsf{Enc2}(m_2) = m_2 R + \mathsf{Enc}(0_{\log q \cdot (N+1) \times 1}) \ \in \mathbb{Z}_q^{\log q \cdot (N+1) \times (N+1)} \quad (10.6)$$

---

[9] In practice, parameters $d \in \mathbb{N}$ and $\nu \in \mathbb{N}$ is chosen such that $q = \nu^d$, which customizes the dimension of the ciphertext $\mathbf{M_2}$ as $\mathbf{M_2} \in \mathbb{Z}_q^{d(N+1) \times (N+1)}$.

[10] The idea of using different encryption method for m_2 was introduced in [11, 12], and is customized for our context in this chapter because of its efficiency and suitability for encryption of dynamic controllers, as will be seen in Sect. 10.3. Other multiplication methods for LWE-based cryptosystems can be found in, e.g., [13].

where $0_{\log q \cdot (N+1) \times 1}$ is the plaintext of zero vector in $[p]^{\log q \cdot (N+1) \times 1}$ and, with the Kronecker product being denoted by $\otimes$,

$$R := [10^0, 10^1, 10^2, \ldots, 10^{\log q - 1}]^T \otimes I_{N+1}$$

where I_{N+1} is the identity matrix of size $N + 1$, so that R is a matrix of $\log q \cdot (N + 1)$ by $(N + 1)$. Note that $\mathbf{O} := \mathsf{Enc}(0_{\log q \cdot (N+1) \times 1})$ is a matrix in $\mathbb{Z}_q^{\log q \cdot (N+1) \times (N+1)}$, each row of which is an encryption of the plaintext 0 (but they are all different due to the randomness of $\mathbf{A}$ and $\mathbf{e}$). This modified encryption $\mathsf{Enc2}$ has the same level of security as Enc because the plaintext m_2 is still hiding in the ciphertext $\mathbf{O}$. Now we note that any vector c in $\mathbb{Z}_q^{1 \times (N+1)}$ can be represented using the radix of 10 as $c = \sum_{i=0}^{\log q - 1} c_i \cdot 10^i$, where $c_i \in \mathbb{Z}_q^{1 \times (N+1)}$, in which each component of the row vector c_i is one of the single digit $0, 1, 2, \ldots, 9$. Therefore, one can define the function $D : \mathbb{Z}_q^{1 \times (N+1)} \to \mathbb{Z}_q^{1 \times \log q \cdot (N+1)}$ that decomposes the argument by its string of digits as

$$D(c) := [c_0, c_1, \ldots, c_{\log q - 1}]. \tag{10.7}$$

An example when $q = 10^2$ and $N = 2$ is:

$$D([40, 35, -27]) = D([40, 35, 73]) = [0, 5, 3, 4, 3, 7]$$

because $[40, 35, 73] = [0, 5, 3] \cdot 10^0 + [4, 3, 7] \cdot 10^1$. As a result, it follows that $c = D(c)R$ for any $c \in \mathbb{Z}_q^{1 \times (N+1)}$. Now, the multiplication of two ciphertexts $\mathbf{m_1} = \mathsf{Enc}(m_1)$ and $\mathbf{m_2} = \mathsf{Enc}(m_2)$ can be done by, with $\mathbf{M_2} = \mathsf{Enc2}(\mathsf{Dec}(\mathbf{m_2}))$,

$$\mathbf{M_2} \times_C \mathbf{m_1} := D(\mathbf{m_1}) \cdot \mathbf{M_2} \quad \in \mathbb{Z}_q^{1 \times (N+1)}$$

where $\cdot$ is the standard matrix multiplication. It should be noted that the operation $\mathsf{Enc2}(\mathsf{Dec}(\mathbf{m_2}))$ requires the secret key sk, and thus, the above operation is more suitable when the multiplier m_2 is encrypted as $\mathbf{M_2} = \mathsf{Enc2}(m_2)$ a priori, and used repeatedly for different $\mathbf{m_1}$'s (which will be the case when we construct dynamic feedback controllers). In this case, the product of the multiplicand $\mathbf{m_1}$ with the multiplier $\mathbf{M_2}$ is simply performed as $D(\mathbf{m_1}) \cdot \mathbf{M_2}$. To see the homomorphic property, we first note that, with the secret key s,

$$\begin{aligned}(\mathbf{M_2} \times_C \mathbf{m_1}) \cdot \mathsf{s} = D(\mathbf{m_1}) \cdot \mathbf{M_2} \cdot \mathsf{s} &= D(\mathbf{m_1}) \cdot (m_2 R + \mathbf{O}) \cdot \mathsf{s} \\ &= m_2 \mathbf{m_1} \cdot \mathsf{s} + D(\mathbf{m_1}) \cdot \mathbf{e_{M_2}}\end{aligned} \tag{10.8}$$

where $\mathbf{e_{M_2}} \in \mathbb{Z}_q^{\log q \cdot (N+1) \times 1}$ is the error vector inside the ciphertext $\mathbf{O}$. From this, we observe that multiplication by $\mathbf{M_2}$ is equivalent to multiplication by the plaintext m_2 plus an error. Then, we have the homomorphic property for multiplication as

$$\text{Dec}(\mathbf{M_2} \times_C \mathbf{m_1}) = \left\lceil \frac{D(\mathbf{m_1}) \cdot \mathbf{M_2} \cdot \mathbf{s} \mod q}{L} \right\rfloor$$

$$= \left\lceil \frac{m_2(Lm_1 + \mathbf{e_1}) + D(\mathbf{m_1}) \cdot \mathbf{e_{M_2}}}{L} \right\rfloor = m_2 m_1, \tag{10.9}$$

as long as $m_2 m_1 \in [p]$ and

$$\left| \frac{m_2 \mathbf{e_1}}{L} + \frac{D(\mathbf{m_1}) \cdot \mathbf{e_{M_2}}}{L} \right| < \frac{1}{2}. \tag{10.10}$$

Note that each component of $D(\mathbf{m_1})$ does not exceed 10 so that the amplification of the error is limited. We will return to this point in Sect. 10.2.2.1.

A sample run and the codes for two operations are as follows:

```
q = env.p*env.L;
c1 = Enc(-2,sk,env);
c2 = Enc(3,sk,env);
c2m = Enc2(3,sk,env);

c_add = c1 + c2;
c_mul = Decomp(c1, q)*c2m;

Dec(c_add,sk,env)
ans =
1

Dec(c_mul,sk,env)
ans =
-6
```

Functions Enc2, Decomp

```
function ciphertext = Enc2(m,sk,env)
q = env.L*env.p;   N = env.N;   lq = log10(q);
R = kron( power(10, [0:1:lq-1]'), eye(N+1) );
ciphertext = Mod(m*R + Enc(zeros(lq*(N+1),1), sk, env), q);
end

function strdigits = Decomp(c, q)
lq = log10(q);
c = mod(c, q);
strdigits = [];
for i=0:lq-1,
```

```
Q = c - mod(c, 10^(lq-1-i));
strdigits = [ Q/10^(lq-1-i), strdigits ];
c = c - Q;
end
end
```

If a ciphertext $\mathbf{m_1} \in \mathbb{Z}_q^{1 \times (N+1)}$ is multiplied with a plaintext $m_2 \in [p]$, then it is simply performed by $\mathbf{m_1} m_2 \pmod{q}$ because this case can be considered as a repeated homomorphic addition.

Finally, we close this section by presenting a code for obtaining the product of the ciphertext of a matrix $F \in [p]^{m \times n}$ and the ciphertext of a column vector $x \in [p]^n$ that is equivalent to the ciphertext of Fx. With $\mathbf{F}_{i,j} = \mathsf{Enc2}(F_{i,j})$ where $F_{i,j}$ is the (i, j)-th element of F, and $\mathbf{x}_j$ being the j-th row of $\mathbf{x} = \mathsf{Enc}(x)$, we define the multiplication as

$$\mathbf{F} \times_C \mathbf{x} := \sum_{j=1}^{n} \begin{bmatrix} D(\mathbf{x}_j) \cdot \mathbf{F}_{1,j} \\ D(\mathbf{x}_j) \cdot \mathbf{F}_{2,j} \\ \vdots \\ D(\mathbf{x}_j) \cdot \mathbf{F}_{n,j} \end{bmatrix}$$

in which, we abuse the notation $\times_C$ that was defined for a scalar product. Then, analogously to (10.8) and (10.9), it follows that

$$(\mathbf{F} \times_C \mathbf{x}) \cdot \mathbf{s} = \sum_{j=1}^{n} \begin{bmatrix} D(\mathbf{x}_j) \cdot \mathbf{F}_{1,j} \cdot \mathbf{s} \\ D(\mathbf{x}_j) \cdot \mathbf{F}_{2,j} \cdot \mathbf{s} \\ \vdots \\ D(\mathbf{x}_j) \cdot \mathbf{F}_{n,j} \cdot \mathbf{s} \end{bmatrix} = \sum_{j=1}^{n} \begin{bmatrix} F_{1,j} \cdot \mathbf{x}_j \cdot \mathbf{s} + D(\mathbf{x}_j) \cdot \mathbf{e}_{\mathbf{F}_{1,j}} \\ F_{2,j} \cdot \mathbf{x}_j \cdot \mathbf{s} + D(\mathbf{x}_j) \cdot \mathbf{e}_{\mathbf{F}_{2,j}} \\ \vdots \\ F_{n,j} \cdot \mathbf{x}_j \cdot \mathbf{s} + D(\mathbf{x}_j) \cdot \mathbf{e}_{\mathbf{F}_{n,j}} \end{bmatrix}$$

$$= F \cdot \mathbf{x} \cdot \mathbf{s} + \sum_{j=1}^{n} \begin{bmatrix} D(\mathbf{x}_j) \cdot \mathbf{e}_{\mathbf{F}_{1,j}} \\ D(\mathbf{x}_j) \cdot \mathbf{e}_{\mathbf{F}_{2,j}} \\ \vdots \\ D(\mathbf{x}_j) \cdot \mathbf{e}_{\mathbf{F}_{n,j}} \end{bmatrix} =: F \cdot \mathbf{x} \cdot \mathbf{s} + \Delta(\mathbf{F}, \mathbf{x})$$

(10.11)

where $\mathbf{e}_{\mathbf{F}_{i,j}}$ is the error inside $\mathbf{F}_{i,j}$, and the homomorphic property is obtained as

$$\mathsf{Dec}(\mathbf{F} \times_C \mathbf{x}) = \left\lceil \frac{1}{L}(F \cdot \mathbf{x} \cdot \mathbf{s} + \Delta(\mathbf{F}, \mathbf{x}) \pmod{q}) \right\rfloor$$

$$= Fx + \left\lceil \frac{1}{L}(F \cdot \mathbf{e}_{\mathbf{x}} + \Delta(\mathbf{F}, \mathbf{x})) \right\rfloor = Fx$$

as long as $Fx \in [p]^m$ and the error $|F \cdot \mathbf{e}_{\mathbf{x}} + \Delta(\mathbf{F}, \mathbf{x})|$ is less than $L/2$.

Therefore, the operation

$$\mathbf{F} \times_C \mathbf{x} + \mathbf{G} \times_C \mathbf{y}$$

where $+$ is the standard addition, corresponds to the plaintext operation $Fx + Gy$ with two matrices F and G, and two vectors x and y. An example code and a sample run are as follows.

```
F = [1, 2; 3, 4];
cF = Enc2Mat(F,sk,env);

x = [1;2];
cx = Enc(x,sk,env);

cFcx = MatMult(cF,cx,env);
Dec(cFcx,sk,env)
ans =
5
11
```

Functions Enc2Mat, MatMult

```
function cA = Enc2Mat(A,sk,env)
q = env.p*env.L;    N = env.N;    [n1,n2] = size(A);
cA = zeros(log10(q)*(N+1), N+1, n1, n2);
for i=1:n1,
  for j=1:n2,
    cA(:,:,i,j) = Enc2(A(i,j),sk,env);
  end
end
end


function Mm = MatMult(M,m,env)
[n1,n2,n3,n4] = size(M);    q = env.p*env.L;
Mm = zeros(n3,env.N+1);
for i=1:n3,
  for j=1:n4,
    Mm(i,:) = Mod(Mm(i,:) + ...
    Decomp(m(j,:),env.p*env.L) * M(:,:,i,j), q);
  end
end
end
```

10.2.2.1 Error Growth Problem Caused by Error Injection

As can be seen in (10.2), a newborn scalar ciphertext $\mathbf{m}$ has its error $\mathbf{e}$ whose size is less than $r/2$; that is, $\|\mathbf{e}\|_\infty \le r/2$ where $\|\cdot\|_\infty$ is the infinity norm of a vector. However, the size of the error inside a ciphertext can grow as the arithmetic operations are performed on the variable. For example, if $\mathsf{MaxError}(\mathbf{m})$ measures the maximum size of the worst-case error in $\mathbf{m}$, then $\mathsf{MaxError}(\mathbf{m_1} + \mathbf{m_2}) = \mathsf{MaxError}(\mathbf{m_1}) + \mathsf{MaxError}(\mathbf{m_2})$ and $\mathsf{MaxError}(\mathbf{m_1}m_2) = m_2\mathsf{MaxError}(\mathbf{m_1})$ for a plaintext m_2, which can be seen from (10.4). The multiplication may cause more increase of the error as can be seen in (10.10). That is, the error in the product $\mathbf{m_{prod}}$ of the multiplicand $\mathbf{m_1}$ and the multiplier $\mathbf{M_2}$ leads to $\mathsf{MaxError}(\mathbf{m_{prod}}) = m_2\mathsf{MaxError}(\mathbf{m_1}) + 9(r/2)\log q$ because each element of $\mathbf{e_{M_2}}$ is the error of the newborn ciphertext so that its absolute value is less than $r/2$ and the component of $D(\mathbf{m_1})$ ranges from 0 to 9. It is noted that the first term $m_2\mathsf{MaxError}(\mathbf{m_1})$ is expected just as in the case of multiplying the plaintext m_2, but the amount of the second term adds more whenever the multiplication is performed.

The discussions so far show that, if the arithmetic operations are performed many times on the ciphertext, then the error may grow unbounded in the worst case, and it may damage the message in the ciphertext. Damage of the message happens when the size of the error in the ciphertext becomes larger than $L/2$. Indeed, the following example shows this phenomenon:

```
c = Enc(1,sk,env);

Dec(3*c,sk,env)
ans =
3

Dec(3000*c,sk,env)
ans =
2999
```

The error growth problem restricts the number of consecutive arithmetic operation on the ciphertext. In order to overcome the restriction, the blueprint of 'bootstrapping' procedure has been developed in [6]. In a nutshell, the grown-up error can be eliminated if the ciphertext is once decrypted and encrypted back again. The bootstrapping algorithm reduces the size of error by performing this process without the knowledge of the secret key. However, the complexity of the bootstrapping process hinders from being used for dynamic feedback controls. We discuss how to overcome this problem without bootstrapping in Sect. 10.4.

10.3 LWE-Based Cryptosystem and Dynamic Feedback Controller

For a comprehensive discussion with dynamic controllers, let us consider a discrete-time single-input-single-output linear time-invariant plant:

$$x_p[t+1] = Ax_p[t] + Bu[t], \quad y[t] = Cx_p[t]. \tag{10.12}$$

To control the plant (10.12), we suppose that a discrete-time linear time-invariant dynamic feedback controller has been designed as

$$x[t+1] = Fx[t] + Gy[t] \tag{10.13a}$$

$$u[t] = Hx[t] + Jy[t] \tag{10.13b}$$

where $x \in \mathbb{R}^n$ is the state of the controller, $y \in \mathbb{R}$ is the controller input, and $u \in \mathbb{R}$ is the controller output. Note that they are real numbers in general, and not yet quantized. In order to implement the controller by a digital computer, we need to quantize the signals y, u, and x, and to use the cryptosystem for the controller, we also need to make them integer values. This procedure is called 'quantization' in this chapter.

Quantization is performed both on the sensor signal $y[t]$, on the control parameters, and finally on the actuator signal $u[t]$. The quantization level for $y[t]$ is often determined by the specification of the sensor under the name of *resolution* R_y. Therefore, we define the quantized integer value of the signal $y[t]$ as

$$y[t] \longrightarrow \bar{y}[t] := \left\lceil \frac{y[t]}{R_y} \right\rceil. \tag{10.14}$$

For example, with $R_y = 0.1$, the signal $y[t] = 12.11$ becomes $\bar{y}[t] = 121$. This procedure is performed at the sensor stage before the encryption. On the other hand, the matrices in (10.13) are composed of real numbers in general. These numbers should be truncated for digital implementation, but it is often the case when the significant digits of them include fractional parts. In this case, we can "scale" the controller (10.13) by taking advantages of the linear system. Before discussing the scaling, we assume that the matrix F consists of integer numbers so that the scaling for F is not necessary. This is an important restriction and we will discuss this issue in detail in Sect. 10.5. Now, take $G = [5.19, 38]^T$ for example. If those numbers are to be kept up to the fraction $1/10 =: S_G$, then the quantized G can be defined as $\bar{G} := \lceil G/S_G \rfloor$ so that $\bar{G} = [52, 380]^T$. By dividing (10.13a) by $R_y S_G$, we obtain the quantized equation as

$$\frac{x[t+1]}{S_G R_y} = F \frac{x[t]}{S_G R_y} + \frac{G}{S_G} \frac{y[t]}{R_y} \xrightarrow{\text{truncation}} \bar{x}[t+1] = F\bar{x}[t] + \bar{G}\bar{y}[t]$$

where $\bar{x}[t] := x[t]/(S_G R_y)$ which becomes integer for all $t > 0$ if the initial condition is set as $\bar{x}[0] = \lceil x[0]/(S_G R_y) \rceil$. Since there may be still some significant fractional numbers in the matrices H or J/S_G in general, the output Eq. (10.13b) is scaled with additional scaling factor S_{HJ} as

$$\frac{u[t]}{S_{HJ} S_G R_y} = \frac{H}{S_{HJ}} \frac{x[t]}{S_G R_y} + \frac{J}{S_{HJ} S_G} \frac{y[t]}{R_y} \xrightarrow{\text{truncation}} \bar{u}[t] = \bar{H}\bar{x}[t] + \bar{J}\bar{y}[t],$$

where $\bar{H} := \lceil H/S_{HJ} \rceil$, $\bar{J} := \lceil J/(S_{HJ} S_G) \rceil$, and $\bar{u}[t] := u[t]/(S_{HJ} S_G R_y)$. Therefore, the *quantized controller*

$$\bar{x}[t + 1] = \bar{F}\bar{x}[t] + \bar{G}\bar{y}[t] \tag{10.15a}$$

$$\bar{u}[t] = \bar{H}\bar{x}[t] + \bar{J}\bar{y}[t] \tag{10.15b}$$

is composed of integer values, and the state $\bar{x}[t]$ evolves on the integer state-space. Finally, the real number input $u[t]$ is obtained by

$$\bar{u}[t] \quad \longrightarrow \quad u[t] = R_u \left\lceil \frac{R_y S_G S_{HJ}}{R_u} \bar{u}[t] \right\rceil \tag{10.16}$$

at the actuator stage, where R_u is the resolution of the actuator. The rounding in (10.16) does nothing if $R_y S_G S_{HJ}/R_u$ is an integer, because $\bar{u}[t]$ is always an integer. It is clear that the digital implementation of (10.13), given by (10.14), (10.15), and (10.16), works well if the truncation error is small.

Since the quantized controller (10.15) consists of all the integer matrices and vectors, it is straightforward to convert it to the homomorphically encrypted controller

$$\mathbf{x}[t + 1] = \mathbf{F} \times_C \mathbf{x}[t] + \mathbf{G} \times_C \mathbf{y}[t] \tag{10.17a}$$

$$\mathbf{u}[t] = \mathbf{H} \times_C \mathbf{x}[t] + \mathbf{J} \times_C \mathbf{y}[t] \tag{10.17b}$$

where the operations on the ciphertexts should be understood as explained in Sect. 10.2.2. Note that $\mathbf{y}[t] = \mathsf{Enc}(\bar{y}[t])$ is always a newborn ciphertext for each t because it is encrypted and transmitted from the sensor stage. Moreover, the ciphertexts $\mathbf{F}$, $\mathbf{G}$, $\mathbf{H}$, and $\mathbf{J}$ can be considered as all newborn ciphertexts because they are generated when the controller is set and not updated by the control operation. The equation (10.17) is solved at each time step with the initial condition $\mathbf{x}[0] = \mathsf{Enc}(\bar{x}[0])$. Under this setting, two new ciphertexts $\mathbf{x}[t + 1]$ and $\mathbf{u}[t]$ are created at each time step, or the system (10.17) is considered to be driven by $\mathbf{y}[t]$ with $\mathbf{x}[0]$. The vector $\mathbf{x}[0]$ also has the newborn error, but the error in $\mathbf{x}[t]$ may grow as time goes on because of the recursion in (10.17a).

As an example, consider a first-order plant given by $x_p[t + 1] = \sqrt{2}x_p[t] + u[t]$ and $y[t] = x_p[t]$, for which a first-order dynamic feedback controller $x[t + 1] = -1 \cdot x[t] + 1 \cdot y[t]$ and $u[t] = -1.414 \cdot x[t] + 0 \cdot y[t]$ stabilizes the closed-loop system.

With the parameters $R_y = 10^{-3}$, $S_G = 1$, $S_{HJ} = 10^{-3}$, and $R_u = R_y S_G S_{HJ} = 10^{-6}$ the simulation can be done for timesteps = 150 as follows.[11]

```
A = sqrt(2); B = 1; C = 1;          % plant
F = -1; G = 1; H = -1.414; J = 0; % controller
Ry = 1e-3; Sg = 1e0; Shj = 1e-3;
G_ = round(G/Sg); H_ = round(H/Shj); J_ = round(J/(Sg*Shj));
cFG = Enc2Mat([F,G_],sk,env);   cHJ = Enc2Mat([H_,J_],sk,env);
xp = -3.4;  x = 4.3;                 % i.c. of plant and ctr
cx = Enc(round(x/(Ry*Sg)), sk, env);

for i = 1:timesteps
    y  = C*xp;                       % Plant output
    cy = Enc(round(y/Ry), sk, env);  % Encryption
    cu = MatMult(cHJ, [cx;cy], env); % Controller output
    u  = Ry*Sg*Shj*Dec(cu, sk, env); % Plant input after Dec
    xp = A*xp + B*u;                 % Plant update
    cx = MatMult(cFG, [cx;cy], env); % Controller update
end
```

10.4 Controlled Error Growth by Closed-Loop Stability

As mentioned previously, the growth of the error in the ciphertext $\mathbf{x}[t]$ is of major concern in this section. The growth should be suppressed in the control operation because, if not, the control signals are damaged by the error signal.

Actually, the source of error growth is the arithmetic operations in (10.17). To see both the message and the error in the state $\mathbf{x}[t]$, let us decrypt the dynamics (10.17) with the secret key $\mathbf{s}$ except the rounding operation; i.e., we define

$$\xi[t] := \frac{(\mathbf{x}[t] \cdot \mathbf{s}) \mod q}{L} \in \mathbb{R}^n$$

so that $\mathsf{Dec}(\mathbf{x}[t]) = \lceil \xi[t] \rfloor$, and see the evolution of ξ-system over real-valued signals, which in turn is equivalent to the operation of (10.17). According to the homomorphic property (10.11), the ξ-system is derived as

---

[11] An example parameter set for this example is env.p = 1e9, env.L = 100, and env.r = 10.

$$\xi[t+1] = F\xi[t] + \bar{G}\left(\bar{y}[t] + \frac{e_{y[t]}}{L}\right) + \frac{\Delta(\mathbf{F}, \mathbf{x}[t])}{L} + \frac{\Delta(\mathbf{G}, \mathbf{y}[t])}{L}$$

$$=: F\xi[t] + \bar{G}\bar{y}[t] + \Delta_1[t], \qquad \xi[0] = \bar{x}[0] + \frac{e_{x[0]}}{L},$$

$$\bar{u}'[t] = \bar{H}\xi[t] + \bar{J}\left(\bar{y}[t] + \frac{e_{y[t]}}{L}\right) + \frac{\Delta(\mathbf{H}, \mathbf{x}[t])}{L} + \frac{\Delta(\mathbf{J}, \mathbf{y}[t])}{L}$$

$$=: \bar{H}\xi[t] + \bar{J}\bar{y}[t] + \Delta_2[t],$$

(10.18)

in which $e_{y[t]}$ and $e_{x[0]}$ are the errors injected to the encryptions $\mathbf{y}[t]$ and $\mathbf{x}[0]$, respectively, $\Delta(\mathbf{F}, \mathbf{x}[t])$, $\Delta(\mathbf{G}, \mathbf{y}[t])$, $\Delta(\mathbf{H}, \mathbf{x}[t])$, and $\Delta(\mathbf{J}, \mathbf{y}[t])$ are the errors caused by ciphertext multiplication, which are defined as the same as in (10.11), and $\bar{u}'[t]$ is defined as $\bar{u}'[t] := (\mathbf{u}[t] \cdot \mathbf{s} \mod q)/L$ so that $\mathsf{Dec}(\mathbf{u}[t]) = \lceil \bar{u}'[t] \rfloor$.

For the comparison with the quantized controller (10.15), the first observation is that if there is no error injected to ciphertexts $\mathbf{y}[t]$, $\mathbf{x}[0]$, and $\{\mathbf{F}, \mathbf{G}, \mathbf{H}, \mathbf{J}\}$ so that $\Delta_1[t]$, $\Delta_2[t]$, and $e_{x[0]}$ are all zero, the operation of (10.18) is exactly the same way as the operation of (10.15). Then, with the control perspective, the signals $\Delta_1[t]$ and $\Delta_2[t]$ can be understood as *external disturbances* injected to the feedback loop, and the quantity $e_{x[0]}$ can be regarded as *perturbation* of the initial condition (see Fig. 10.2). Here, the sizes of $\Delta_1[t]$, $\Delta_2[t]$, and $e_{x[0]}$ can be made arbitrarily small by increasing the parameter L for the encryption. This is because $\|e_{y[t]}/L\|_\infty$ and $\|e_{x[0]}/L\|_\infty$ are less than $r/(2L)$, and the disturbance caused by multiplication of $\{\mathbf{x}[t], \mathbf{y}[t]\}$ by $\{\mathbf{F}, \mathbf{G}, \mathbf{H}, \mathbf{J}\}$ can also be made arbitrary small with the choice of L; for example, as seen in (10.11), the size of signal $\Delta(\mathbf{F}, \mathbf{x}[t])/L$ is bounded as

$$\left\|\frac{\Delta(\mathbf{F}, \mathbf{x}[t])}{L}\right\|_\infty \le \frac{1}{L}\sum_{j=1}^{n}\left\|\begin{bmatrix} D(\mathbf{x}_j[t]) \cdot e_{\mathbf{F}_{1,j}} \\ D(\mathbf{x}_j[t]) \cdot e_{\mathbf{F}_{2,j}} \\ \vdots \\ D(\mathbf{x}_j[t]) \cdot e_{\mathbf{F}_{n,j}} \end{bmatrix}\right\|_\infty$$

$$\le \frac{9nr\log q}{2L}$$

$$= \frac{9nr(\log p + \log L)}{2L}.$$

Now, in terms of the error growth problem of the controller state, the difference $\xi[t] - \bar{x}[t]$ corresponds to the error of our concern. One might expect that the size of $\xi[t] - \bar{x}[t]$ can be made arbitrarily small by increasing L, but it is not true due to the rounding operations in the sensor and actuator; for example, if the difference $\xi[t] - \bar{x}[t]$ is so small that the difference of actuator inputs is less than the size of input resolution, it is truncated and the difference is not compensated in the closed-loop stability. As a result, the error eventually grows up to the resolution range, but is controlled not to grow more than that. Therefore, the damage of the message in the ciphertexts $\mathbf{x}[t]$ is inevitable, but it can be limited up to the last a few digits. Motivated by this fact, one may intentionally enhance the resolutions by a few more digits in order to preserve the significant figures. In this way, as long as the injected

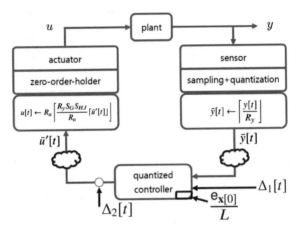

Fig. 10.2 This figure describes the closed-loop system with the controller (10.18). In other words, the behavior of the closed-loop system with the encrypted controller (10.17) is the behavior of the closed-loop with the quantized controller (10.15) with the norm-bounded external disturbances $\Delta_1[t]$ and $\Delta_2[t]$, and the perturbation $\mathbf{e}_{\mathbf{x}[0]}/L$ on the initial condition of the controller

errors $\Delta_1[t]$, $\Delta_2[t]$, and $\mathbf{e}_{\mathbf{x}[0]}/L$ are sufficiently small, the error (i.e., the difference $\xi[t] - \bar{x}[t]$) is controlled not to grow unbounded by the closed-loop stability. See a simulation result in Fig. 10.3.

In the rest of this section, we briefly analyze the control performance in terms of the encryption as well as the quantization. For this, let us recall that $\mathsf{Dec}(\mathbf{u}[t]) = \lceil \bar{u}'[t] \rfloor$. This leads to, by (10.16),

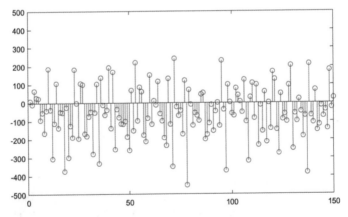

Fig. 10.3 The error in the ciphertext $\mathbf{x}[t]$ from a sample run of the simulation in Sect. 10.3. This is the plot of $L(\xi[t] - \bar{x}[t])$ where $\bar{x}[t]$ is the state of the quantized controller and $\xi[t]$ is the state of (10.18), with the parameters $L = 100$, $1/R_y S_G = 1000$, and $r = 10$. It is seen that the error goes beyond L, but is suppressed within one digit in the resolution range. This means that the message is damaged but only for one last digit

$$u = R_u \left\lceil \frac{R_y S_G S_{HJ}}{R_u} \lceil \bar{u}' \rfloor \right\rfloor = R_y S_G S_{HJ} \bar{u}' + \Delta_{\text{Dec}} + \Delta_u$$

where

$$\Delta_{\text{Dec}} := R_y S_G S_{HJ} \left(\lceil \bar{u}' \rfloor - \bar{u}' \right), \quad \Delta_u := R_u \left\lceil \frac{R_y S_G S_{HJ}}{R_u} \lceil \bar{u}' \rfloor \right\rfloor - R_y S_G S_{HJ} \lceil \bar{u}' \rfloor$$

in which, Δ_{Dec} implies the error caused by the rounding in the decryption, and Δ_u implies the error by the quantization of the input stage. Now, for the sake of simplicity, let us assume that $G = S_G \bar{G}$, $H = S_{HJ} \bar{H}$, and $J = S_{HJ} S_G \bar{J}$, which means there is no error due to the scaling of the matrices. By defining $\xi' := R_y S_G \xi$, we obtain from (10.18) that

$$R_y S_G S_{HJ} \bar{u}' = H\xi' + Jy + J\Delta_y + R_y S_G S_{HJ} \Delta_2 \quad \text{where} \quad \Delta_y := R_y \left\lceil \frac{y}{R_y} \right\rfloor - y$$

in which, Δ_y is the error caused by the quantization at the sensor stage. Putting together, the closed-loop system of the plant (10.12) and the controller (10.18) is equivalently described by

$$x_p[t+1] = Ax_p[t] + B(H\xi'[t] + JCx_p[t])$$
$$+ \left\{ B(J\Delta_y[t] + R_y S_G S_{HJ} \Delta_2[t] + \Delta_{\text{Dec}}[t] + \Delta_u[t]) \right\}$$
$$\xi'[t+1] = F\xi'[t] + GCx_p[t] + \left\{ G\Delta_y[t] + S_G S_{HJ} \Delta_1[t] \right\}$$

with the initial condition of the controller is set to be

$$\xi'[0] = x[0] + \left\{ R_y S_G \left\lceil \frac{x[0]}{R_y S_G} \right\rfloor - x[0] + \frac{R_y S_G \mathbf{e}_{x[0]}}{L} \right\}$$

where $x[0]$ is the initial condition of (10.13). Note that all the braced terms (i.e., errors) can be made arbitrarily small with sufficiently small R_y and R_u and with sufficiently large L. Moreover, with all these errors being zero, the above system is nothing but the closed-loop system of the plant (10.12) and the controller (10.13), which is supposed to be asymptotically stable. Therefore, it is seen that the control performance with the encrypted controller (10.17) can be made arbitrarily close to the nominal control performance with the linear controller (10.13).

10.5 Conclusion and Need for Integer System Matrix

In this chapter, with the use of fully homomorphic encryption, we have seen a method as well as an illustrative example to implement a dynamic feedback controller over encrypted data. Exploiting both additively and multiplicatively homomorphic properties of LWE-based scheme, all the operations in the controller are performed over encrypted parameters and signals. Once the designed controller (10.13) is converted to the dynamical system (10.15) over integers, it can be directly encrypted as (10.17). From the nature of fully homomorphic encryption schemes, the error injected to the encryption $\mathbf{y}[t]$ may be accumulated in the controller state $\mathbf{x}[t]$ under the recursive state update and may affect the message. However, from the control perspective, it has been seen that the effect of error is controlled and suppressed by the stability of the closed-loop system.

For the concluding remark, let us revisit that the encryption scheme for the dynamic controller (10.13) is based on the assumption that all entries of the system matrix F are *integers*. To see the necessity of this assumption, let us suppose the matrix F consists of non-integer real numbers. One may attempt the scaling of F as $\lceil F/S_F \rfloor$ with the scaling factor $1/S_F > 1$ in order to keep the fractional part of F, but this scaling is hopeless because it results in recursive multiplication by $1/S_F$ for each update of the controller. Indeed, for this case, it can be checked that the state $\bar{x}[t]$ of the quantized controller (10.15a) is multiplied by $\lceil F/S_F \rfloor$ (instead of F) for each time step, so (10.15a) should be remodeled as the form

$$\bar{x}[t+1] = \left\lceil \frac{F}{S_F} \right\rfloor \bar{x}[t] + \left\lceil \frac{G}{S_F^{t+1} S_G} \right\rfloor \bar{y}[t] \tag{10.19}$$

with the relation $\bar{x}[t] = x[t]/(S_F^t S_G R_y)$. However, encryption of (10.19) is hopeless, because in this case the message of the encrypted state is unbounded due to the term $1/S_F^t$. It will lose its value when it eventually go beyond the bound $\pm p/2$ of the plaintext space $[p]$ represented as (10.1), unless the state is reset to eliminate the accumulated scaling factor.

This problem, which is from the constraint that encrypted variables can be multiplied by scaled real numbers only a finite number of times, is in fact one of the main difficulties of encrypting dynamic controllers having non-integer system matrix.[12] In this respect, one may find potential benefits of using proportional-integral-derivative (PID) controllers or finite-impulse-response (FIR) filters for the design of encrypted control system, because they can be realized with the matrix F being integer as follows:

- Given an FIR filter written as $C(z) = \sum_{i=0}^{n} b_{n-i} z^{-i}$, the dynamic feedback controller can be realized as

---

[12]The problem is the same for encrypted controllers based on additively homomorphic encryption schemes. See [14] for the details.

$$x[t+1] = \begin{bmatrix} 0 & \cdots & 0 & 0 \\ 1 & \cdots & 0 & 0 \\ \vdots & \ddots & \vdots & \vdots \\ 0 & \cdots & 1 & 0 \end{bmatrix} x[t] + \begin{bmatrix} 1 \\ 0 \\ \vdots \\ 0 \end{bmatrix} y[t],$$

$$u[t] = \begin{bmatrix} b_{n-1} & \cdots & b_1 & b_0 \end{bmatrix} x[t] + b_n y[t].$$

- A discrete PID controller in the parallel form is given by

$$C(z) = k_p + \frac{k_i T_s}{z-1} + \frac{k_d}{\frac{T_s}{N_d} + \frac{T_s}{z-1}}$$

where k_p, k_i, and k_d are the proportional, integral, and derivative gains, respectively, T_s is the sampling time, and $N_d \in \mathbb{N}$ is the parameter for the derivative filter. This controller can be realized as

$$x[t+1] = \begin{bmatrix} 2-N_d & N_d - 1 \\ 1 & 0 \end{bmatrix} x[t] + \begin{bmatrix} 1 \\ 0 \end{bmatrix} y[t],$$

$$u[t] = \begin{bmatrix} b_1 & b_0 \end{bmatrix} x[t] + b_2 y[t],$$

where $b_1 = k_i T_s - k_d N_d^2/T_s$, $b_0 = k_i T_s N_d - k_i T_s + k_d N_d^2/T_s$, and $b_2 = k_p + k_d N_d/T_s$.

Another idea of approximating the effect of non-integer real numbers of F has been presented in [14] by using stable pole-zero cancellation. However, it was done at the cost of increased steady-state error in control performance. Further research is called for in this direction.

Acknowledgements The authors are grateful to Prof. Jung Hee Cheon and Dr. Yongsoo Song, Department of Mathematical Sciences, Seoul National University, for helpful discussions. This work was supported by National Research Foundation of Korea (NRF) grant funded by the Korea government (Ministry of Science and ICT) (No. NRF-2017R1E1A1A03070342).

References

1. Kogiso K, Fujita T (2015) Cyber-security enhancement of networked control systems using homomorphic encryption. In: Proceedings of IEEE conference on decision and control (CDC)
2. Farokhi F, Shames I, Batterham N (2016) Secure and private cloud-based control using semi-homomorphic encryption. In: Proceedings of IFAC workshop on distributed estimation and control in networked systems (NecSys), pp 163–168
3. Kim J, Lee C, Shim H, Cheon JH, Kim A, Kim M, Song Y (2016) Encrypting controller using fully homomorphic encryption for security of cyber-physical systems. IFAC-PapersOnLine 49(22):175–180
4. El Gamal T (1985) A public key cryptosystem and a signature scheme based on discrete logarithms. In: Proceedings of CRYPTO 84 on advances in cryptology. Springer, New York, pp 10–18

5. Paillier P (1999) Public-key cryptosystems based on composite degree residuosity classes. In: Annual international conference on the theory and applications of cryptographic techniques. Springer, pp 223–238
6. Gentry C (2009) Fully homomorphic encryption using ideal lattices. STOC 9:169–178
7. Rivest RL, Adleman L, Dertouzos ML (1978) On data banks and privacy homomorphisms. Found Secur Comput 4(11):169–180
8. Lindner R, Peikert C (2011) Better key sizes (and attacks) for LWE-based encryption. In: Cryptographers' track at the RSA conference. Springer, pp 319–339
9. Regev O (2009) On lattices, learning with errors, random linear codes, and cryptography. J ACM (JACM) 56(6):34
10. Chen L, Jordan S, Liu Y-K, Moody D, Peralta R, Perlner R, Smith-Tone D (2016) Report on post-quantum cryptography. US Department of Commerce, National Institute of Standards and Technology
11. Gentry C, Sahai A, Waters B (2013) Homomorphic encryption from learning with errors: conceptually-simpler, asymptotically-faster, attribute-based. In: Advances in cryptology–CRYPTO 2013. Springer, pp 75–92
12. Chillotti I, Gama N, Georgieva M, Izabachène M (2016) Faster fully homomorphic encryption: bootstrapping in less than 0.1 seconds. In: Advances in cryptology–ASIACRYPT 2016: 22nd international conference on the theory and application of cryptology and information security. Springer, pp 3–33
13. Lyubashevsky V, Peikert C, Regev O (2010) On ideal lattices and learning with errors over rings. In: Annual international conference on the theory and applications of cryptographic techniques. Springer, pp 1–23
14. Cheon JH, Han K, Kim H, Kim J, Shim H (2018) Need for controllers having integer coefficients in homomorphically encrypted dynamic system. In: IEEE 57th conference on decision and control (CDC), pp 5020–5025

Chapter 11
Encrypted Model Predictive Control in the Cloud

Moritz Schulze Darup

Abstract In this chapter, we focus on encrypted model predictive control (MPC) implemented in a single cloud. In general, encrypted control enables confidential controller evaluations in networked control systems. Technically, an encrypted controller is a modified control algorithm that is capable of computing encrypted control actions based on encrypted system states without intermediate decryptions. Encrypted control can, for example, be realized using homomorphic encryption that allows simple mathematical operations to be carried out on encrypted data. However, encrypting optimization-based control schemes such as MPC is non-trivial. Against this background, the contribution of the chapter is twofold. First, we summarize and unify two existing encrypted MPCs using the additively homomorphic Paillier cryptosystem. Second, we present a novel encrypted MPC based on real-time iterations of the alternating direction method of multipliers (ADMM). We theoretically and experimentally compare the three approaches and highlight unique features of the new scheme.

11.1 Introduction and Problem Statement

Cloud-computing and distributed computing become more and more essential for modern control systems such as smart grids, building automation, robot swarms, or intelligent transportation systems. While cloud-based and distributed control schemes are powerful, flexible, and efficient, they also increase the risk of cyber-attacks such as spying or sabotage. In fact, the involved communication of sensitive data via public networks and the data processing on third-party platforms promote eavesdropping and manipulation. Future control schemes should counteract those threats and ensure confidentiality and integrity of the involved data.

Realizing secure cloud-based control is non-trivial. In fact, simply encrypting the communication with the cloud using standard techniques (such as the advanced

M. Schulze Darup (✉)
Automatic Control Group, Department of Electrical Engineering and Information Technology,
Universität Paderborn, Paderborn, Germany
e-mail: moritz.schulzedarup@rub.de

© Springer Nature Singapore Pte Ltd. 2020
F. Farokhi (ed.), *Privacy in Dynamical Systems*,
https://doi.org/10.1007/978-981-15-0493-8_11

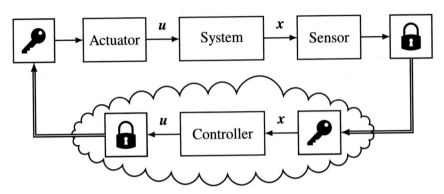

Fig. 11.1 Illustration of a cloud-based control scheme with encrypted communications but insecure controller evaluation. Double-arrows highlight encrypted data transmission

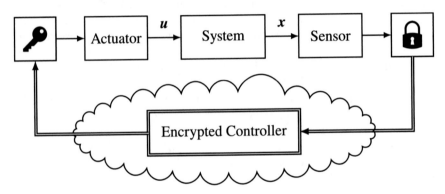

Fig. 11.2 Illustration of a cloud-based control scheme with encrypted communications and encrypted controller evaluation. Double-arrows and double-framing highlight encrypted data transmission and encrypted data processing, respectively

encryption standard (AES)) is not sufficient since a curious cloud provider then has access to the unsecured data during the controller evaluation (see Fig. 11.1). Protection against eavesdropping throughout the whole control loop can, however, be realized using encrypted control schemes as in Fig. 11.2. This setup differs from the scheme in Fig. 11.1 in that plaintext states x and inputs u are not available to the cloud. To achieve this confidentiality, the original control algorithm has to be modified in such a way that it computes encrypted control actions based on encrypted system states without intermediate decryptions. At first glance, such a modification seems hard to realize. However, encrypted controllers have already been implemented successfully (see, for instance, [1, 2, 8, 9, 15, 16, 30–32]). Regarding the underlying technology, two (occasionally overlapping) types of realizations can be distinguished. The first group of encrypted control schemes (represented by, e.g., [8, 9, 15, 16, 31, 32]) is primarily based on homomorphic encryption (HE). HE (see, e.g., [7, 10, 21]) is a natural technological choice for encrypted control since it enables elementary mathematical operations on encrypted data. The second group (represented by [1, 2, 30]) primarily uses secret sharing in combination with secure multi-party computation. Secret sharing [33] divides secret data into multiple shares in such a way that the individual shareholders learn nothing about the secret but the secret can be reconstructed by combining a certain number of shares. Secure multi-party computation (see, e.g., [23, 35]) provides protocols to perform computations on the secret data based on the previously introduced shares. The two types of encrypted control schemes do not only vary with respect to the dominating cryptosystem but also regarding the network architecture. In fact, schemes in the first group usually involve a single cloud whereas controllers (or observers) in the second group make use of multiple clouds or servers.

In this chapter, we focus on the first group, i.e., on encrypted cloud-based controllers based on HE. More precisely, we concentrate on encrypted model predictive control (MPC) implemented in a single cloud. MPC (see, e.g., [24] for an overview) is an optimization-based control strategy tailored for systems with input or state constraints. Since solving the underlying optimal control problems (OCPs) is numerically demanding, MPC can benefit from cloud-based or distributed implementations. Hence, encrypted MPC has been considered shortly after the seminal works on encrypted control in [8, 16]. Encrypted MPC has already been realized using various optimization techniques. In fact, multi-parametric programming (MPP) builds the basis for the implementation in [32], a projected gradient scheme (PGS) is applied in [1], and real-time projected gradient iterations are considered in [31]. MPP (see [3]) solves the OCP of interest for all relevant system states offline, i.e., before runtime of the controller. In contrast, PGSs (see, e.g., [20, 25]) provide structurally simple online solvers. While structurally simple, the numerical effort for running PGSs depends on the complexity of the involved projection step. In the framework of MPC, those projections can usually not be implemented efficiently if state constraints are present. Hence, solely input constraints are considered in [1, 31] whereas state (and input) constraints can be included in [32]. This advantage of the encrypted MPC in [32] comes with three drawbacks. First, applying MPP for MPC requires simple prediction models with only a few states and short prediction horizons. Second, the offline solution of the optimization leads to a large catalogue of controller segments that needs to be stored and accessed online. Third, in order to realize the encrypted implementation in [32], the selection of the "active" segment is carried out at the sensor, i.e., outside the cloud.

In view of the previous discussion, the contribution of the chapter is twofold. First, we provide a unifying summary of the encrypted MPCs in [31, 32], where we specify the pros and cons briefly addressed above. We do not further comment on the two-party-based approach in [1] since our focus is on implementations in a single cloud. Our second contribution is a novel homomorphically encrypted controller that realizes online MPC for systems with state and input constraints. Similar to [31], the new approach uses real-time iterations for optimization, i.e., a single solver iteration per sampling instant. However, in contrast to the PGS-based scheme in [31], the iterations originate from the alternating direction method of multipliers (ADMM). ADMM (see [5] for an overview) provides efficient solvers for the minimization of composite objective functions. In the context of MPC, the alternating minimization of the decomposed objectives allows to consider state and input constraints. The closed-loop dynamics of an MPC based on real-time ADMM iterations have recently been studied in [28]. Here, we present an encrypted implementation of this novel controller in a cloud.

The chapter is organized as follows. Background on MPC and HE is given in Sect. 11.2. The existing encrypted MPC schemes [31, 32] are summarized and analyzed in Sects. 11.3 and 11.4. The novel encrypted MPC based on real-time ADMM iterations is presented in Sect. 11.5. The three approaches are compared and applied to a simple numerical example in Sect. 11.6. Conclusions and an outlook are given in Sect. 11.7.

11.1.1 Notation and Acronyms

We denote the sets of real, integer, and natural numbers by $\mathbb{R}$, $\mathbb{Z}$, and $\mathbb{N}$, respectively. The sets $\mathbb{Z}_P$ and $\mathbb{Z}_P^*$ refer to the additive and multiplicative group of integers modulo $P \in \mathbb{N}$, respectively. The standard representatives of $\mathbb{Z}_P$ are collected in the set $\mathbb{N}_P := \{0, 1, \ldots, P - 1\}$. We use bold-face letters whenever we refer to matrices or vectors. In particular, we denote the identity matrix in $\mathbb{R}^{n \times n}$ by $\boldsymbol{I}_n$. Moreover, with $\boldsymbol{1}_m, \boldsymbol{0}_m$, and $\boldsymbol{0}_{m \times n}$, we refer to the vector in $\mathbb{R}^m$ full of ones, the vector full of zeros, and the matrix in $\mathbb{R}^{m \times n}$ full of zeros, respectively. Finally, vector-valued inequalities are understood element-wise. We complete the notation section with a list of frequently used acronyms in Table 11.1.

Table 11.1 List of Acronyms

Acronym	Meaning
ADMM	Alternating direction method of multiplies
HE	Homomorpihic encryption
LQR	Linear quadratic regulator
MPC	Model predictive control
MPP	Multi-parametric programming
OCP	Optimal control problem
PGS	Projected gradient scheme
QP	Quadratic program

11.2 Background and Preliminaries

In this section, we provide background on MPC for linear systems with constraints and a brief introduction to HE. In particular, we recall the additively homomorphic Paillier encryption that is applied in many encrypted controllers. As most encryption schemes, the Paillier cryptosystem acts on the set of integers. Hence, encrypted control usually requires quantization and we provide a suitable framework.

11.2.1 Model Predictive Control for Linear Constrained Systems

Throughout the chapter, we consider linear discrete-time systems

$$x(k+1) = Ax(k) + Bu(k), \qquad x(0) := x_0, \tag{11.1}$$

that are subject to state and input constraints of the form

$$x(k) \in \mathcal{X} \subseteq \mathbb{R}^n \quad \text{and} \quad u(k) \in \mathcal{U} := \{u \in \mathbb{R}^m \mid u_{\min} \leq u \leq u_{\max}\}. \tag{11.2}$$

We further assume that the pair (A, B) is stabilizable, that the constraints $\mathcal{X}$ are convex and polyhedral with $\mathbf{0}_n \in \text{int}(\mathcal{X})$, and that $u_{\min} < \mathbf{0}_m < u_{\max}$. MPC then typically builds on solving the OCP

$$V(x) := \min_{\substack{\tilde{x}(0),\ldots,\tilde{x}(N) \\ \tilde{u}(0),\ldots,\tilde{u}(N-1)}} \tilde{x}(N)^\top P \tilde{x}(N) + \sum_{\kappa=0}^{N-1} \tilde{x}(\kappa)^\top Q \tilde{x}(\kappa) + \tilde{u}(\kappa)^\top R \tilde{u}(\kappa) \tag{11.3a}$$

$$\text{s.t.} \quad \tilde{x}(0) = x, \tag{11.3b}$$

$$\tilde{x}(\kappa+1) = A\tilde{x}(\kappa) + B\tilde{u}(\kappa), \qquad \forall \kappa \in \mathbb{N}_N \tag{11.3c}$$

$$\tilde{x}(\kappa) \in \mathcal{X}, \qquad \forall \kappa \in \mathbb{N}_N \tag{11.3d}$$

$$\tilde{u}(\kappa) \in \mathcal{U}, \qquad \forall \kappa \in \mathbb{N}_N \tag{11.3e}$$

$$\tilde{x}(N) \in \mathcal{T} \tag{11.3f}$$

in every time step for the current state $x = x(k)$. The weighting matrices P, Q, and R are design parameters that specify the objective. For simplicity, we here assume that P, Q, and R are all positive definite. The terminal set $\mathcal{T}$ can be used to enforce stability of the closed-loop system and feasibility of the MPC (see [17, Sect. 3.3] for details). However, stability and feasibility can also be obtained for $\mathcal{T} = \mathcal{X}$ and sufficiently long prediction horizons N (see, e.g., [4, Theorem 13] or [29, Theorem 3]). In this context, the closed-loop dynamics are determined by the feedback

$$u(k) = \tilde{u}^*(0), \tag{11.4}$$

where $\tilde{u}^*(0), \ldots, \tilde{u}^*(N-1), \tilde{x}^*(0), \ldots, \tilde{x}^*(N)$ refer to the optimizers for (11.3). In order to compute those optimizers numerically, the OCP is usually rewritten such that standard solvers can be applied (see Sects. 11.3–11.5). Here, the central observation is that (11.3) is a quadratic program (QP) for polyhedral constraints $\mathcal{X}, \mathcal{U},$ and $\mathcal{T}$.

11.2.2 Basics on Homomorphic Encryption and the Paillier Cryptosystem

We later use HE (see, e.g., [10]) or, more precisely, the Paillier cryptosystem [21] to implement encrypted control schemes. In general, HE stands for a special family of cryptosystems that enables certain mathematical operations to be carried out on encrypted data. More precisely, we call a cryptosystem additively homomorphic if there exists an operation "$\oplus$" such that

$$z_1 + z_2 = \mathrm{Dec}\left(\mathrm{Enc}(z_1) \oplus \mathrm{Enc}(z_2)\right) \tag{11.5}$$

holds, where z_1 and z_2 are two arbitrary numbers in the message space of the cryptosystem and where the functions Enc and Dec refer to the encryption and decryption procedure, respectively. Clearly, relation (11.5) allows to evaluate encrypted additions. Analogously, cryptosystems are called multiplicatively homomorphic if an operation "$\otimes$" exist such that encrypted multiplications can be realized using

$$z_1 z_2 = \mathrm{Dec}\left(\mathrm{Enc}(z_1) \otimes \mathrm{Enc}(z_2)\right). \tag{11.6}$$

A popular additively HE scheme is the Paillier cryptosystem that is detailed in the following. Multiplicatively homomorphic encryption is often implemented using ElGamal [7]. Encryption schemes that are both additively and multiplicatively homomorphic are called fully homomorphic. In principle, fully HE schemes can be used to encrypt arbitrary functions [10].

Most existing encrypted controllers (such as [1, 2, 8, 9, 15, 31, 32]) make use of the Paillier cryptosystem [21]. In this asymmetric encryption scheme, the encryption is carried out based on a public key P and the decryption requires the private key S. The key generation for the simplified Paillier scheme (see [14, Sect. 13.2]) builds on two primes $p_1, p_2 \in [2^{l-1}, 2^l - 1]$ of the same "length" $l \in \mathbb{N}$. The public key then is $P = p_1 p_2$ and the private key evaluates to $S = (p_1 - 1)(p_2 - 1)$. Now, the encryption of a number z from the message space $\mathbb{Z}_P$ is realized by

$$\mathrm{Enc}(z, r) := (P+1)^z r^P \bmod P^2, \tag{11.7}$$

where r is a random number that is picked from $\mathbb{Z}_P^*$ for every single encryption. The resulting ciphertext c lies in $\mathbb{Z}_{P^2}^*$. The decryption is carried out by computing

$$\text{Dec}(c) := \frac{(c^S \bmod P^2) - 1}{P} S^{-1} \bmod P, \tag{11.8}$$

where $S^{-1} \bmod P$ refers to the multiplicative inverse of S modulo P. In the following, we will restrict ourselves to messages from $\mathbb{N}_P$ i.e, the set of standard representatives of $\mathbb{Z}_P$. We then obtain $\text{Dec}(\text{Enc}(z, r)) = z$ for every $z \in \mathbb{N}_P$ and every $r \in \mathbb{Z}_P^*$, i.e., reversibility of the encryption. As already mentioned, the Paillier cryptosystem is additively homomorphic. In fact, we have

$$z_1 + z_2 = \text{Dec}\left(\text{Enc}(z_1, r_1)\,\text{Enc}(z_2, r_2) \bmod P^2\right). \tag{11.9}$$

for all $z_1, z_2 \in \mathbb{N}_P$ such that $z_1 + z_2 \in \mathbb{N}_P$ and all $r_1, r_2 \in \mathbb{Z}_P^*$. In addition, the Paillier cryptosystems supports multiplications with one encrypted factor as apparent from the relation

$$z_1 z_2 = \text{Dec}\left(\text{Enc}(z_1, r)^{z_2} \bmod P^2\right) \tag{11.10}$$

that holds for all $z_1, z_2 \in \mathbb{N}_P$ such that $z_1 z_2 \in \mathbb{N}_P$ and all $r \in \mathbb{Z}_P^*$. Relations (11.9) and (11.10) build the basis for many existing encrypted controllers and the schemes to be presented in Sects. 11.3–11.5.

11.2.3 Quantization via Fixed-Point Numbers

The previously presented Paillier cryptosystem is designed for integer messages. Hence, applying Paillier encryption to realize encrypted control requires to map the states x to the corresponding integer message space. This mapping usually starts with a quantization of the real-valued states (and controller parameters). Here, we approximate the states $x_j \in \mathbb{R}$ with fixed-point numbers from the set

$$\mathbb{Q}_{b,\gamma,\delta} := \left\{ -b^\gamma, -b^\gamma + b^{-\delta}, \ldots, b^\gamma - 2b^{-\delta}, b^\gamma - b^{-\delta} \right\},$$

where $b, \gamma, \delta \in \mathbb{N}$ with $b \geq 1$. The three parameters can be understood as the basis, the magnitude, and the resolution of the fixed-point numbers, respectively. For example, for $b = 10$ and $\gamma = \delta = 1$, we obtain $\mathbb{Q}_{b,\gamma,\delta} = \{-10, -9.9, \ldots, 9.8, 9.9\}$. Now, different user-defined mappings $h : \mathbb{R} \to \mathbb{Q}_{b,\gamma,\delta}$ can be used to compute fixed-point approximations of the form $\hat{x}_j := h(x_j)$. Rounding down can, for example, be realized by

$$h(x) := \max\left(\{-b^\gamma\} \cup \left\{ q \in \mathbb{Q}_{b,\gamma,\delta} \mid q \leq x \right\}\right).$$

However, we leave the specification of h to the user apart from the restriction that

$$|h(x) - x| \leq b^{-\delta} \quad \text{for all} \quad x \in [-b^\gamma, b^\gamma].$$

In other words, the quantization error should be limited by the resolution for real-valued data in the range of $\mathbb{Q}_{b,\gamma,\delta}$. It remains to address the mapping from $\mathbb{Q}_{b,\gamma,\delta}$ to the message space $\mathbb{N}_P$ of the Paillier cryptosystem. In this context, we first note that

$$b^\delta \, \mathbb{Q}_{b,\gamma,\delta} = \{-b^{\gamma+\delta}, -b^{\gamma+\delta} + 1, \dots, b^{\gamma+\delta} - 1\} \subset \mathbb{Z}.$$

Hence, scaling with b^δ maps $\mathbb{Q}_{b,\gamma,\delta}$ onto a subset of $\mathbb{Z}$. In order to obtain a mapping onto a subset of $\mathbb{N}_P$, we choose a number $Q \in \mathbb{N}_P$ with $Q \geq 2$ and define the function $f : \mathbb{Z} \to \mathbb{N}_Q$ as

$$f(z) := z \bmod Q = z - Q \left\lfloor \frac{z}{Q} \right\rfloor. \tag{11.11}$$

The combination of the quantization via h, the scaling with b^δ, and the mapping f leads to $f(b^\delta h(\boldsymbol{x}_j)) \in \mathbb{N}_Q \subseteq \mathbb{N}_P$. The resulting numbers can then be processed in the Paillier cryptosystem. In this context, it turns out to be useful that the relations

$$f(z_1 + z_2) = f(z_1) + f(z_2) \bmod Q \quad \text{and} \tag{11.12a}$$
$$f(z_1 z_2) = f(z_1) f(z_2) \bmod Q. \tag{11.12b}$$

hold for every $z_1, z_2 \in \mathbb{Z}$. Moreover, the function $\varphi : \mathbb{Z} \to \mathbb{Z}$ with

$$\varphi(z) := \begin{cases} z - Q & \text{if } z \geq \frac{Q}{2}, \\ z & \text{otherwise} \end{cases}$$

is a partial inverse of f. In fact, we have $z = \varphi(f(z))$ for every

$$z \in \mathbb{Z}_Q^\varphi := \left\{ -\left\lfloor \frac{Q}{2} \right\rfloor, -\left\lfloor \frac{Q}{2} \right\rfloor + 1, \dots, \left\lceil \frac{Q}{2} \right\rceil - 1 \right\} = \mathbb{N}_Q - \left\lfloor \frac{Q}{2} \right\rfloor.$$

Invertibility of f is required for correctness of encrypted control schemes. In this context, it is interesting to note that $|\mathbb{Z}_Q^\varphi| = Q$. Hence, the choice of Q determines the cardinality (or size) of the set where invertibility of f is given.

11.3 Encrypted MPC via Multi-parametric Programming

In this section, we summarize the encrypted MPC introduced in [32]. The approach is currently the only encrypted controller that can handle arbitrary polyhedral state constraints of the form

$$\mathcal{X} = \left\{ \boldsymbol{x} \in \mathbb{R}^n \mid \boldsymbol{\Phi}^{(x)} \boldsymbol{x} \leq \mathbf{1}_{q_x} \right\} \quad \text{and} \quad \mathcal{T} = \left\{ \boldsymbol{x} \in \mathbb{R}^n \mid \boldsymbol{\Phi}^{(t)} \boldsymbol{x} \leq \mathbf{1}_{q_t} \right\}, \tag{11.13}$$

where the matrices $\boldsymbol{\Phi}^{(x)} \in \mathbb{R}^{q_x \times n}$ and $\boldsymbol{\Phi}^{(t)} \in \mathbb{R}^{q_t \times n}$ reflect the hyperplanes characterizing $\mathcal{X}$ and $\mathcal{T}$. The OCP (11.3) with the constraint specifications (11.13) can be rewritten as the QP

$$z^*(x) = \arg\min_z \frac{1}{2} z^\top H z + x^\top F^\top z \tag{11.14a}$$

$$\text{s.t.} \quad Gz \le Ex + \mathbf{1}_{q_z} \tag{11.14b}$$

with the decision variables

$$z := \begin{pmatrix} \tilde{u}(0) \\ \vdots \\ \tilde{u}(N-1) \end{pmatrix} \in \mathbb{R}^{Nm} \tag{11.15}$$

and $q_z := N(q_x + 2m) + q_t$ linear inequality constraints. We refer to [3, Sect. 2.1] for details on the computation of the matrices E, F, G, and H. We note, however, that their computation is determined by the elimination of the state variables $\tilde{x}(0), \ldots, \tilde{x}(N)$ in (11.3). In fact, the remaining decision variables (11.15) solely reflect the predicted input sequence.

11.3.1 Offline Optimization Using Multi-parametric Programming

The QP (11.3) is parametrized by the current state $x = x(k)$. Since $x \in \mathbb{R}^n$ is, in general, a multi-dimensional vector, (11.3) is a so-called multi-parametric QP. Multi-parametric QPs of the form (11.3) have interesting properties. In fact, it is well-known that the optimizer $z^*(x)$ is a piecewise affine function of the parameter x. More precisely, $z^*(x)$ has the structure

$$z^*(x) = \begin{cases} L^{(1)}x + \beta^{(1)} & \text{if } x \in \mathcal{P}^{(1)}, \\ \vdots & \vdots \\ L^{(s)}x + \beta^{(s)} & \text{if } x \in \mathcal{P}^{(s)}, \end{cases} \tag{11.16}$$

where the individual affine segments are specified by the matrices $L^{(\sigma)} \in \mathbb{R}^{Nm \times n}$, the vectors $\beta b^{(\sigma)} \in \mathbb{R}^{Nm}$, and the polytopic domains $\mathcal{P}^{(\sigma)} \subseteq \mathcal{X}$. The number of segments s largely depends on the problem at hand but is bounded above by (the typically quite conservative bound) 2^{q_z} (see [3, Sect. 4.4]). Now, from a practical point of view, the structure (11.16) is particularly interesting since it can, in principle, be computed offline using MPP. A suitable procedure, which can be understood as a multi-parametric active-set solver, is described in [3, Sect. 4.3]. Having computed and stored the structure (11.16) offline, MPC can online be implemented as summarized in the following algorithm.

Algorithm 11.3.1: MPC using offline computed piecewise affine optimizers

(i) For the current state $x(k)$, the "active" segment σ is identified by localizing $x(k) \in \mathcal{P}^{(\sigma)}$.

(ii) The corresponding affine control law is applied by evaluating

$$u(k) = Cz^*(x(k)) = CL^{(\sigma)}x(k) + C\beta^{(\sigma)} \quad \text{with} \quad C := \begin{pmatrix} I_m & 0_{m \times (N-1)m} \end{pmatrix}.$$
(11.17)

Algorithm 11.3.1 provides an attractive implementation of MPC since online optimization is avoided. However, computing the piecewise affine structure (11.16) offline can only be carried out efficiently for a relatively small number of inequality constraints q_z in (11.14b). Moreover, storing the data associated with the s affine segments can be demanding depending on the hardware at hand. However, as apparent from (11.17), it is not required to completely store $L^{(\sigma)}$ and $\beta^{(\sigma)}$ for every segment. In fact, it is sufficient to store the entries relevant for the actual control action, i.e.,

$$K^{(\sigma)} := CL^{(\sigma)} \in \mathbb{R}^{m \times n} \quad \text{and} \quad b^{(\sigma)} := C\beta^{(\sigma)} \in \mathbb{R}^m.$$

Finally, identifying the currently active segment σ can be exhausting for large s. Yet, an efficient localization of $\mathcal{P}^{(\sigma)}$ can often be realized using binary search trees [34].

11.3.2 Encrypted Implementation in the Cloud

The encrypted implementation of the previously discussed MPC mainly builds on the observation that the second step in Algorithm 11.3.1 results in the evaluation of an affine control law of the form

$$u(k) = K^{(\sigma)}x(k) + b^{(\sigma)}.$$
(11.18)

Affine operations can be easily encrypted using the homomorphisms (11.9) and (11.10) of the Paillier cryptosystem. To see this, we first consider a quantized version of the i-th equation in (11.18), i.e.,

$$\breve{u}_i(k) := \hat{b}_i^{(\sigma)} + \sum_{j=1}^{n} \widehat{K}_{ij}^{(\sigma)} \hat{x}_j(k),$$
(11.19)

where $\hat{x}_j(k) := h(x_j(k))$, $\widehat{K}_{ij}^{(\sigma)} := h(K_{ij}^{(\sigma)})$, and $\hat{b}_i^{(\sigma)} := h(b_i^{(\sigma)})$. We note that $\breve{u}_i(k)$ is defined as the result of the computation on the right-hand side of (11.19) which is, in general, unequal to $h(u_i(k))$ with $u(k)$ from (11.18). Now, an encrypted evaluation of the quantized control law (11.19) can be realized as follows. At the sensor, the quantized states are encrypted by evaluating

$$c_j(k) := \text{Enc}\left(f\left(b^\delta \hat{x}_j(k)\right), r_j(k)\right) \tag{11.20}$$

for every $j \in \{1, \dots, n\}$ according to (11.7) using random numbers $r_j(k) \in \mathbb{Z}_p^*$. The n ciphertexts $c_j(k) \in \mathbb{Z}_{P^2}^*$ are then sent to the cloud, where the ciphertext

$$v_i^{(\sigma)}(k) := e_i^{(\sigma)} \prod_{j=1}^{n} c_j(k)^{Z_{ij}^{(\sigma)}} \bmod P^2 \tag{11.21}$$

is computed for every $i \in \{1, \dots, m\}$ with

$$Z_{ij}^{(\sigma)} := f\left(b^\delta \widehat{K}_{ij}^{(\sigma)}\right) \quad \text{and} \quad e_i^{(\sigma)} := \text{Enc}\left(f\left(b^{2\delta}\hat{b}_i^{(\sigma)}\right), r_i^{(\sigma)}\right)$$

using $r_i^{(\sigma)} \in \mathbb{Z}_P^*$. Finally, the m ciphertexts $v_i^{(\sigma)}(k)$ are transmitted to the actuator, where the decryption is carried out by computing

$$\check{u}_i(k) = b^{-2\delta} \varphi\left(\text{Dec}\left(v_i^{(\sigma)}(k)\right) \bmod Q\right) \tag{11.22}$$

with Dec as in (11.8). Now, by exploiting the reversibility of the encryption, the homomorphisms (11.9) and (11.10), and the invertibility of the mapping f by φ, it is straightforward to show that the computed inputs $\check{u}_i(k)$ in (11.19) and (11.22) are identical presupposed that the underlying conditions for reversibility, the homomorphisms and invertibility hold. As detailed in [32, Sect. 4] or [26, Sect. 3.2], these conditions are satisfied if

$$f\left(b^{2\delta}\hat{b}_i^{(\sigma)}\right) + \sum_{j=1}^{n} f\left(b^\delta \hat{x}_j(k)\right) Z_{ij}^{(\sigma)} \in \mathbb{N}_P \quad \text{and} \tag{11.23a}$$

$$b^{2\delta}\left(\hat{b}_i^{(\sigma)} + \sum_{j=1}^{n} \widehat{K}_{ij}^{(\sigma)} \hat{x}_j(k)\right) \in \mathbb{Z}_Q^\varphi \tag{11.23b}$$

for every $i \in \{1, \dots, m\}$. The conditions (11.23) obviously depend on the current state $x(k)$ or, more precisely, on its quantization. It is, however, possible to choose P, $\mathbb{Q}_{b,\gamma,\delta}$, and Q such that (11.23) holds for all (feasible) states $x(k)$. For example, taking into account that f maps onto $\mathbb{N}_Q$, a sufficient condition for (11.23a) is

$$(Q-1)\left(1 + \left\|\mathbf{Z}^{(\sigma)}\right\|_\infty\right) \le P - 1.$$

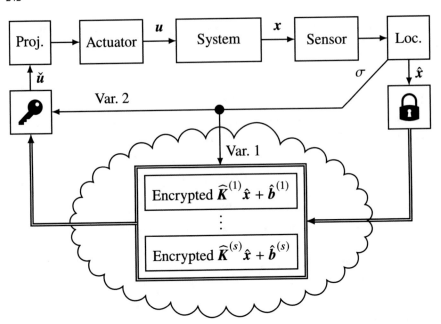

Fig. 11.3 Encrypted MPC via MPP. The piecewise affine MPC law is computed offline and implemented in an encrypted fashion. The localization (Loc.) of the current affine controller segment is carried out at the sensor. The projection (Proj.) of the computed quantized control action onto the input constraints $\mathcal{U}$ takes place at the actuator. Two variants of the encrypted controller can be distinguished based on the recipient of the identified segment σ

A similar time-invariant condition can be derived for (11.23b) when $x(k) \in \mathcal{X}$ is considered (cf. [26, Eq. (21)]).

Summing up, the encrypted evaluation of the quantized control law (11.19) can be realized by computing (11.20), (11.21), and (11.22) at the sensor, in the cloud, and at the actuator, respectively. In principle, these computations allows us to encrypt the second step in Algorithm 11.3.1. However, encrypting the first step of that algorithm is challenging and has not been realized yet. In [32], the following workaround is proposed. The localization $x(k) \in \mathcal{P}^{(\sigma)}$ is carried out at the sensor before the current state is quantized, encrypted, and sent to the cloud. The identified segment index σ is then either sent to the cloud (variant 1) or to the actuator (variant 2). In the first variant, the cloud evaluates the encrypted affine control law corresponding to the received index σ. In the second variant, the cloud evaluates the encrypted control laws for all segments and sends the resulting encrypted control actions to the actuator. There, only the encrypted input related to segment σ is decrypted and applied. The two variants of the encrypted MPC scheme are illustrated in Fig. 11.3.

Both variants include projections of the computed inputs $\check{u}(k)$ onto the input constraints $\mathcal{U}$. These projections are occasionally required since the quantizations may lead to $\check{u}(k) \notin \mathcal{U}$ although $u(k) = Cz^*(x(k)) \in \mathcal{U}$ is guaranteed for every feasible $x(k)$. In principle, the projections on $\mathcal{U}$ are easy to evaluate. In fact, due to the box-shaped input constraints $\mathcal{U}$ as in (11.2),

$$\text{proj}_{\mathcal{U}}(\check{u}) = \min \left\{ \max\{\check{u}, u_{\min}\}, u_{\max} \right\}. \tag{11.24}$$

However, an encrypted implementation of (11.24) is demanding. Hence, the projections are carried out at the actuator before applying the control actions (see Fig. 11.3). The potential differences between $\check{u}(k)$ and $u(k)$ not only require projections. In fact, they further cause deviations from the optimal control strategy. These deviations can be interpreted as disturbances of the optimally controlled system. More formally, the dynamics of the resulting closed-loop system can be described as

$$x(k+1) = Ax(k) + BCz^*\big(x(k)\big) + d(k), \tag{11.25}$$

where the first part refers to the dynamics of the optimally controlled system and where the additive disturbances

$$d(k) := B\left(\text{proj}_{\mathcal{U}}(\check{u}(k)) - Cz^*\big(x(k)\big)\right)$$

reflect the quantization errors. This interpretation is useful since the disturbances $d(k)$ can be kept arbitrarily small for suitable quantizations (in terms of $\mathbb{Q}_{b,\gamma,\delta}$ and h) under the assumption that $\mathcal{X}$ is bounded. The resulting bounds on $d(k)$ have two interesting applications. First, assuming asymptotic stability of the optimally controlled system, bounded disturbances allow to certify a posteriori robust stability of the quantized and encrypted control loop. Second, bounds on $d(k)$ can a priori be used to design robust MPC laws with stability and feasibility guarantees. We refer to [32] for technical details on a systematic approach for guaranteeing robustness.

11.4 Encrypted MPC via Real-Time Projected Gradient Scheme

This section summarizes the encrypted MPC introduced in [31]. The approach is tailored for systems with input constraints only. More precisely, we here consider the input constraints $\mathcal{U}$ from (11.2) and $\mathcal{X} = \mathcal{T} = \mathbb{R}^n$ (i.e., no state constraints). The OCP (11.3) can then be rewritten as the QP

$$z^*(x) = \arg\min_z \frac{1}{2} z^\top H z + x^\top F^\top z \tag{11.26a}$$

$$\text{s.t.} \quad z_{\min} \leq z \leq z_{\max} \tag{11.26b}$$

with z, H, and F as in (11.14) and

$$z_{\min} := \begin{pmatrix} u_{\min} \\ \vdots \\ u_{\min} \end{pmatrix} \in \mathbb{R}^{Nm} \quad \text{and} \quad z_{\max} := \begin{pmatrix} u_{\max} \\ \vdots \\ u_{\max} \end{pmatrix} \in \mathbb{R}^{Nm}.$$

Now, the box-constrained QP (11.26) can be easily solved with a PGS (see, e.g., [19, 20, 25]). The procedure is detailed in the next section.

11.4.1   Online Optimization Using Projected Gradients

Solving (11.26) using a PGS can be realized by recurringly evaluating the iterations

$$z^{(j+1)} := \text{proj}_{\mathcal{Z}} \left(z^{(j)} - \rho \left(H z^{(j)} + F x \right) \right), \tag{11.27}$$

where $H z^{(j)} + F x$ is the gradient of the objective function in (11.26a), where ρ is the step size, and where the projection onto

$$\mathcal{Z} := \{ z \in \mathbb{R}^{Nm} \mid z_{\min} \leq z \leq z_{\max} \}$$

reflects the constraints (11.26b). The choice of the parameter ρ affects the convergence (or divergence) of the scheme. Convergence of the iterates $z^{(j)}$ to the optimizer $z^*(x)$ is guaranteed for every $\rho \in \left(0, 2\lambda_{\max}^{-1}(H) \right)$ [20]. Obviously, the iterations (11.27) can only be implemented efficiently if the projection is numerically easy to evaluate. Analogously to (11.24), we here find

$$\text{proj}_{\mathcal{Z}}(\zeta) = \min \{ \max\{\zeta, z_{\min}\}, z_{\max} \}.$$

due to the box constraints $\mathcal{Z}$. Hence, the PGS (11.27) can indeed be implemented efficiently and used to numerically solve (11.26).

11.4.2   Dynamics for Real-Time Iterations

In general, multiple iterations (11.27) are required to approximate $z^*(x)$ with a certain accuracy. The required number of iterations may, in addition, depend on the current state x. For some applications, ensuring real-time capability may require to a priori fix a small number of iterations independent of x. In this context, an extreme case has been considered recently in [22]. In fact, in the proposed predictive control scheme, only a single iteration (11.27) is applied at every sampling instant. At first sight, such a coarse optimization seems to be doomed to fail. However, in the framework of optimization-based control, a single iteration per time-step can be sufficient since additional iterations follow at future sampling instances. In the resulting setup, solver iterations are coupled with time-triggered controller iterations and thus called real-time iterations. MPC via real-time iterations has not only been proposed in [22] but also in other works (see, e.g., [6]). However, among these real-time iteration schemes, [22] stands out in that closed-loop stability under (input) constraints has been guaranteed for the first time.

In the following, we briefly summarize the stability analysis in [22]. At first, the real-time iterations derived from (11.27) can be written as

$$z^{(1)}(k) = \text{proj}_{\mathcal{Z}}\left(z^{(0)}(k) - \rho\left(Hz^{(0)}(k) + Fx(k)\right)\right), \qquad (11.28)$$

which expresses that, in every time-step k, a single iteration is evaluated that delivers $z^{(1)}(k)$ based on $z^{(0)}(k)$ and the current state $x(k)$. Taking the definition of z from (11.15) and the optimal feedback (11.4) into account, we then apply the input

$$u(k) = Cz^{(1)}(k) \qquad (11.29)$$

with C as in (11.17). At the next sampling instant, we measure the state $x(k+1)$ and repeat the procedure. As apparent from (11.28), the computation of $z^{(1)}(k)$ and $z^{(1)}(k+1)$ in step k and $k+1$ requires the specification of $z^{(0)}(k)$ and $z^{(0)}(k+1)$, respectively. In principle, the "initial guesses" $z^{(0)}(k)$ and $z^{(0)}(k+1)$ can be chosen freely. However, it turns out to be useful to utilize the result $z^{(1)}(k)$ from step k for the specification of $z^{(0)}(k+1)$. This coupling can be understood as a warmstart of the solver iterations. We here restrict ourselves to warmstarts of the form

$$z^{(0)}(k+1) = Dz^{(1)}(k), \qquad (11.30)$$

i.e., linear dependencies between $z^{(0)}(k+1)$ and $z^{(1)}(k)$. The closed-loop dynamics can then be described based on the augmented system

$$\xi(k+1) = \mathcal{A}\xi(k) + \mathcal{B}\,\text{proj}_{\mathcal{Z}}\left(\mathcal{K}\xi(k)\right), \qquad \xi(0) := \xi_0 \qquad (11.31)$$

with the augmented states

$$\xi := \begin{pmatrix} x \\ z^{(0)} \end{pmatrix} \in \mathbb{R}^{n+p}, \qquad (11.32)$$

the augmented systems matrices

$$\mathcal{A} := \begin{pmatrix} A & 0_{n\times p} \\ 0_{p\times n} & 0_{p\times p} \end{pmatrix}, \quad \mathcal{B} := \begin{pmatrix} BC \\ D \end{pmatrix}, \quad \text{and} \quad \mathcal{K} := \begin{pmatrix} -\rho F & I_p - \rho H \end{pmatrix}, \quad (11.33)$$

and $p := Nm$. Apart from the MPC parameters N, P, Q, and R, the resulting dynamics can be influenced by the choice of the step size ρ, the update matrix D, and the initialization $z_0^{(0)}$. In [22], stabilizing parameter choices are provided for open-loop stable systems (11.1). In fact, under the assumption that the system matrix A is (Schur) stable, choosing P as the solution of the Lyapunov equation

$$A^\top P A - P + Q = \mathbf{0}_{n \times n}, \tag{11.34}$$

choosing any $\rho \in \left(0, 2\lambda_{\max}^{-1}(H)\right)$, and choosing the update matrix

$$D := \begin{pmatrix} \mathbf{0}_{(p-m) \times m} & I_{p-m} \\ \mathbf{0}_{m \times m} & \mathbf{0}_{m \times (p-m)} \end{pmatrix} \tag{11.35}$$

guarantees asymptotic stability of the closed-loop system for arbitrary (positive) prediction horizons N, (positive definite) weighting matrices Q and R, and arbitrary initializations $z_0^{(0)}$ (see [22, Sect. IV.C]). The stability proof builds on the observation that the objective function (11.3a) of the original OCP, that can be expressed as

$$W(\xi) := \frac{1}{2}\left(z^{(0)}\right)^\top H z^{(0)} + x^\top F^\top z^{(0)}$$
$$+ x^\top \left(\left(A^N\right)^\top P A^N + \sum_{\kappa=0}^{N-1} \left(A^\kappa\right)^\top Q A^\kappa\right) x,$$

is a Lypunov function for the augmented system (11.31). In fact, as apparent from

$$W(\xi(k+1)) - W(\xi(k)) + x(k)^\top Q x(k) \le -\alpha \left\| z^{(0)}(k) - z^*(x(k)) \right\|_2^2 \tag{11.36}$$

with $\alpha > 0$, f is decreasing along every system trajectory (see [22, Sect. IV.B] for details). Relation (11.36) is similar to standard Lyapunov conditions for exact MPC (see, e.g., [17, Eq. (3.1)]). However, the term on the right-hand side in (11.36) reflects the inexact optimization applied here. More precisely, this term reflects the objective function decrease resulting from a single PGS iteration (11.27) that is analyzed in [22, Sect. III.C] in more detail. In the following, we concentrate on the encrypted implementation of the scheme. As preparation, we summarize the proposed control law in the following algorithm.

Algorithm 11.4.1: MPC using real-time PGS iterations

(i) For the current state $x(k)$ and the current $z^{(0)}(k)$, iteration (11.28) is evaluated, i.e,

$$z^{(1)}(k) = \text{proj}_{\mathcal{Z}}\left(-\rho F x(k) + (I_p - \rho H)z^{(0)}(k)\right) = \text{proj}_{\mathcal{Z}}\left(\mathcal{K}\xi(k)\right). \tag{11.37}$$

(ii) The control action (11.29) is applied and the update (11.30) is performed.

11.4.3   Encrypted Implementation in the Cloud

We observed in Sect. 11.3.2 that affine operations can be easily encrypted while projections are hard to encrypt. Obviously, the derived real-time iteration (11.37) is composed of a linear mapping and a projection. Hence, it is hard to completely encrypt (11.37). In [31] an encrypted implementation is thus realized as follows. In the cloud, the linear mapping $\zeta(k) = \mathcal{K}\xi(k)$ or, more precisely, its quantized version

$$\check{\zeta}_i(k) := \sum_{j=1}^{p} \widehat{\mathcal{K}}_{ij}\hat{\xi}_j(k) \tag{11.38}$$

is evaluated in an encrypted fashion, where $\widehat{\mathcal{K}}_{ij} := h(\mathcal{K}_{ij})$ and $\hat{\xi}_j(k) := h(\xi_j(k))$. The encrypted result $\check{\zeta}(k)$ is then send to the actuator, where it is decrypted and where the projection

$$\check{z}^{(1)}(k) := \mathrm{proj}_{\mathcal{Z}}\left(\check{\zeta}(k)\right) \tag{11.39}$$

is carried out. Next, the input $u(k) = C\check{z}^{(1)}(k)$ is applied, the update $z^{(0)}(k+1) = D\check{z}^{(1)}(k)$ is executed, and $z^{(0)}(k+1)$ is send to the sensor. There, the next state $x(k+1)$ is measured and the augmented state $\xi(k+1)$ is formed according to (11.32). Finally, the augmented state is quantized, encrypted, and send to the cloud, where the procedure starts anew. The resulting control loop is illustrated in Fig. 11.4.

It remains to briefly comment on the encryption details. Encrypting the augmented states is realized analogously to (11.20). In fact, at the sensor, we compute

$$c_j(k) := \mathrm{Enc}\left(f\left(b^\delta \hat{\xi}_j(k)\right), r_j(k)\right) \tag{11.40}$$

for every $j \in \{1, \ldots, n+p\}$ using random numbers $r_j(k) \in \mathbb{Z}_p^*$. In the cloud, the ciphertexts

$$v_i(k) := \prod_{j=1}^{n+p} c_j(k)^{Z_{ij}} \bmod P^2 \tag{11.41}$$

are evaluated for every $i \in \{1, \ldots, p\}$ with $Z_{ij} := f\left(b^\delta \widehat{\mathcal{K}}_{ij}\right)$. Finally, the encrypted vector $v_i(k)$ is send to the actuator, where $\check{\zeta}(k)$ is recovered by computing

$$\check{\zeta}_i(k) = b^{-2\delta}\varphi\left(\mathrm{Dec}(v_i(k)) \bmod Q\right) \tag{11.42}$$

for every $i \in \{1, \ldots, p\}$. Similar to (11.23), correctness of the encrypted control scheme is guaranteed if the conditions

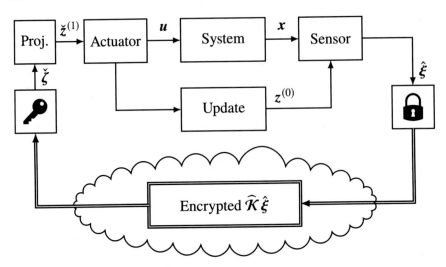

Fig. 11.4 Encrypted MPC via real-time PGS iterations. The linear part of the iteration (11.37) is implemented in an encrypted fashion. The projection (Proj.) of the intermediate results $\check{\zeta}$ onto the constraints $\mathcal{Z}$ is carried out at the actuator. The resulting approximation $\check{z}^{(1)}$ of the optimizer $z^*(x)$ is used to define the input u and the initialization $z^{(0)}$ for the following step. At the sensor, the augmented state is formed, encrypted, and send to the cloud

$$\sum_{j=1}^{n+p} f\left(b^\delta \hat{\xi}_j(k)\right) Z_{ij} \in \mathbb{N}_P \qquad \text{and} \qquad (11.43a)$$

$$b^{2\delta} \sum_{j=1}^{n+p} \widehat{\mathcal{K}}_{ij} \hat{\xi}_j(k) \in \mathbb{Z}_Q^\varphi \qquad\qquad (11.43b)$$

hold for every $i \in \{1, \ldots, p\}$ (cf. [31, Eq. (25)]). Similar to (11.25), the robustness of the scheme can be investigated based on the closed-loop system

$$\xi(k+1) = \mathcal{A}\xi(k) + \mathcal{B}\operatorname{proj}_{\mathcal{Z}}\left(\mathcal{K}\xi(k)\right) + d(k) \qquad (11.44)$$

with the additive disturbances

$$d(k) := \mathcal{B}\left(\operatorname{proj}_{\mathcal{Z}}\left(\widehat{\mathcal{K}}\hat{\xi}(k)\right) - \operatorname{proj}_{\mathcal{Z}}\left(\mathcal{K}\xi(k)\right)\right).$$

A detailed analysis of the disturbed system can be found in [31].

11.5 Encrypted MPC via Real-Time ADMM

In this section, we present a novel encrypted MPC that can be seen as an extension of the scheme from Sect. 11.4 to systems with state and input constraints. In fact, in addition to the input constraints from (11.2), we here consider state constraints of the form

$$\mathcal{X} = \mathcal{T} = \left\{ x \in \mathbb{R}^n \mid x_{\min} \le x \le x_{\max} \right\} \tag{11.45}$$

with $x_{\min} < \mathbf{0}_n < x_{\max}$. In general, the presence of state constraints precludes rewriting the OCP (11.3) as in (11.26). Hence, the PGS from the previous section cannot be applied efficiently. The OCP (11.3) with $\mathcal{X}$ as in (11.45) can, however, be solved easily using ADMM. Applying ADMM requires to rewrite (11.3) in terms of a decomposed optimization problem. Different decompositions have been discussed in the literature (see, e.g., [11, 13]). Here, we follow the approach from [13] that is based on reformulating (11.3) in terms of the QP

$$z^*(x) = \arg\min_z \frac{1}{2} z^\top H z \tag{11.46a}$$

$$\text{s.t.} \quad z_{\min} \le z \le z_{\max}, \tag{11.46b}$$

$$Gz = Ex \tag{11.46c}$$

with the decision variables and bounds

$$z := \begin{pmatrix} \tilde{u}(0) \\ \vdots \\ \tilde{u}(N-1) \\ \tilde{x}(1) \\ \vdots \\ \tilde{x}(N) \end{pmatrix} \in \mathbb{R}^p, \quad z_{\min} := \begin{pmatrix} u_{\min} \\ \vdots \\ u_{\min} \\ x_{\min} \\ \vdots \\ x_{\min} \end{pmatrix} \in \mathbb{R}^p, \quad \text{and} \quad z := \begin{pmatrix} u_{\max} \\ \vdots \\ u_{\max} \\ x_{\max} \\ \vdots \\ x_{\max} \end{pmatrix} \in \mathbb{R}^p,$$

$$\tag{11.47}$$

respectively, where $p := Nm + Nn$. We refer to [27, Eq. 11] for details on the matrices H, G, and E. We further note that (11.46) differs from (11.26) in particular by means of the equality constraints (11.46c). A suitable decomposition of the QP (11.46) is detailed in the next section.

11.5.1 Online Optimization Using ADMM

To prepare the application of ADMM, we first rewrite (11.46) in terms of an unconstrained optimization problem. With the help of indicator functions, we find

$$z^*(x) = \arg \min_z \frac{1}{2} z^\top H z + \mathcal{I}_{\mathcal{Z}}(z) + \mathcal{I}_{\mathcal{E}(x)}(z),$$

where the involved sets are defined as

$$\mathcal{Z} := \{z \in \mathbb{R}^p \mid z_{\min} \leq z \leq z_{\max}\} \quad \text{and} \quad \mathcal{E}(x) := \{z \in \mathbb{R}^p \mid Gz = Ex\}.$$

Following the approach in [13], we then introduce the "copy" y of z in order to derive the decomposed optimization problem

$$\min_{y,z} \frac{1}{2} y^\top H z + \mathcal{I}_{\mathcal{Z}}(z) + \mathcal{I}_{\mathcal{E}(x)}(y) \tag{11.48a}$$

$$\text{s.t.} \qquad y = z. \tag{11.48b}$$

Now, ADMM builds on the augmented Lagrangian

$$L_\rho(y, z, \mu) := \frac{1}{2} y^\top H y + \mathcal{I}_{\mathcal{Z}}(z) + \mathcal{I}_{\mathcal{E}(x)}(y) + \mu^\top (y - z) + \frac{\rho}{2} \|y - z\|_2^2$$

of (11.48) and consists of the three iterations

$$y^{(j+1)} := \arg \min_y L_\rho\left(y, z^{(j)}, \mu^{(j)}\right)$$

$$= \arg \min_y \frac{1}{2} y^\top (H + \rho I_p) y + (\mu^{(j)} - \rho z^{(j)})^\top y + \mathcal{I}_{\mathcal{E}(x)}(y) \tag{11.49a}$$

$$z^{(j+1)} := \arg \min_z L_\rho\left(y^{(j+1)}, z, \mu^{(j)}\right)$$

$$= \arg \min_z \mathcal{I}_{\mathcal{Z}}(z) + \frac{\rho}{2} \left\| y^{(j+1)} + \frac{1}{\rho} \mu^{(j)} - z \right\|_2^2 \tag{11.49b}$$

$$\mu^{(j+1)} := \mu^{(j)} + \rho\left(y^{(j+1)} - z^{(j+1)}\right) \tag{11.49c}$$

related to the three variables y, z, and μ [5, Sect. 3.1]. The first iteration (11.49a) can be written as an equality-constrained QP. Hence, the optimizer $y^{(j+1)}$ can be inferred from the equation

$$\begin{pmatrix} H + \rho I_p & G^\top \\ G & 0_{q \times q} \end{pmatrix} \begin{pmatrix} y^{(j+1)} \\ * \end{pmatrix} = \begin{pmatrix} \rho z^{(j)} - \mu^{(j)} \\ Ex \end{pmatrix}.$$

Due to H being positive definite, ρ being positive, and G having full rank, the inverse

$$\Gamma := \begin{pmatrix} \Gamma_{11} & \Gamma_{12} \\ \Gamma_{12}^\top & \Gamma_{22} \end{pmatrix} = \begin{pmatrix} H + \rho I_p & G^\top \\ G & 0_{q \times q} \end{pmatrix}^{-1}$$

is well-defined and we obtain

$$y^{(j+1)} = \Gamma_{11}(\rho z^{(j)} - \mu^{(j)}) + \Gamma_{12} Ex. \tag{11.50}$$

The second iteration (11.49b) refers to the proximal operator for the indicator function $\mathcal{I}_{\mathcal{Z}}$. Thus,

$$z^{(j+1)} = \text{prox}_{\mathcal{I}_{\mathcal{Z}},\rho}\left(y^{(j+1)} + \frac{1}{\rho}\mu^{(j)}\right) = \text{proj}_{\mathcal{Z}}\left(y^{(j+1)} + \frac{1}{\rho}\mu^{(j)}\right). \tag{11.51}$$

Interestingly, the resulting iterations do not depend on $y^{(j)}$. Hence, we can eliminate the copy y by substituting $y^{(j+1)}$ from (11.50) in (11.51) and (11.49c). The two remaining iterations then are

$$z^{(j+1)} = \text{proj}_{\mathcal{Z}}\left(\Gamma_{11}(\rho z^{(j)} - \mu^{(j)}) + \Gamma_{12} Ex + \frac{1}{\rho}\mu^{(j)}\right), \tag{11.52a}$$

$$\mu^{(j+1)} = \mu^{(j)} + \rho(\Gamma_{11}(\rho z^{(j)} - \mu^{(j)}) + \Gamma_{12} Ex - z^{(j+1)}). \tag{11.52b}$$

The iterations (11.52) are known to converge to the optimum of (11.46) for any $\rho > 0$ (see [13, Sect. III.B] and references therein). Hence, MPC with state and input constraints can be implemented based on (11.52).

11.5.2 Dynamics for Real-Time Iterations

As for the PGS iterations (11.27), we usually require multiple ADMM iterations (11.52) to approximate $z^*(x)$ with a certain accuracy. However, analogously to Sect. 11.4.2, we here consider a single iteration (11.52) per time-step. The dynamics of the resulting closed-loop system have recently been analyzed in [28]. In the following, we summarize the central results from [28] to prepare the encrypted implementation of the real-time ADMM iterations in Sect. 11.5.3.

At first, we specify the real-time iterations

$$z^{(1)}(k) = \text{proj}_{\mathcal{Z}}\left(\Gamma_{11}(\rho z^{(0)}(k) - \mu^{(0)}(k)) + \Gamma_{12} Ex(k) + \frac{1}{\rho}\mu^{(0)}(k)\right), \tag{11.53a}$$

$$\mu^{(1)}(k) = \mu^{(0)}(k) + \rho(\Gamma_{11}(\rho z^{(0)}(k) - \mu^{(0)}(k)) + \Gamma_{12} Ex(k) - z^{(1)}(k)). \tag{11.53b}$$

analogously to (11.28). Similar to (11.17) and (11.29), we then apply the input

$$u(k) = Cz^{(1)}(k) \quad \text{with} \quad C := \begin{pmatrix} I_m & 0_{m \times (p-m)}. \end{pmatrix} \tag{11.54}$$

We note that the C-matrices in (11.17) and (11.54) differ in terms of the number of columns. Clearly, this difference results from the different dimensions of the decision variables z in (11.15) and (11.47). Now, as in (11.30), the initializations $z^{(0)}(k)$ and $\mu^{(0)}(k)$ for every real-time iteration can, in principle, be chosen freely. However, inspired by (11.30), we consider warmstarts of the form

$$z^{(0)}(k+1) = D_z z^{(1)}(k), \tag{11.55a}$$

$$\mu^{(0)}(k+1) = D_\mu \mu^{(1)}(k). \tag{11.55b}$$

The resulting closed-loop dynamics can again be described by an augmented system with the structure (11.31). Yet, the corresponding augmented states and the augmented systems matrices differ from (11.32) and (11.33), respectively. In fact, we here find the augmented states

$$\xi := \begin{pmatrix} x \\ z^{(0)} \\ \mu^{(0)} \end{pmatrix} \in \mathbb{R}^{n+2p}$$

and the augmented system matrices

$$\mathcal{A} := \begin{pmatrix} \begin{pmatrix} A & 0_{n\times 2p} \end{pmatrix} \\ 0_{p\times(n+2p)} \\ \rho D_\mu \mathcal{K} \end{pmatrix}, \quad \mathcal{B} := \begin{pmatrix} BC_u \\ D_z \\ -\rho D_\mu \end{pmatrix}, \quad \mathcal{K} := \begin{pmatrix} \Gamma_{12}E & \rho\Gamma_{11} & \frac{1}{\rho}I_p - \Gamma_{11} \end{pmatrix}.$$

Due to the state constraints, the stability analysis for the augmented system here is more difficult than in Sect. 11.4.2 respectively [22]. Nevertheless, asymptotic stability can be guaranteed locally around the augmented origin. To see this, we first note that the origin 0_{n+2p} is an interior point of $\mathcal{Z}$. As a consequence, the product $\mathcal{K}\xi$ will be contained in $\mathcal{Z}$ for small augmented states ξ. Hence, for those states, we find $\text{proj}_{\mathcal{Z}}(\mathcal{K}\xi) = \mathcal{K}\xi$. As a result, the dynamics of the augmented system (11.31) around the origin become linear, i.e.,

$$\xi(k+1) = (\mathcal{A} + \mathcal{B}\mathcal{K})\xi(k). \tag{11.56}$$

The linear dynamics are asymptotically stable if and only if the matrix $\mathcal{S} := \mathcal{A} + \mathcal{B}\mathcal{K}$ is (Schur) stable. It can be shown that there always exists a suitable parameter $\rho > 0$ such that $\mathcal{S}$ is stable [27, Proposition 6]. Interestingly, this result holds independent of the choices for D_z and D_μ. It is further interesting that the region where the local stability result holds can be specified. In fact, it is easy to see that the positively invariant set

$$\mathcal{R} := \{\xi \in \mathbb{R}^{n+2p} \mid \mathcal{K}\mathcal{S}^k\xi \in \mathcal{Z}, \forall k \in \mathbb{N}\}$$

represents the subset of the domain of attraction where the linear dynamics (11.56) apply. While stability of $\mathcal{S}$ and non-emptiness of $\mathcal{R}$ hold for a suitable ρ and arbitrary D_z and D_μ, the dynamics of the closed-loop system and the volume of $\mathcal{R}$ significantly change with the choice of the parameters ρ, D_z, and D_μ. We refer to [28] for the numerical evaluation of different parameter sets. For the numerical example in Sect. 11.6.2, we here choose $\rho := 1$,

$$D_z := \begin{pmatrix} \mathbf{0}_{(N-1)m \times m} & I_{(N-1)m} & \mathbf{0}_{(N-1)m \times n} & \mathbf{0}_{(N-1)m \times (N-2)n} & \mathbf{0}_{(N-1)m \times n} \\ \mathbf{0}_{m \times m} & \mathbf{0}_{m \times (N-1)m} & \mathbf{0}_{m \times n} & \mathbf{0}_{m \times (N-2)n} & K^* \\ \mathbf{0}_{(N-2)n \times m} & \mathbf{0}_{(N-2)n \times (N-2)m} & \mathbf{0}_{(N-2)n \times n} & I_{(N-2)n} & \mathbf{0}_{(N-2)n \times n} \\ \mathbf{0}_{n \times m} & \mathbf{0}_{n \times (N-2)m} & \mathbf{0}_{n \times n} & \mathbf{0}_{n \times (N-2)n} & I_n \\ \mathbf{0}_{n \times m} & \mathbf{0}_{n \times (N-2)m} & \mathbf{0}_{n \times n} & \mathbf{0}_{n \times (N-2)n} & A + BK^* \end{pmatrix},$$

$$(11.57)$$

$$D_\mu := \begin{pmatrix} \mathbf{0}_{(N-1)m \times m} & I_{(N-1)m} & \mathbf{0}_{(N-1)m \times n} & \mathbf{0}_{(N-1)m \times (N-2)n} & \mathbf{0}_{(N-1)m \times n} \\ \mathbf{0}_{m \times m} & \mathbf{0}_{m \times (N-1)m} & \mathbf{0}_{m \times n} & \mathbf{0}_{m \times (N-2)n} & \mathbf{0}_{m \times n} \\ \mathbf{0}_{(N-2)n \times m} & \mathbf{0}_{(N-2)n \times (N-2)m} & \mathbf{0}_{(N-2)n \times n} & I_{(N-2)n} & \mathbf{0}_{(N-2)n \times n} \\ \mathbf{0}_{n \times m} & \mathbf{0}_{n \times (N-2)m} & \mathbf{0}_{n \times n} & \mathbf{0}_{n \times (N-2)n} & I_n \\ \mathbf{0}_{n \times m} & \mathbf{0}_{n \times (N-2)m} & \mathbf{0}_{n \times n} & \mathbf{0}_{n \times (N-2)n} & \mathbf{0}_{n \times n} \end{pmatrix},$$

$$(11.58)$$

where K^* refers to the controller gain of the linear quadratic regulator (LQR). In other words, to obtain $z^{(0)}(k+1)$, we eliminate the elements associated with $\tilde{u}(0)$ and $\tilde{x}(1)$ from $z^{(1)}(k)$, we shift the remaining input and state sequences by one time step, and we add an LQR step to complete $z^{(0)}(k+1)$. A similar shift is applied to derive $\mu^{(0)}(k+1)$ from $\mu^{(1)}(k)$. However, we complete the set of Lagrange multipliers $\mu^{(0)}(k+1)$ by adding zero entries. The resulting real-time ADMM-based MPC is summarized in the following algorithm.

Algorithm 11.5.1: MPC using real-time ADMM iterations

(i) For the current state $x(k)$ and the current variables $z^{(0)}(k)$ and $\mu^{(0)}(k)$, the iterations (11.53) are evaluated, i.e.,

$$z^{(1)}(k) = \text{proj}_{\mathcal{Z}} \left(\Gamma_{12} Ex(k) + \rho \Gamma_{11} z^{(0)}(k) + \left(\frac{1}{\rho} I_p - \Gamma_{11} \right) \mu^{(0)}(k) \right)$$
$$= \text{proj}_{\mathcal{Z}} \left(\mathcal{K} \xi(k) \right), \qquad (11.59a)$$
$$\mu^{(1)}(k) = \rho \Gamma_{12} Ex(k) + \left(I_p - \rho \Gamma_{11} \right) \mu^{(0)}(k) + \rho^2 \Gamma_{11} z^{(0)}(k) - \rho z^{(1)}(k)$$
$$= \rho \left(\mathcal{K} \xi(k) - z^{(1)}(k) \right). \qquad (11.59b)$$

(ii) The control action (11.54) is applied and the updates (11.55) are performed.

11.5.3 Encrypted Implementation in the Cloud

In principle, the encrypted and cloud-based implementation of ADMM-based MPC can be realized analogously to the encryption of the PGS in Sect. 11.4.3. An efficient implementation should, however, reflect the two following characteristics of Algorithm 11.5.2. First, a projection is only required for the evaluation of $z^{(1)}(k)$ (and not for $\mu^{(1)}(k)$). Second, $\mu^{(1)}(k)$ is not required to compute the input $u(k)$. As a consequence, it turns out to be sufficient that the cloud sends $\zeta(k) := \mathcal{K}\xi(k)$ to the actuator (and not $\mu^{(1)}(k)$). Moreover, it will not be required to send the full augmented state $\xi(k)$ from the sensor to the actuator.

In order to detail the proposed implementation, we initially focus on the cloud-based realization. The encryption will be addressed afterwards. As in Sect. 11.4.3, the cloud mainly computes the argument $\zeta(k)$ of the projection required for the computation of $z^{(1)}(k)$. In this context, the key observation is that $\zeta(k)$ can not only be obtained by calculating $\mathcal{K}\xi(k)$. In fact,

$$\zeta(k) = \mathcal{L} \begin{pmatrix} x(k) \\ z^{(1)}(k-1) \\ \zeta(k-1) \end{pmatrix} \quad \text{with} \quad \mathcal{L} := \mathcal{K} \begin{pmatrix} I_n & 0_{n\times p} & 0_{n\times p} \\ 0_{p\times n} & D_z & 0_{p\times p} \\ 0_{p\times n} & -\rho D_\mu & \rho D_\mu \end{pmatrix}. \quad (11.60)$$

To see this, we note that

$$\mathcal{K}\xi(k) = \mathcal{K} \begin{pmatrix} x(k) \\ z^{(0)}(k) \\ \mu^{(0)}(k) \end{pmatrix} = \mathcal{K} \begin{pmatrix} x(k) \\ D_z z^{(1)}(k-1) \\ D_\mu \mu^{(1)}(k-1) \end{pmatrix}$$

$$= \mathcal{K} \begin{pmatrix} x(k) \\ D_z z^{(1)}(k-1) \\ \rho D_\mu \left(\zeta(k-1) - z^{(1)}(k-1) \right) \end{pmatrix}$$

according to (11.59) and by definition of ζ. Now, relation (11.60) enables the following efficient implementation of Algorithm 11.5.2. At time step k, the cloud computes $\zeta(k)$ according to (11.60). The result is submitted to the actuator but also stored in the cloud. At the actuator, the projection of $\zeta(k)$ onto $\mathcal{Z}$ is evaluated. The resulting $z^{(1)}(k)$ is used to derive $u(k)$. In addition, $z^{(1)}(k)$ is additionally forwarded to the sensor. At step $k + 1$, the sensor measures the current state $x(k + 1)$ and sends it to the cloud together with $z^{(1)}(k)$ from the previous step. In the cloud, the controller evaluation starts anew with the computation of $\zeta(k + 1)$ based on $x(k + 1)$, $z^{(1)}(k)$, and $\zeta(k)$. An illustration of the control loop is given in Fig. 11.5.

It remains to comment on the initialization of the control scheme. At the initial step $k = 0$, we assume that no information on $\zeta(-1)$ and $z^{(1)}(-1)$ is available to the cloud and the sensor, respectively. The cloud will consequently compute $\zeta(0)$ based on the original relation $\mathcal{K}\xi(0)$. To this end, the sensor measures and sends $x(0)$. Moreover, it provides the user-defined initialization $z^{(0)}(0)$ (instead of $z^{(1)}(-1)$). Setting $\mu^{(0)}(0) = 0_p$, the cloud then computes

$$\zeta(0) = \mathcal{K}\xi(0) = \Gamma_{12} E x(0) + \rho \Gamma_{11} z^{(0)}(0).$$

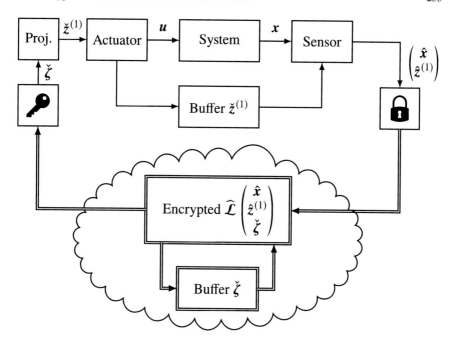

Fig. 11.5 Encrypted MPC via real-time ADMM iterations. Relation (11.60) is implemented in an encrypted fashion. The projection (Proj.) of the intermediate results $\check{\zeta}$ onto the constraints $\mathcal{Z}$ is carried out at the actuator. The resulting approximation $\hat{z}^{(1)}$ of the optimizer $z^*(x)$ is used to define the input u and additionally forwarded to the sensor. At the sensor, the current state x is measured and augmented with the previously buffered $\hat{z}^{(1)}$. The resulting vector is encrypted and send to the cloud

We next address the encrypted implementation of the proposed scheme. We focus on the steps $k > 0$ and omit the initialization at $k = 0$ for brevity. Again, we first quantize the central controller operation. Here, quantizing (11.60) yields

$$\check{\zeta}_i(k) := \sum_{j=1}^{n} \widehat{\mathcal{L}}_{ij} \hat{x}_j(k) + \sum_{j=n+1}^{n+p} \widehat{\mathcal{L}}_{ij} \hat{z}_{j-n}^{(1)}(k-1) + \sum_{j=n+p+1}^{n+2p} \widehat{\mathcal{L}}_{ij} \check{\zeta}_{j-n-p}(k-1) \quad (11.61)$$

for every $i \in \{1, \ldots, p\}$, where $\widehat{\mathcal{L}}_{ij} := h(\mathcal{L}_{ij})$, $\hat{x}_j(k) := h(x_j(k))$, and $\hat{z}_j^{(1)}$ $(k-1) := h(z_j^{(1)}(k-1))$ are standard quantizations. The terms $\check{\zeta}_i(k)$ and $\check{\zeta}_i(k-1)$ are slightly nonstandard. In fact, the computed value $\check{\zeta}_i(k)$ may or may not be contained in $\mathbb{Q}_{b,\gamma,\delta}$. A similar observation has already been made with respect to (11.19). Here, however, the observation is more important since $\check{\zeta}_i(k)$ is reused in the next time step. As a consequence, we have to analyze $\check{\zeta}_i(k)$ more carefully. In principle, studying the products $\widehat{\mathcal{L}}_{ij} \hat{x}_j(k)$ is sufficient for understanding the nature of $\check{\zeta}_i(k)$. In fact, multiplying two numbers from a quantization grid with a resolution of $b^{-\delta}$, i.e., $\widehat{\mathcal{L}}_{ij}, \hat{x}_j(k) \in \mathbb{Q}_{b,\gamma,\delta}$, yields a product that may refer to a quantization with a resolution of $b^{-2\delta}$. Roughly speaking, this observation is the reason for the factor $b^{-2\delta}$ in (11.22) and (11.42). Now, if the intermediate result $\check{\zeta}_i(k)$ requires a resolution of $b^{-2\delta}$, the reutilization of $\check{\zeta}_i(k)$ in the subsequent time step may demand a resolution of $b^{-3\delta}$ for $\check{\zeta}_i(k+1)$. Hence, for $k \to \infty$, the required resolution will be infinitely small and thus impractical. In order to avoid this situation, we refresh the quantization of $\check{\zeta}_i(k)$ every $\Delta k \geq 1$ time steps. The role of Δk will specified in the following.

Analogously to Sects. 11.3.2 and 11.4.3, the encrypted evaluation of (11.61) starts at the sensor with the encryption of (the augmented) states. For time steps $k > 0$ with $k \bmod \Delta k \neq 0$, we compute the ciphertexts

$$c_j(k) := \begin{cases} \mathrm{Enc}\left(f\left(b^{(1+(k \bmod \Delta k))\delta}\hat{x}_j(k)\right), r_j(k)\right) & \text{if } j \leq n, \\ \mathrm{Enc}\left(f\left(b^{(1+(k \bmod \Delta k))\delta}\hat{z}^{(1)}_{j-n}(k-1)\right), r_j(k)\right) & \text{otherwise} \end{cases} \tag{11.62}$$

for every $j \in \{1, \dots, n+p\}$ using random numbers $r_j(k) \in \mathbb{Z}_P^*$. The encrypted vector $c(k)$ is then send to the cloud, where the ciphertexts

$$v_i(k) := \left(\prod_{j=1}^{n+p} c_j(k)^{Z_{ij}}\right)\left(\prod_{j=n+p+1}^{n+2p} v_{j-n-p}(k-1)^{Z_{ij}}\right) \bmod P^2 \tag{11.63}$$

are evaluated and stored for every $i \in \{1, \dots, p\}$ with $Z_{ij} := f\left(b^\delta \widehat{\mathcal{L}}_{ij}\right)$. Next, the encrypted vector $v_i(k)$ is transmitted to the actuator, where $\check{\zeta}(k)$ is recovered from

$$\check{\zeta}_i(k) = b^{-(2+(k \bmod \Delta k))\delta}\varphi\left(\mathrm{Dec}(v_i(k)) \bmod Q\right) \tag{11.64}$$

and where the projection onto $\mathcal{Z}$ is carried out analogously to (11.39). Finally, the input $u(k) = C\check{z}^{(1)}(k)$ is applied and $\check{z}^{(1)}(k)$ is forwarded to the sensor. Obviously, Δk affects the encryption in (11.62) and the decryption in (11.64). For $\Delta k = 1$, we obtain similar statements as in Sects. 11.3.2 and 11.4.3. However, for $\Delta k > 1$, the exponents of the base b vary with k. Now, every Δk time-steps, we refresh the quantization of the encrypted $\check{\zeta}(k)$ as follows. Before a time-step k satisfying $k \bmod \Delta k = 0$, the actuator forwards not only $\check{z}^{(1)}(k-1)$ to the sensor but also $\check{\zeta}(k-1)$. At such a time-step, the sensor quantizes $\check{\zeta}(k-1)$, encrypts it by evaluating

$$v_{j-n-p}(k-1) := \mathrm{Enc}\left(f\left(b^\delta \hat{\zeta}^{(1)}_j(k-1)\right), r_j(k)\right)$$

for every $j \in \{n+p+1, \dots, n+2p\}$ using random numbers $r_j(k) \in \mathbb{Z}_P^*$, and sends it to the cloud in addition to (11.62). The cloud uses the received ciphertexts $v_{j-n-p}(k-1)$ instead of the internally buffered ones to compute $v(k)$ as in (11.63).

The modifications during the encryptions (11.62) and the decryptions (11.64) also affect the conditions for correctness of the encrypted operations. Nevertheless, the resulting requirements

$$\sum_{j=1}^{n} f\left(b^{(1+(k \bmod \Delta k))\delta}\hat{x}_j(k)\right)Z_{ij}$$

$$+ \sum_{j=n+1}^{n+p} f\left(b^{(1+(k \bmod \Delta k))\delta}\hat{z}_{j-n}^{(1)}(k-1)\right)Z_{ij}$$

$$+ \sum_{j=n+p+1}^{n+2p} f\left(b^{(1+(k \bmod \Delta k))\delta}\check{\zeta}_{j-n-p}(k-1)\right)Z_{ij} \in \mathbb{N}_P, \tag{11.65}$$

and

$$b^{(2+(k \bmod \Delta k))\delta}\left(\sum_{j=1}^{n} \widehat{\mathcal{L}}_{ij}\hat{x}_j(k) + \sum_{j=n+1}^{n+p} \widehat{\mathcal{L}}_{ij}\hat{z}_{j-n}^{(1)}(k-1) \right.$$

$$\left. + \sum_{j=n+p+1}^{n+2p} \widehat{\mathcal{L}}_{ij}\check{\zeta}_{j-n-p}(k-1) \right) \in \mathbb{Z}_Q^\varphi, \tag{11.66}$$

still show many similarities to the conditions (11.23) and (11.43). Moreover, robustness against quantization effects can, in principle, be analyzed with a disturbed model similar to (11.44). However, a solid analysis requires stronger stability results for the unencrypted control loop.

11.6 Comparison and Benchmark

The three discussed encrypted MPC schemes show many similarities. In fact, the computations carried out at the sensor, in the cloud, and at the actuator are structurally akin to each other. In the following, we specify those similarities but we also point out differences between the three schemes. In particular, by analyzing a numerical example, we show that the closed-loop behavior can vary significantly for the three controllers.

11.6.1 Structural Similarities and Differences

In all three schemes, the sensor is responsible for the encryption of the (augmented) states, the cloud implements the encrypted controller evaluation, and the actuator carries out the decryption. However, comparing the actual computations (11.20),

(11.40), and (11.62) at the sensor, we recognize that the number of encryptions per time step varies significantly. In fact, only n encryptions are carried out for the MPC based on MPP while $n + 2(Nm + Nn)$ encryptions are required for the real-time ADMM-based MPC during the refresh of the quantization of $\check{\zeta}$ (see second column in Tab. 11.2). Similarly, the cloud computations (11.21), (11.41), and (11.63) differ in terms of the numbers of required operations (see third and fourth columns in Tab. 11.2). It is, however, remarkable that the cloud implements only two types of operations. In fact, it solely evaluates modular exponentiations (Exp) and modular multiplications (Mul) referring to partially encrypted multiplications (enabled by (11.10)) and fully encrypted additions (enabled by (11.9)), respectively. Finally, the number of decryptions (11.22), (11.42), and (11.64) at the actuator differs as apparent from the fifth column in Table 11.2.

The listed operations in Table 11.2 allow to estimate the numerical effort associated with each scheme. To this end, we simply measure the effort for one elementary operation and extrapolate the total effort based on the operation counts in Table 11.2. We measure the effort for the elementary operations based on computation times resulting from a Python implementation running on a 2.7 GHz Intel Core i7-7500U. Not surprisingly, the effort for the various operations changes significantly with the "size" of the divisor P, i.e., the public key of the Paillier cryptosystem. As specified in Sect. 11.2.2, P results from the multiplication of the primes p_1 and p_2.

In Table 11.3 (that is taken from [26]), computations times for the four elementary operations Enc, Exp, Mul, and Dec are given for different lengths of the primes p_1

Table 11.2 Number of elementary operations performed at the sensor, in the cloud, and at the actuator for the discussed encrypted control schemes. In the cloud, modular exponentiations (Exp) and modular multiplications (Mul) are carried out. The dimensions of the primal decision variables are $p_I := Nm$ for PGS (and MPP) and $p_{II} := Nm + Nn$ for ADMM, respectively

Encrypted control scheme	Sensor	Cloud		Actuator
	Enc	Exp	Mul	Dec
MPC via MPP (variant 1)	n	mn	mn	m
MPC via MPP (variant 2)	n	mns	mns	m
MPC via real-time PGS	$n + p_I$	$p_I(n + p_I)$	$p_I(n + p_I - 1)$	p_I
MPC via real-time ADMM	$n + p_{II}$	$p_{II}(n + 2p_{II})$	$p_{II}(n + 2p_{II} - 1)$	p_{II}
MPC via real-time ADMM (refresh)	$n + 2p_{II}$	$p_{II}(n + 2p_{II})$	$p_{II}(n + 2p_{II} - 1)$	p_{II}

Table 11.3 Average computation times (in milliseconds from [26]) for the elementary operations in Table 11.2 and different lengths of the primes p_1 and p_2 defining the Paillier encryption (see Sect. 11.2.2)

Length l	Enc	Exp	Mul	Dec
512	19.976	1.139	0.016	18.982
1024	150.981	4.629	0.057	143.198
1536	406.075	8.105	0.099	397.140

and p_2. The listed bit-lengths l reflect past (512), current (1024), and future (1536) recommendations to ensure security for factoring-based encryption schemes like Paillier (see [18, Table 2]). Now, the combination of the data in Tables 11.2 and 11.3 allows to estimate the numerical effort for each scheme. This is illustrated for the numerical example in the following section.

11.6.2 Numerical Example

We apply all three schemes to the double-integrator system with the matrices

$$A = \begin{pmatrix} 1 & 1 \\ 0 & 1 \end{pmatrix} \quad \text{and} \quad B := \begin{pmatrix} 0.5 \\ 1 \end{pmatrix}.$$

and the constraints $\mathcal{X} := \{x \in \mathbb{R}^2 \mid |x_1| \leq 25, |x_2| \leq 5\}$ and $\mathcal{U} := [-1, 1]$. We consider the weighting matrices $Q = I_2$ and $R = 0.1$. The prediction horizon is set to $N = 12$. For the MPP-based approach, we choose the terminal weighting P as the solution of the (discrete-time) algebraic Riccati equation

$$A^\top (P - P B (R + B^\top P B)^{-1} B^\top P) A - P + Q = 0_{n \times n}. \quad (11.67)$$

Moreover, in addition to $\mathcal{X}$ and $\mathcal{U}$, we consider the standard terminal set (see [12])

$$\mathcal{T} := \{x \in \mathbb{R}^2 \mid (A + BK^*)^k x \in \mathcal{X}, \ K^*(A + BK^*)^k x \in \mathcal{U}, \ \forall k \in \mathbb{N}\}, \quad (11.68)$$

where $K^* := (R + B^\top P B)^{-1} B^\top P A$ reflects LQR. For the ADMM-based approach, we consider $\mathcal{X}$ and $\mathcal{U}$ since they can be written as box constraints with

$$x_{\min} := \begin{pmatrix} -25 \\ -5 \end{pmatrix}, \quad x_{\max} := \begin{pmatrix} 25 \\ 5 \end{pmatrix}, \quad u_{\min} = -1, \quad \text{and} \quad u_{\max} = 1.$$

Furthermore, we also choose P from (11.67). However, we ignore the terminal set (11.68) and set $\mathcal{T} = \mathcal{X}$. Regarding the ADMM iterations, we choose $\rho = 1$, D_z and D_μ as in Sect. 11.5.2, and $\mu^{(0)}(0) := 0_p$. Moreover, we use the unconstrained LQR input and state sequences

$$z^{(0)}(0) := \begin{pmatrix} K^*(A + BK^*)^0 \\ \vdots \\ K^*(A + BK^*)^{N-1} \\ (A + BK^*)^1 \\ \vdots \\ (A + BK^*)^N \end{pmatrix} x_0$$

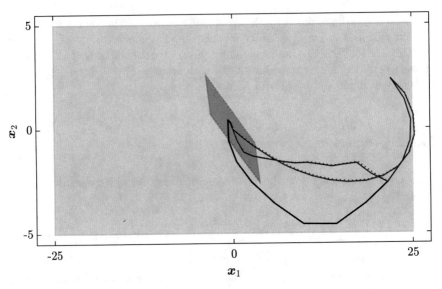

Fig. 11.6 Illustration of the state trajectories for the MPCs based on MPP (black), PGS (red), and ADMM (blue). The solid trajectories refer to the encrypted versions whereas dotted trajectories reflect unencrypted implementations. The light gray set in the background illustrates the state constraints $\mathcal{X}$. The dark gray set is the terminal set $\mathcal{T}$ from (11.68)

as a state-dependent initialization for $z^{(0)}(0)$. Finally, for the PGS-based approach, we consider the input constraints $\mathcal{U}$ only and set $\mathcal{X} = \mathcal{T} = \mathbb{R}^2$. Regarding the choice of P, we take (11.34) into account. Strictly speaking, (11.34) cannot be applied here since A has eigenvalues on the unit circle. Hence, we choose P from (11.34) but we set $R = 10^{-4}$ (only) during the computation of P. Concerning the PGS iterations, we consider the update matrix D from (11.35) and $z^{(0)} := 0_p$. For the remaining choice of ρ, we need to account for $\rho \in (0, 2\lambda_{\max}^{-1}(H))$. Based on the maximal eigenvalue of H, we infer the upper bound $2\lambda_{\max}^{-1} = 3.7593 \cdot 10^{-4}$ and choose $\rho = 10^{-4}$.

The parameters above configure the unencrypted versions of all three control schemes. In order to complete the configuration for the encrypted controllers, we select the parameters $b = 2$, $\gamma = 5$, and $\delta = 10$ for the quantization. In other words, we consider signed 16-bit quantizations with 5 integer bits and 10 fractional bits. We further choose $Q = 2^{128}$ and prime numbers p_1 and p_2 of length $l = 1024$. On the one hand, these choices ensure that the conditions (11.23), (11.43), and (11.65), (11.66) are satisfied during the simulation. On the other hand, the choice of l reflects current security recommendations (see, e.g., [18]). Regarding the ADMM-based approach, we refresh the quantization of $\check{\zeta}$ every $\Delta k = 5$ time steps. We next simulate the closed-loop behavior for all three controllers and their unencrypted and encrypted versions. For every scenario, the simulation starts with the initial state

$$x_0 := \begin{pmatrix} 22 \\ 2.4 \end{pmatrix}.$$

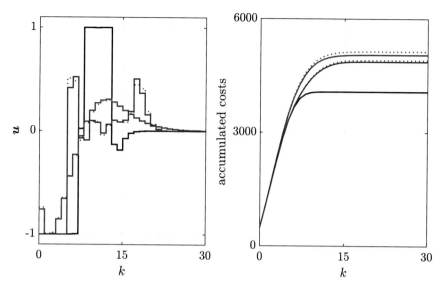

Fig. 11.7 Illustration of the input sequences (left) and the accumulated stage cost (right) for the MPCs based on MPP (black), PGS (red), and ADMM (blue). The solid curves refer to the encrypted versions whereas dotted curves reflect unencrypted implementations

The six resulting state trajectories are illustrated in Fig. 11.6. Apparently, all trajectories are converging to the origin. This is interesting since asymptotic stability is only guaranteed for the unencrypted MPC based on MPP. In fact, for the encrypted MPC via MPP, quantization errors have not been considered during the design phase. For the MPC via PGS, existing stability guarantees do not cover systems with unstable (i.e., not Schur stable) A matrices. For the MPC via ADMM, convergence is only guaranteed in a neighborhood of the origin. Apart from convergence, the six trajectories show major and minor deviations from each other. Major deviations occur for the three different control schemes. Minor deviations reflect the quantization-based differences between the unencrypted and encrypted versions of each scheme. Not surprisingly, similar deviations also arise for the input sequences and the accumulated stage costs $x(k)^\top Qx(k) + u(k)^\top Ru(k)$ in Fig. 11.7.

Studying the state, input, and cost sequences in more detail, three interesting observations can be made. First, since all three control schemes take $\mathcal{U}$ explicitly into account, input constraints are never violated (see Fig. 11.7). Second, constraints on states are only considered by the approaches based on MPP and ADMM. In contrast, the PGS-based scheme neglects state constraints. This explains why the red trajectories in Fig. 11.6 temporarily leave $\mathcal{X}$ while the black and blue trajectories stay inside the set. It is, however, important to note that state constraints satisfaction is guaranteed only for the MPP-based approach. In fact, the suboptimal real-time ADMM iterations currently come without such guarantees. Third, the accumulated stage costs in Fig. 11.7 suggest that the MPP-based approach performs best followed by the ADMM-based scheme and the MPC via PGS. Clearly, the superior

Table 11.4 Number of elementary operations and accumulated computation times (Δt in milliseconds) for primes with $l = 1024$ at the sensor, in the cloud, and at the actuator for the numerical example

Encrypted control scheme	Sensor		Cloud			Actuator	
	Enc	Δt	Exp	Mul	Δt	Dec	Δt
MPC via MPP (variant 1)	2	301.962	2	2	9.372	1	143.198
MPC via MPP (variant 2)	2	301.962	458	458	2146.188	1	143.198
MPC via real-time PGS	14	2113.734	168	156	786.564	12	1718.376
MPC via real-time ADMM	38	5737.278	2664	2628	12481.452	36	5155.128
MPC via real-time ADMM (refresh)	74	11172.594	2664	2628	12481.452	36	5155.128

performance of the MPP-based MPC is expectable since the QP (11.14) is solved to optimality.

It remains to comment on the numerical effort of the three encrypted controllers. To this end, we first count the number of elementary operations according to Table 11.2. For the present example, we have $n = 2$ states and $m = 1$ input. Moreover, the horizon length has been chosen as $N = 12$. Finally, solving (11.14) using MPP provides a piecewise affine solution (11.16) with $s = 229$ segments. Based on these dimensions, we find the operation counts in Table 11.4. Combining these counts with the computation times from Table 11.3 yield the listed efforts at the sensor, in the cloud, and at the actuator. As apparent from the computation times, the ADMM-based approach involves the highest numerical effort. In fact, based on the extrapolated times, up to 11.2, 12.5, and 5.2 s are required to perform the computations at the sensor, in the cloud, and at the actuator, respectively. We stress, however, that the extrapolated computation times do not necessarily reflect actual time spans required to perform the computations. In fact, most of the computations are highly parallelizable which enables more efficient implementations.

11.7   Conclusion and Outlook

In this chapter, we summarized and unified two existing encrypted predictive controllers and introduced a novel encrypted MPC based on real-time ADMM iterations (i.e., a single iteration per time step). The novel approach provides the first encrypted MPC that considers state and input constraints, that is based on online optimization, and that runs in a single cloud. These unique features come at a price. In fact, the numerical benchmark in Sect. 11.6.2 showed that the numerical effort for evaluating the novel scheme is higher than for the two existing encrypted MPCs from [31, 32].

This conclusion and some intermediate results of the chapter provide interesting starting points for future research. It is clearly of great interest to develop a more efficient implementation of an encrypted MPC with the above features. In this context, we note that the high numerical effort of the novel scheme mainly results from the high dimension of the decision variable z that reflects the predicted state and input sequences. One intuitive way to reduce this dimension is to consider an ADMM implementation where z reflects the predicted inputs only. Suitable ADMM-based MPC formulations exist (see, e.g., [11]). However, for those schemes, it is not straightforward to design appropriate warmstarts as required for efficient real-time iterations. Clearly, optimal control based on real-time solver iterations has useful applications beyond encrypted control. In this context, rigorous stability guarantees would significantly increase the applicability of real-time iterations. In turn, those guarantees would serve as a basis for certifying robustness of encrypted controllers against quantization effects. Finally, the computation times in Table 11.4 do not only reveal that the novel encrypted MPC is numerically demanding. In fact, the listed times also indicate that cloud-based implementations of encrypted control can be somewhat artificial. In particular, for (the first variant of) the MPP-based MPC from [32], the computations at the sensor and at the actuator are significantly more demanding than the computations in the cloud. Hence, including the cloud becomes questionable (although there might be reasons beyond computing performance). To avoid such effects, more efficient cryptosystems and more flexible encrypted operations are required. It seems that secure multi-party computation [23, 35] provides a powerful tool to satisfy this demand. Inspired by the promising results in [1, 2], future research on encrypted optimization-based control should increasingly focus on multi-cloud implementations.

Acknowledgements Support by the German Research Foundation (DFG) under the grant SCHU 2940/4-1 is gratefully acknowledged.

References

1. Alexandru AB, Morari M, Pappas GJ (2018) Cloud-based MPC with encrypted data. In: Proceedings of the 57th conference on decision and control, pp 5014–5019
2. Alexandru AB, Pappas GJ (2019) Encrypted LQG using labeled homomorphic encryption. In: Proceedings of the 10th ACM/IEEE international conference on cyber-physical systems, pp 129–140
3. Bemporad A, Morari M, Dua V, Pistikopoulos EN (2002) The explicit linear quadratic regulator for constrained systems. Automatica 38(1):3–20
4. Boccia A, Grüne L, Worthmann K (2014) Stability and feasibility of state constrained MPC without stabilizing terminal constraints. Syst Control Lett 72:14–21
5. Boyd S, Parikh N, Chu E, Peleato B, Eckstein J (2011) Distributed optimization and statistical learning via the alternating direction method of multipliers. Found Trends Mach Learn 3(1):1–122
6. Diehl M, Bock HG, Schlöder JP (2005) A real-time iteration scheme for nonlinear optimization in optimal feedback control. SIAM J Control Optim 43(5):1714–1736
7. ElGamal T (1985) A public key cryptosystem and a signature scheme based on discrete logarithms. IEEE Trans Inf Theory 31(4):469–472
8. Farokhi F, Shames I, Batterham N (2016) Secure and private cloud-based control using semi-homomorphic encryption. In: Proceedings of 6th IFAC workshop on distributed estimation and control in networked system

9. Farokhi F, Shames I, Batterham N (2017) Secure and private control using semi-homomorphic encryption. Control Eng Pract 67:13–20
10. Gentry C (2010) Computing arbitrary functions of encrypted data. Commun ACM 22(11):612–613
11. Ghadimi E, Teixeira A, Shames I, Johansson M (2015) Optimal parameter selection for the alternating direction method of multipliers (ADMM): quadratic problems. IEEE Trans Autom Control 60(3):644–658
12. Gilbert EG, Tan KT (1991) Linear systems with state and control constraints: the theory and application of maximal output admissible sets. IEEE Trans Autom Control 36(9):1008–1020
13. Jerez JL, Goulart PJ, Richter S, Constantinides GA, Kerrigan EC, Morari M (2014) Embedded online optimization for model predictive control at megahertz rates. IEEE Trans Autom Control 59(12):3238–3251
14. Katz J, Lindell Y (2014) Introduction to modern cryptography, 2nd edn. CRC Press
15. Kim J, Lee C, Shim H, Cheon JH, Kim A, Kim M, Song Y (2016) Encrypting controller using fully homomorphic encryption for security of cyber-physical systems. In: Proceedings of the 6th IFAC workshop on distributed estimation and control in networked systems, pp 175–180
16. Kogiso K, Fujita T (2015) Cyber-security enhancement of networked control systems using homomorphic encryption. In: Proceedings of the 54th conference on decision and control, pp 6836–6843
17. Mayne DQ, Rawlings JB, Rao C, Scokaert POM (2000) Constrained model predictive control: stability and optimality. Automatica 36:789–814
18. Barker E (2016) Recommendation for key management Part 1. NIST Spec Publ 800(57). National Institute of Standards and Technology
19. Nesterov Y (2004) Introductory lectures on convex optimization: a basic course, applied optimization, vol 87. Kluwer Academic Publishers
20. Nesterov Y (2013) Gradient methods for minimizing composite functions. Math Program 140(1):125–161
21. Paillier P (1999) Public-key cryptosystems based on composite degree residuosity classes. In: Advances in cryptology-eurocrypt '99. Lecture notes in computer science, vol 1592. Springer, pp 223–238
22. van Parys R, Pipeleers G (2018) Real-time proximal gradient method for linear MPC. In: Proceedings of the 2018 European control conference, pp 1142–1147
23. Pinkas B, Schneider T, Smart N, Williams S (2009) Secure two-party computation is practical. In: Advances in cryptology-asiacrypt 2009. Lecture notes in computer science, vol 5912. Springer, pp 250–267
24. Rawlings JB, Mayne DQ, Diehl MM (2017) Model predictive control: theory, computation, and design, 2nd edn. Nob Hill Publishing
25. Richter S, Jones CN, Morari M (2009) Real-time input-constrained MPC using fast gradient methods. In: Proceedings of the 48th IEEE conference on decision and control, pp 7387–7392
26. Schulze Darup M (2019) Verschlüsselte Regelung in der Cloud - Stand der Technik und offene Probleme. at - Automatisierungstechnik 67(8): 668–681
27. Schulze Darup M, Book G (2019) arXiv:1911.02641 [math.OC]
28. Schulze Darup M, Book G (2019) Towards real-time ADMM for linear MPC. In: Proceedings of the 2019 European control conference, pp 4276–4282
29. Schulze Darup M, Cannon M (2016) Some observations on the activity of terminal constraints in linear MPC. In: Proceedings of the 2016 European control conference, pp 4977–4983
30. Schulze Darup M, Jager T (2019) Encrypted cloud-based control using secret sharing with one-time pads. In: Proceedings of the 58th conference on decision and control
31. Schulze Darup M, Redder A, Quevedo DE (2018) Encrypted cloud-based MPC for linear systems with input constraints. In: Proceedings of 6th IFAC nonlinear model predictive control conference, pp 635–642
32. Schulze Darup M, Redder A, Shames I, Farokhi F, Quevedo D (2018) Towards encrypted MPC for linear constrained systems. IEEE Control Syst Lett 2(2):195–200
33. Shamir A (1979) How to share a secret. Commun ACM 53(3):97–105

34. Tøndel P, Johansen TA, Bemporad A (2002) Computation and approximation of piecewise affine control laws via binary search trees. In: Proceedings of the 41st conference on decision and control, pp 3144–3149
35. Yao AC (1982) Protocols for secure computations. In: Proceedings of the 23rd annual symposium on foundations of computer science, SFCS '82. IEEE Computer Society, pp 160–164

Chapter 12
Encrypted Control Using Multiplicative Homomorphic Encryption

Kiminao Kogiso

Abstract This chapter introduces the concept of controller encryption for enhancing the cybersecurity of networked control systems and presents how to encrypt a linear controller using a homomorphic public key ElGamal encryption system. A remarkable advantage of controller encryption is that it can conceal several pieces of information processed inside the controller device, such as controller parameters, references (recipes), measurements, control commands, and parameters of plant models in the internal model principle while maintaining the original function of the controller. A numerical example confirmed that only the scrambled parameters and signals can be seen in the controller device of the security-enhanced control system.

12.1 Introduction

A controller is the heart of an automatic control system and needs to operate safely and reliably. During operation, the controller processes sensor measurement (feedback) and reference (recipe) signals to yield adequate control inputs; controller parameters such as the proportional–integral–derivative (PID) gain are used as well. Thus, the controller is a device that gathers and contains all of the signals and parameters of the control system and stores essential and confidential production information for organizations and companies. For people who have access to the controller device and who can read the processed input/output signals and controller parameters, technical knowledge of control engineering makes guessing and identifying the dynamics of a plant and process not very complicated. Therefore, the contents of the controller

K. Kogiso (✉)
The University of Electro-Communications, 1-5-1 Chofugaoka, Chofu, Tokyo 1828585, Japan
e-mail: kogiso@uec.ac.jp

© Springer Nature Singapore Pte Ltd. 2020
F. Farokhi (ed.), *Privacy in Dynamical Systems*,
https://doi.org/10.1007/978-981-15-0493-8_12

must be protected from unauthorized accesses and be secluded from the view of many and unspecified persons, including adversaries. Companies need to keep the controller private. There have been some cyberattacks against controllers, such as falsifying a controller parameter to excite a natural frequency of a plant with Stuxnet [17] and switching off the Ukraine power grid [6].

These incidents have motivated research into the prevention and detection of cyberattacks against networked control systems. For example, several studies have addressed cyberattacks from a control theoretical viewpoint: modeling attacks for analysis and to consider potential countermeasures [25], model-based detection methods [18, 19, 24, 26], fallback and recovery control [23], and secure state estimation and resilient control upon the occurrence of a cyberattack [2, 10, 20]. Since 2015, studies have proposed fusing cryptography with a control system [1, 3–5, 8, 9, 11–16]; the present author's work [16] is a pioneer in the control engineering field. The primary purpose of these studies has been to conceal the feedback and control input signals as well as the controller parameters; the idea is to employ the homomorphism of public key encryption schemes into the controller's processing. This approach provides an interesting and significant property that the decryption process is not required to be inside the controller. This means that the controller does not need to keep any private keys to calculate the control input. Therefore, cryptography fusion can help enhance a control system's cybersecurity. Such a control system with built-in encryption is called an "encrypted control system." The issue of how to encrypt the controller is called "controller encryption."

This chapter focuses on encrypting a linear controller with multiplicative homomorphic encryption, which is an ElGamal encryption scheme. First, fundamental knowledge about the homomorphic encryption scheme is introduced. A remarkable feature of the encrypted control is that the controller calculates the encrypted output (i.e., control input) directly from the encrypted input with the encrypted controller parameters without any decryption processes. The resulting controller has only ciphertext, which means that the controller is kept private against unauthorized persons and adversaries. The numerical example in this chapter provides some hints on how to make codes to simulate encrypted control systems and illustrates how the encrypted controller works in a networked control system. Another concern is the computation cost of the encryption and decryption processes. Thus, the measured computation time are also presented for every security parameter from $2^5 = 32$ to $2^{14} = 16,384$ bits; this range include more security parameters than the practical ones (i.e., 1024 and 2048 bits).

This chapter uses the following notation:. $\mathbb{R}$ is the set of real numbers, $\mathbb{Z}^+$ is the set of non-negative integers, $\mathbb{Z}_n$ is a reduced residue system modulo n (i.e., $\mathbb{Z}_n := \{0, 1, \ldots, n-1\}$), $\mathcal{M}$ is a message space, $\mathcal{C}$ is a ciphertext space, $\mathbb{Z}_n^\times$ is the set of integers co-prime to n that belongs to $\mathbb{Z}_n$ (or is an multiplicative group of residues mudulo n), and 1^k is a binary string of length k.

12.2 Multiplicative Homomorphic Encryption Scheme

This section introduces the public key encryption scheme and fundamental information for concealing controllers. The general framework of the public key encryption scheme consists of three algorithms: key generation (Gen), encryption (Enc), and decryption (Dec).

- Gen: on input 1^k, generate the private key sk and the public key pk:

$$\mathsf{Gen}\left(1^k\right) = (\mathrm{pk}, \mathrm{sk}).$$

Determine a plaintext space (or message space) $\mathcal{M}$.
- Enc: on input of the public key pk and plaintext $m \in \mathcal{M}$, compute the ciphertext c:

$$\mathsf{Enc}(\mathrm{pk}, m) = c.$$

- Dec: on input of the private key sk (or both of pk and sk) and cihpertext $c \in \mathcal{C}$, compute the plaintext m':

$$\mathsf{Dec}(\mathrm{sk}, c) = m'.$$

Here, the cihpertext encrypted by Enc must be decrypted to be the same as the original plaintext.

$$m' = \mathsf{Dec}\,(\mathrm{sk}, \mathsf{Enc}(\mathrm{pk}, m)) = m, \ \forall m \in \mathcal{M}$$

For convenience, a public key encryption scheme is defined as $\mathcal{E} = (\mathsf{Gen}, \mathsf{Enc}, \mathsf{Dec})$ of the algorithms Gen, Enc, and Dec. The parameter k, which is called a security parameter or key length, denotes the length in bits of positive integers such as m and c that is admissible for the encryption scheme and provides a way of measuring the difficulty for an adversary to break the encryption scheme. Some public key encryption schemes preserve a map between the sets of plaintext and cihpertext (i.e., $\mathcal{M}$ and $\mathcal{C}$, respectively). Such an encryption scheme is called homomorphic.

The homomorphism of the public key encryption plays a significant role in this chapter, so it is defined below.

Definition 12.1 An encryption scheme $\mathcal{E} = (\mathsf{Gen}, \mathsf{Enc}, \mathsf{Dec})$ is *homomorphic* if the three following conditions are fulfilled:

1. The set $\mathcal{M}$ together with operation $\bullet$ and the set $\mathcal{C}$ together with operation $*$ form a group, respectively.
2. Any plaintext m in $\mathcal{M}$ is mapped into $\mathcal{C}$:

$$\mathsf{Enc}(\mathrm{pk}, m) \in \mathcal{C}, \ \forall m \in \mathcal{M}.$$

3. If, for any plaintext m_1 and $m_2 \in \mathcal{M}$ the corresponding ciphertext is written as

$$c_1 = \mathsf{Enc}(\mathsf{pk}, m_1) \in \mathcal{C} \text{ and } c_2 = \mathsf{Enc}(\mathsf{pk}, m_2) \in \mathcal{C},$$

then the following equation holds:

$$\mathsf{Enc}(\mathsf{pk}, m_1 \bullet m_2) = c_1 * c_2. \tag{12.1}$$

Definition 12.1 can be used to identify some types of homomorphic encryption schemes. If a homomorphic encryption scheme has an appropriate operation $*$ over the ciphertext such that $\bullet$ is a multiplicative operation over the plaintext, then the scheme is multiplicative homomorphic encryption. Examples include the RSA encryption [22] and the ElGamal encyrption [7]. If $\bullet$ is an additive operation, the scheme is an additive homomorphic encryption, such as Paillier encryption [21]. If $\bullet$ is both of addition and multiplication, the scheme is a fully homomorphic encryption. The difference in homomorphism leads to several types of encrypted control methods. This approach employs the multiplicative homomorphic encryption scheme because the resulting encrypted control methods are not complicated for implementation in real control systems.

The next section introduces the ElGamal encryption scheme; it uses a random variable and thus is capable of concealing signals and control parameters such as noise.

12.2.1 ElGamal Encryption

The ElGamal encryption scheme is defined in the form of the algorithms, $\mathcal{E}_E = (\mathsf{Gen}, \mathsf{Enc}, \mathsf{Dec})$.

- **Gen**: Generate the public key $\mathsf{pk} = (\mathbb{G}, q, g, h)$ and private key $\mathsf{sk} = s$, where q is the k-bit binary of a prime integer, $\mathbb{G} \subset \mathbb{Z}_p^\times$ is a cyclic group of the order q modulo prime p satisfying $p - 1 \bmod q = 0$, $g \in \mathbb{G}$ is a generator of $\mathbb{G}$, s is a randomly chosen element of $\mathbb{Z}_q$ so that $h = g^s \bmod p$, and the message space is $\mathcal{M} = \mathbb{G}$.
- **Enc**: Encrypt the plaintext $m \in \mathbb{G}$ with the public key $\mathsf{pk} = (\mathbb{G}, q, g, h)$ to compute the ciphertext C:

$$\begin{aligned} \mathsf{Enc}(\mathsf{pk}, m) &:= \left(g^r \bmod p, \ mh^r \bmod p\right) \\ &= (c_1, c_2) = C, \end{aligned} \tag{12.2}$$

where r is a random value uniformly chosen from $\mathbb{Z}_q$.
- **Dec**: Decrypt the ciphertext $C = (c_1, c_2)$ with the private key $\mathsf{sk} = s$ and public key $\mathsf{pk} = (\mathbb{G}, q, g, h)$ to compute the plaintext m':

$$\mathsf{Dec}(\mathrm{sk}, C) := c_2 \left(c_1^s\right)^{-1} \bmod p = m', \tag{12.3}$$

where $\left(c_1^s\right)^{-1}$ is a modular multiplicative inverse of the integer c_1^s modulo p. If the ciphertext C is encrypted by (12.2), then $m' = m$ holds because of the following:

$$m' = c_2 \left(c_1^s\right)^{-1} \bmod p = mh^r \left(g^{rs}\right)^{-1} \bmod p = mg^{rs}g^{-rs} \bmod p = m.$$

The properties of the ElGamal encryption are as follows. The encryption scheme $\mathcal{E}_E$ is not deterministic; two ciphertexts computed by $m_1 = m_2$ are not always the same because the random value r is used in the Enc algorithm. The message space $\mathcal{M} = \mathbb{G}$ generated by the above algorithms is a proper subgroup of the group $\mathbb{Z}_p^{\times}$. Therefore, the message space includes only a part of the elements of $\mathbb{Z}_p^{\times}$, and arbitrary integers more than 1 and less than p cannot be encrypted.

Multiplicative Homomorphism The ElGamal encryption allows homomorphic computation of the multiplication on the ciphertext. For the two plain texts m_1 and $m_2 \in \mathbb{G}$ and the corresponding ciphertexts in (12.2):

$$C_1 = \mathsf{Enc}(\mathrm{pk}, m_1) = \left(g^{r_1} \bmod p, \ m_1 h^{r_1} \bmod p\right),$$
$$C_2 = \mathsf{Enc}(\mathrm{pk}, m_2) = \left(g^{r_2} \bmod p, \ m_2 h^{r_2} \bmod p\right).$$

Then, the ciphertext of $m_1 m_2$, which is a multiplication of m_1 and m_2 with the random value $r = r_1 + r_2$, results in

$$\begin{aligned}
\mathsf{Enc}(\mathrm{pk}, m_1 m_2) &= \left(g^r \bmod p, \ m_1 m_2 h^r \bmod p\right) \\
&= \left(g^{r_1} g^{r_2} \bmod p, \ m_1 h^{r_1} m_2 h^{r_2} \bmod p\right) \\
&= C_1 \times_e C_2 \bmod p, \tag{12.4}
\end{aligned}$$

where the notation $\times_e$ in (12.4) denotes an element-wise (Hadamard) product of the C modulo p, and the relationship between the right and the left hand sides is illustrated in Fig. 12.1. This means that, in (12.1), the operation $\bullet$ over the set $\mathcal{M}$ is multiplication, and the operation $*$ over the set $\mathcal{C}$ is the Hadamard product.

Example

Based on the key generation algorithm, whose details are introduced in Sect. 12.4, g is set to 3. This results in the public key sk $= (\mathbb{G}, q, g, h)$:

$$\mathbb{G} = \{1, 3, 4, 7, 9, 10, 11, 12, 16, 17, 21, 23, 25, 26, 27, 28, 29, 30, 31, 33, 36, 37,$$
$$38, 40, 41, 44, 48, 49, 51, 59, 61, 63, 64, 65, 68, 69, 70, 75, 77, 78, 81\}$$
$$q = 41, \quad g = 3, \quad h = 9,$$

where $p = 83$ and the corresponding private key sk is found at $s = 2$.

Fig. 12.1 Two different paths to reach the encrypted $m_1 m_2$ from two plain messages m_1 and m_2

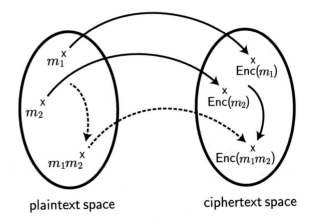

plaintext space   ciphertext space

For the plaintext $m = 21$ with the randomly chosen $r = 40$, the corresponding ciphertext is given by

$$\mathsf{Enc}(k_p, 21) = (3^{40}\mathrm{mod}\ 83,\ 21 \times 9^{40}\mathrm{mod}\ 83) = (40, 68).$$

Then, from the decryption (12.3) with the private-key $s = 2$, the following holds:

$$\mathsf{Dec}(k_s, (40, 68)) = 68\left(40^2\right)^{-1} \mathrm{mod}\ 83 = 21.$$

This confirms that a value travels around to be an original number.

The multiplicative homomorphism can be shown with $\mathsf{Enc}(\mathrm{pk}, 21) = (40, 68) := C_1$ and $\mathsf{Enc}(\mathrm{pk}, 3) = (16, 21) := C_2$. From the Hadamard product in (12.4) the following is obtained:

$$\mathsf{Enc}(\mathrm{pk}, 21 \times 3) = (40 \times 16\ \mathrm{mod}\ 83,\ 68 \times 21\ \mathrm{mod}\ 83) = (59, 17).$$

Then, the following holds:

$$\mathsf{Dec}(\mathrm{sk}, (59, 17)) = 63 = 21 \times 3.$$

This confirms that a multiplication exists that looks different but is essentially the same in the plaintext and ciphertext spaces, as illustrated in Fig. 12.1.

The homomorphism helps conceal linear operations in controllers. Its application is presented in the following section.

12.3 Encrypted Control

This section presents the incorporation of the previously discussed ElGamal encryption scheme into a control system to keep the controller's information secret. The approach of controller encryption is to conceal the input/output signals of the controller, such as feedback measurements and control inputs, and controller parameters such as the PID gain.

12.3.1 Controller Encryption

Consider the situation of a discrete-time linear controller being designed for a plant to achieve appropriate and desired control specifications of the networked control system:

$$f : \begin{cases} x(t+1) = Ax(t) + Bv(t) \\ u(t) = Cx(t) + Dv(t) \end{cases} \tag{12.5}$$

where t is a step; A, B, C, and D are parameters of the controller; $x \in \Re^{n_c}$ is a state of the controller; $v \in \Re^{m_c}$ is an input to the controller (e.g., output of the plant and command from an operator); and $u \in \Re^{l_c}$ is an output of the controller. Then, (12.5) can be rewritten as follows:

$$\begin{bmatrix} x(t+1) \\ u(t) \end{bmatrix} = f(\Phi, \xi(t)) = \Phi\xi(t). \tag{12.6}$$

Here, the parameter Φ and the state and the input ξ are as follows;

$$\Phi := \begin{bmatrix} A & B \\ C & D \end{bmatrix} \in \Re^{\alpha \times \beta}, \quad \xi := \begin{bmatrix} x \\ v \end{bmatrix} \in \Re^{\beta},$$

where $\alpha := n_c + l_c$ and $\beta := n_c + m_c$.

The parameters and the signals Φ and ξ are processed inside a control device or a particular computer to yield the output u. If someone gains unauthorized access to the control device, Φ and ξ may be monitored and recorded to inflict damage on the control system or to steal the information of the control operation based on the recorded ξ and u for the purpose of reverse engineering. Consider the problem of controller encryption, where all of the information appearing in the controller is encrypted. Then, the encrypted output is directly calculated from the encrypted input with the encrypted controller parameters without any decryption processes required inside the controller. The resulting controller has only ciphertext.

The concept of controller encryption is given below [16].

Definition 12.2 Assume that, given a linear controller f in (12.5) for a control system, the controller input v and output u are encrypted by the encryption scheme $\mathcal{E} = (\text{Gen}, \text{Enc}, \text{Dec})$. If there exists a map $f_{\mathcal{E}}$ such that the following holds:

$$f_{\mathcal{E}}\left(\text{Enc}(\text{pk}, \overline{\Phi}), \text{Enc}(\text{pk}, \overline{\xi})\right) = \text{Enc}\left(\text{pk}, \overline{f}(\Phi, \xi)\right), \qquad (12.7)$$

then $f_{\mathcal{E}}$ is an *encrypted controller* to f. Here, $\overline{\Phi} \in \mathcal{M}^{\alpha \times \beta}, \overline{\xi} \in \mathcal{M}^{\beta}$, and $\overline{f}(\cdot) \in \mathcal{M}^{\alpha}$ are the plaintexts obtained from $\Phi \in \Re^{\alpha \times \beta}, \xi \in \Re^{\beta}$, and $f(\cdot) \in \Re^{\alpha}$, respectively.

Above, (12.7) shows that the controller's encrypted output on the right-hand side is equivalent to the output calculated by the controller's encrypted parameters and encrypted state and input through $f_{\mathcal{E}}$. Note that no decryption processes are performed to calculate the encrypted output. Thus, this concept does not require the control device to keep any private keys to recover the encrypted signals. This is the main idea for keeping the controller's information secret.

This approach requires rounding the real number to the nearest integer in the plaintext space $\mathcal{M}$ because the plaintexts in (12.7), such as $\overline{\Phi}$ and $\overline{\xi}$, are whole numbers. Given the plaintext space $\mathcal{M}$ and the real value $\zeta \gamma \in \Re$, the following rounding operation is given:

$$\overline{\zeta} = \lceil (\gamma \zeta)^{+} \rceil_{\mathcal{M}}$$

where γ is a gain (scaling parameter) and $\lceil \cdot \rceil_{\mathcal{M}}$ is the nearest integer in $\mathcal{M}$ rounded from the representation. ζ^{+} denotes a map from the real number ζ to the nonnegative real number ζ^{+}:

$$\zeta^{+} = \begin{cases} \zeta & \text{if } \zeta \geq 0 \\ p + \zeta & \text{if } \zeta < 0 \end{cases},$$

where $\zeta \in \Re$ and $\zeta^{+} \in \Re^{+}$. The above map satisfies that the following equation:

$$\zeta + (-\zeta) \bmod p = \zeta + (p - \zeta) \bmod p = 0,$$

where $-\zeta$ is an addition inverse element and can be replaced with $p - \zeta$. This requirement gives rise to the quantization error:

$$\zeta - \text{Dec}(\text{sk}, \text{Enc}(\text{pk}, \overline{\zeta}))/\gamma.$$

Quantization errors lead to performance degradation. The relationship between the performance degradation and security parameter was revealed in [14]: as $k \to \infty$, the performance degradation converges to zero. Intuitively, this is because $|\mathbb{G}|$ increases with k, so the quantization error becomes approximately zero. Additionally, in the ElGamal encryption, the plaintext space of the cyclic group $\mathbb{G}$ is intermittent. Thus, care needs to be taken when choosing γ.

Here, it was assumed that there are no transmission delays and data losses over the communication links because the focus is on how the controller is encrypted.

12.3.2   Realization of the Encrypted Controller

To realize the encrypted controller $f_{\mathcal{E}}$, the homomorphism (12.1) for multiplication is used. However, the operation of the controller (12.6) consists of multiplication and addition. Thereby, the operation f is considered to be a composite product of addition f^{+} and multiplication $f^{\times}$:

$$f = f^{+} \circ f^{\times}.$$

Then, multiplication is defined as

$$f^{\times}(\varPhi, \xi) := \begin{bmatrix} \varPhi_1 \xi_1 & \varPhi_2 \xi_2 & \cdots & \varPhi_\beta \xi_\beta \end{bmatrix} =: \varPsi,$$

from a linear combination of the matrix $\varPhi$ and vector ξ:

$$\varPhi\xi = \varPhi_1 \xi_1 + \varPhi_2 \xi_2 + \cdots + \varPhi_\beta \xi_\beta = \sum_{l=1}^{\beta} \varPhi_l \xi_l.$$

The addition is defined as

$$f^{+}(\varPsi) := \sum_{l=1}^{\beta} \varPsi_l,$$

where ξ_l denotes the l-th component of the column vector ξ, $\varPhi_l$ denotes the l-th row vector of the matrix $\varPhi$, $\varPsi_l$ denotes the l-th row vector of the matrix $\varPsi$, and β is the maximum row number. Therefore, the multiplication $f^{\times}$ can be replaced with the homomorphism (12.1) to incorporate the ElGamal encryption into the calculation inside the controller device [16].

The encrypted controller is realized as follows. In the controller device, the multiplication $f^{\times}$ of the parameters $\varPhi$ and signals ξ, which can be concealed by the homomorphism, is processed. The controller device then outputs the encrypted matrix $\varPsi$ to the plant. In the decryption device just before the plant, the decryption of $\varPsi$ and the addition f^{+} of the decrypted $\varPsi$ are processed. The addition is separated from the original linear controller (12.5), so the decryption process Dec needs to be modified. The modified encrypted scheme $\mathcal{E}^{+} = (\mathsf{Gen}, \mathsf{Enc}, \mathsf{Dec}^{+})$ is introduced below:

Gen :   On input $1^k \rightarrow$ (pk, sk), generate a public key pk and private key sk similar to $\mathcal{E}$.

Enc : $\mathcal{M}^{\alpha} \rightarrow \mathcal{C}^{\alpha}$ This is the same as that in $\mathcal{E}$, and the plaintext (vector) $m = [m_1, m_2, \ldots, m_\alpha]^{\top} \in \mathcal{M}^{\alpha}$ is encrypted component-wise to compute

$$\mathsf{Enc}(\mathsf{pk}, m) = \begin{bmatrix} \mathsf{Enc}(\mathsf{pk}, m_1) \\ \mathsf{Enc}(\mathsf{pk}, m_2) \\ \vdots \\ \mathsf{Enc}(\mathsf{pk}, m_\alpha) \end{bmatrix}.$$

Here, a matrix version of $\mathsf{Enc}(\cdot, M)$ with $M \in \mathcal{M}^{\alpha \times \beta}$ that returns $C \in \mathcal{C}^{\alpha \times \beta}$ is also defined.

$\mathsf{Dec}^+ : \quad \mathcal{C}^{\alpha \times \beta} \to \mathcal{M}^\alpha$ the ciphertext (matrix):

$$C = \begin{bmatrix} c_{11} & c_{12} & \cdots & c_{1\beta} \\ c_{21} & c_{22} & \cdots & c_{2\beta} \\ \vdots & \vdots & & \vdots \\ c_{\alpha 1} & c_{\alpha 2} & \cdots & c_{\alpha\beta} \end{bmatrix},$$

is decrypted for each component by $\mathsf{Dec}(\mathsf{sk}, \cdot)$. Then, sum each row vector:

$$\mathsf{Dec}^+(\mathsf{sk}, C) = \sum_{i=1}^{\beta} \begin{bmatrix} \mathsf{Dec}(\mathsf{sk}, c_{1i}) \\ \mathsf{Dec}(\mathsf{sk}, c_{2i}) \\ \vdots \\ \mathsf{Dec}(\mathsf{sk}, c_{\alpha i}) \end{bmatrix} =: (f^+ \circ \mathsf{Dec})(\mathsf{sk}, C).$$

where $\mathsf{Dec}(\mathsf{sk}, C)$ gives the plaintext so that the last term is not strictly accurate. Here, $M = \overline{M}(= \mathsf{Dec}(\mathsf{pk}, C))$ with the plaintext (matrix) $\overline{M}$ is assumed to correspond to the real matrix M.

Consequently, the modified encryption scheme $\mathcal{E}^+$ can be used to encrypt the linear controller (12.5).

Theorem 12.1 *Assume that the quantization errors between Φ and $\overline{\Phi}$ and between ξ and $\overline{\xi}$ are approximately zero. For the linear controller f in (12.6), the controller $f_{\mathcal{E}^+}^\times : \mathcal{C}^{\alpha \times \beta} \times \mathcal{C}^\beta \to \mathcal{C}^{\alpha \times \beta}$ encrypted by the encryption scheme $\mathcal{E}^+ = (\mathsf{Gen}, \mathsf{Enc}, \mathsf{Dec}^+)$ is formulated as*

$$f_{\mathcal{E}^+}^\times(\mathsf{Enc}(\mathsf{pk}, \overline{\Phi}), \mathsf{Enc}(\mathsf{pk}, \overline{\xi}))$$
$$= \begin{bmatrix} \mathsf{Enc}(\mathsf{pk}, \overline{\Phi}_{11}) * \mathsf{Enc}(\mathsf{pk}, \overline{\xi}_1) & \cdots & \mathsf{Enc}(\mathsf{pk}, \overline{\Phi}_{1\beta}) * \mathsf{Enc}(\mathsf{pk}, \overline{\xi}_\beta) \\ \mathsf{Enc}(\mathsf{pk}, \overline{\Phi}_{21}) * \mathsf{Enc}(\mathsf{pk}, \overline{\xi}_1) & \cdots & \mathsf{Enc}(\mathsf{pk}, \overline{\Phi}_{2\beta}) * \mathsf{Enc}(\mathsf{pk}, \overline{\xi}_\beta) \\ \vdots & & \vdots \\ \mathsf{Enc}(\mathsf{pk}, \overline{\Phi}_{\alpha 1}) * \mathsf{Enc}(\mathsf{pk}, \overline{\xi}_1) & \cdots & \mathsf{Enc}(\mathsf{pk}, \overline{\Phi}_{\alpha\beta}) * \mathsf{Enc}(\mathsf{pk}, \overline{\xi}_\beta) \end{bmatrix},$$
$$\tag{12.8}$$

where $\overline{\Phi} \in \mathcal{M}^{\alpha \times \beta}$ and $\overline{\xi} \in \mathcal{M}^\beta$ are a plaintext matrix and vector, respectively, corresponding to $\Phi \in \mathfrak{R}^{\alpha \times \beta}$ and $\xi \in \mathfrak{R}^\beta$, and $\overline{\Phi}_{ij}$ denotes the ij-th component of the *matrix* $\overline{\Phi}$. *In particular, the case of* $f_{\mathcal{E}^+}^{\times} = f_{\mathcal{E}_E^+}^{\times}$ *indicates that the encrypted controller is based on the modified scheme of the ElGamal encryption.*

Proof The multiplicative homomorphism in $\mathcal{E}^+$ and the l-th row vector of (12.8) can be rewritten as follows:

$$
\begin{bmatrix}
\mathsf{Enc}(\mathsf{pk}, \overline{\Phi}_{1l}) * \mathsf{Enc}(\mathsf{pk}, \overline{\xi}_l) \\
\mathsf{Enc}(\mathsf{pk}, \overline{\Phi}_{2l}) * \mathsf{Enc}(\mathsf{pk}, \overline{\xi}_l) \\
\vdots \\
\mathsf{Enc}(\mathsf{pk}, \overline{\Phi}_{al}) * \mathsf{Enc}(\mathsf{pk}, \overline{\xi}_l)
\end{bmatrix}
=
\begin{bmatrix}
\mathsf{Enc}(\mathsf{pk}, \overline{\Phi}_{1l}\overline{\xi}_l) \\
\mathsf{Enc}(\mathsf{pk}, \overline{\Phi}_{2l}\overline{\xi}_l) \\
\vdots \\
\mathsf{Enc}(\mathsf{pk}, \overline{\Phi}_{al}\overline{\xi}_l)
\end{bmatrix}.
$$

Therefore, the left-hand side on the above equation is equivalent to the result of the encryption of the components of the multiplication $f^{\times}(\Phi, \xi)$ by $\mathsf{Enc}(\mathsf{pk}, \cdot)$. From the definition of Dec^+ and quantization errors of approximately zero with the sufficient large security parameter k, the following holds:

$$
\mathsf{Dec}^+(\mathsf{sk}, f_{\mathcal{E}^+}^{\times}(\mathsf{Enc}(\mathsf{pk}, \overline{\Phi}), \mathsf{Enc}(\mathsf{pk}, \overline{\xi}))) = (f^+ \circ f_{\mathcal{E}^+}^{\times})(\overline{\Phi}, \overline{\xi}) = f(\Phi, \xi),
$$
$$
\Leftrightarrow f_{\mathcal{E}^+}^{\times}(\mathsf{Enc}(\mathsf{pk}, \overline{\Phi}), \mathsf{Enc}(\mathsf{pk}, \overline{\xi})) = \mathsf{Enc}(\mathsf{pk}, \overline{f}(\Phi, \xi)).
$$

The operation of (12.8) does not require the private key sk, so the $\mathcal{E}^+$-encrypted controller $f_{\mathcal{E}^+}^{\times}$ satisfies the definition of the encrypted controller. □

As stated in Theorem 12.1, the modified encryption scheme can be used to encrypt a general class of linear controllers. Note that the proposed scheme does not allow addition for the state transition inside the controller device so that the updated state of the controller $x(t + 1)$ in (12.5) can be seen for the first time just after the decryption process in front of the plant. Therefore, the state needs to be fed back to the controller device in order to update the controller state.

12.4 Simulation of Encrypted Control Systems

This section illustrates the behaviors of encrypted control systems with the ElGamal encryption algorithms. To obtain the behaviors, arithmetic operations were performed on integers of arbitrary size. This is because the cryptography-based control system requires arithmetic operations over larger digits in integers than 64 bits in the standard C library environment. In this study, the bignum library of OpenSSL API's Crypto Library[1] was used. The library offers several functions for arbitrary-precision arithmetic operations that work on numbers and whose digits of precision are limited only by the available memory of the used computer. This helps generate codes of encrypted control systems to numerically investigate the relationship between the quantization error and security parameter and to measure the computation burden.

---

[1]Documents of the OpenSSL API are available at https://www.openssl.org.

12.4.1 Simulation Codes

This section introduces pseudocodes of the ElGamal encryption algorithms Gen, Enc, and Dec and of the encrypted control simulation.

The **Gen** algorithm in Algorithm 12.2 outputs the encryption keys pk and sk and also computes the safe prime p and width d around zero in the plaintext space that are used in a rounding operation. In the algorithm, "is_prime()" on lines 6 and 8 is a function that determines whether a given number is prime. In the case of the bignum library, the function is replaced with "BN_is_prime_ex()," and "Rand()" on line 24 returns a random number that can be replaced with "BN_rand_range()."

Algorithm 12.2 Gen

Require: $k \in \mathbb{N}$
Ensure: pk $= \{\mathbb{G}, q, g, h\}$, sk $= s$, p, d
1: $min \leftarrow 2^k + 1$
2: $mid \leftarrow 0$
3: $max \leftarrow 2^{k+1}$
4: # search Sophie Germain prime
5: **while** $min < max$ **do**
6:     **if** is_prime(min) $== 1$ **then**
7:         $mid \leftarrow 2min + 1$
8:         **if** is_prime(mid) $== 1$ **then**
9:             $q \leftarrow min$, $p \leftarrow mid$
10:             **break**
11:         **end if**
12:     **end if**
13:     $min \leftarrow min + 2$
14: **end while**
15: # search generator $g \in \mathbb{G}$
16: $g \leftarrow 2$
17: **while** $g < p$ **do**
18:     **if** $g^q \bmod p == 1$ **then**
19:         **break**
20:     **end if**
21:     $g \leftarrow g + 1$
22: **end while**
23: # compute s and h
24: $s \leftarrow$ Rand(q)
25: $h \leftarrow g^s \bmod p$
26: # search width $d \in \mathbb{Z}$
27: $d \leftarrow p - 1$
28: **while** $g < d$ **do**
29:     **if** $d^q \bmod p == 1$ **then**
30:         $d \leftarrow p - d$
31:         **break**
32:     **end if**
33:     $d \leftarrow d - 1$
34: **end while** return pk, sk, p, d

Algorithms 12.3 and 12.4 are the encryption and decryption algorithms, respectively, in Sect. 12.2.1. In the setting of the encrypted control systems, the feedback signals and controller parameters are quantized before the encryption. This means that they are rounded to the nearest integer. Thus, the rounding function is also needed, which is in the **Appendix** of this chapter.

The encrypted controls can be simulated by following the pseudocode of Algorithm 12.5. The processes from line 1 to line 7 encrypt the controller parameters Φ and install the encrypted parameters to the controller. The processes from line 13 to line 24 are done on the plant side and include reading a sensor's value, decrypting the received information Ψ with Dec^+, actuating the plant, and encrypting the signals ξ to send the controller. The processes at lines 26 and 27 run in the controller to perform the multiplicative homomorphic operation and send the encrypted Ψ to the plant. After these processes are repeatedly computed, the time responses of the encrypted control systems can be obtained to check their behavior. To investigate the real-time computation for the established encrypted control system, the execution time of processes within the loop (i.e., lines 13–27) should be measured.

Algorithm 12.3 Enc

Require: $m \in \mathbb{G}$, pk $= \{\mathbb{G}, q, g, h\}$, p
Ensure: $C \in \mathbb{G} \times \mathbb{G}$
1: $r \leftarrow \mathrm{Rand}(q)$
2: $c_1 \leftarrow g^r \bmod p$
3: $c_2 \leftarrow m(h^r \bmod p) \bmod p$
4: **return** $C \leftarrow (c_1, c_2)$

Algorithm 12.4 Dec

Require: $C \in \mathbb{G} \times \mathbb{G}$, pk $= \{\mathbb{G}, q, g, h\}$, sk $= s$, p
Ensure: $m' \in \mathbb{G}$
1: $m' \leftarrow c_2\{(c_1^s \bmod p)^{-1} \bmod p\} \bmod p$
2: **if** $m' > q$ **then**
3: $\quad m' \leftarrow m' - p$
4: **end if**
5: **return** m'

The next section illustrates the encrypted control simulation results obtained from using the codes in a C language environment on a MacBook Air (1.6 GHz Intel Core i5 with 8 GB memory).

12.4.2 Numerical Example

For this numerical simulation, a PID feedback control system is considered. A plant is the following two-dimensional linear time-invariant system:

$$\dot{x}_p(\tau) = \begin{bmatrix} 0 & 2 \\ -2 & -3 \end{bmatrix} x_p(\tau) + \begin{bmatrix} 0 \\ 1 \end{bmatrix} u(\tau),$$

$$y(\tau) = \begin{bmatrix} 1 & 0 \end{bmatrix} x_p(\tau),$$

where τ is the time, x_p is the state of the plant, u is the input, and y is the output. For the simulation, the plant model is discretized by a zero-order hold with a sampling period of 10 ms. A PID controller is designed with the gains pk $= 8$, $K_I = 0.1$, and $K_D = 0.01$, and the controller is discretized by the zero-order hold with the same sampling period. This results in the following controller:

$$\begin{bmatrix} x(t+1) \\ u(t) \end{bmatrix} = \begin{bmatrix} 1 & 0.0063 & 0 \\ 0 & 0.3678 & 0.0063 \\ 10 & -99.90 & 3 \end{bmatrix} \begin{bmatrix} x(t) \\ -y(t) \end{bmatrix} := \Phi\xi(t),$$

where t is the step, x is the state of the controller, Φ is the controller parameter, and ξ is considered an input to the controller.

For example, let the security parameter be set to 1024 bits. In this case, the public and private keys obtained by Algorithm 12.2 result in $g = 3$:

Algorithm 12.5 Encrypted Control Simulation

Require: pk, sk, p, d, Φ, $\gamma_p \geq 0$, $\gamma_c \geq 0$
1: **for** $i = 1$ to α **do**
2:     **for** $j = 1$ to β **do**
3:         enc_Φ_{ij} $\leftarrow$ Enc(roundG($\gamma_c \Phi_{ij}$, pk, p, d), pk, p)
4:     **end for**
5: **end for**
6: # Controller side
7: Set pk, p, enc_Φ, and ξ_0 $\leftarrow$ $(x(0),\ y(0))$.
8: **for** $i = 1$ to β **do**
9:     enc_ξ_{0i} $\leftarrow$ Enc(roundG($\gamma_p \xi_{0i}$, pk, p, d), pk, p)
10: **end for**
11: Send enc_Ψ $\leftarrow$ enc_Φ $\times$ enc_ξ_0 mod p to the plant.
12: **loop**
13:     # Plant side
14:     $y \leftarrow$ SensorRead()
15:     Receive enc_Ψ from the controller.
16:     **for** $i = 1$ to α **do**
17:         $(x,\ u)_i \leftarrow \Sigma_{j=1}^{\beta}$Dec(enc_$\Psi_{ij}$, pk, sk, p)/$(\gamma_p \gamma_c)$
18:     **end for**
19:     Actuate(u)
20:     $\xi \leftarrow (x,\ y)$
21:     **for** $i = 1$ to β **do**
22:         enc_ξ_i $\leftarrow$ Enc(roundG($\gamma_p \xi_i$, pk, p, d), pk, p)
23:     **end for**
24:     Send enc_ξ to the controller.
25:     # Controller side
26:     enc_Ψ $\leftarrow$ enc_Φ $\cdot$ enc_ξ mod p
27:     Send enc_Ψ to the plant.
28: **end loop**

$q = 80000000000000000000000000000000000000000000000000000000000000000$
$00000000000000000000000000000000000000000000000000000000000000000$
$00000000000000000000000000000000000000000000000000000000000000000$
$00000000000000000000000000000000000000000000000000000000000000000$
$0000000CA605,$

$h = 1E243CE692432B4C5B3E6072E73A9B38F61270434E373CFD21E3160$
$A633113D13EB2B6805C20052074C444F343F48EE7EAEC9688311477$
$E657F40AFA97EE6A1CC817E3D74BAA5963D9A454CA03C9D5B8AC$
$764D361D0BD9AAE2889403FB595EAC0A4C1D864DD2DF26067D6E$
$A2BAF3ABCADD17F7B411FD8067E3991EA4827E01E4,$

and

$s = 335CF6903F8A593384689C8E70AD33BAA4B45B8AEA8FFCF630E7$
$B53E1A0DA0BBE275802080FE0E395A5044EDB34A044EDFD3A9F$
$041EE5EC2C17FC409C419F78E5E8171FC5EB0BD09F1290138A1D$
$736210FD68E09D4CDE1239A19BED52F67771DB63DB347827D662$
$5870A078BE1C2FAB3529E321C0C5790C6BB2CBFACA69EC42F,$

where the Sophie Germain prime p is

$p = 10000000000000000000000000000000000000000000000000000000000000000$
$00000000000000000000000000000000000000000000000000000000000000000$
$00000000000000000000000000000000000000000000000000000000000000000$
$00000000000000000000000000000000000000000000000000000000000000000$
$0000000194C0B.$

Additionally, $d = 2$ and the gains of the quantizers are set to $\gamma_c = \gamma_p = 2^{500}$ in an empirical manner.

The simulation results of the encrypted control system are shown in Fig. 12.2. Figure 12.2a, b show the time responses of the encrypted input and output, respectively; these are signal behaviors inside the controller and over the communication links between the plan and controller. They look exactly like noise; this is because those signals and the controller parameters are expressed in ciphertext; each follows an almost uniform distribution. Figure 12.2c, d show the time responses of the control signal and sensor measurement, respectively. The time responses of the normal PID control system are also plotted for comparison. Figure 12.2e, f show the input and output errors between the normal and encrypted PID control systems. These figures numerically confirm that the information about the controller that includes its parameters and the signals is concealed, and the impact of encrypting the controller on the control system is negligible. However, there is a technical concern about real-time computation for the practical use of this security parameter.

Real-time computation is required for practical control application. From the viewpoint of information security, a practical security parameter should be 1024, 2048, or more bits. Thus, such practical security parameters were considered to evaluate whether real-time computation could be achieved. The developed simulation

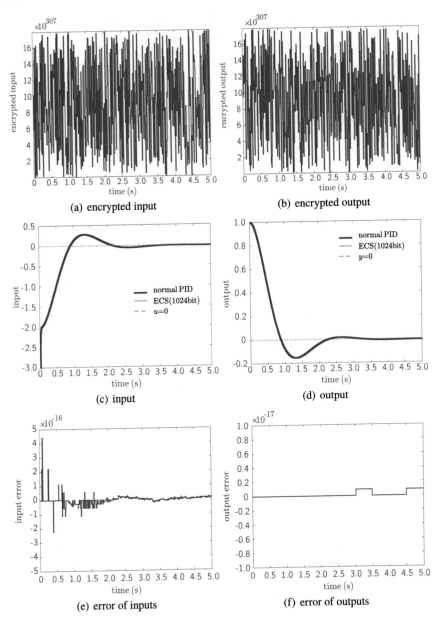

Fig. 12.2 Simulation results for the case with 1024 bits

codes were used to measure the computation time for security parameters from 2^5 to 2^{14} (i.e., from 32 bits to 16,384 bits). Fig. 12.3a shows the computation time from line 14 to line 27 in Algorithm 12.5 for the security parameters. Real-time computation could be achieved with a security parameter of less than 128 bits because the

Fig. 12.3 Processing time
for the encrypted control
system using a MacBook Air
(1.6 GHz Intel Core i5 with 8
GB memory)

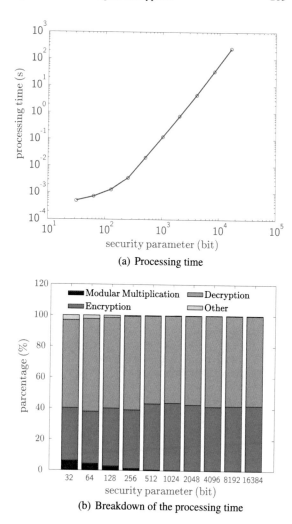

(a) Processing time

(b) Breakdown of the processing time

measurement time was 6.2 ms on average, which was within the sampling period.
The percentage that each of the four processes took up in a control period was investigated: decryption, encryption, multiplicative homomorphic operation, and other
processes. The results are shown in Fig. 12.3b. Most of the total time was used to
execute the encryption and decryption processes, especially for security parameters with more than 1024 bits. These results should motivate the development of
an efficient and lightweight encryption algorithm or fast processing device that can
reducing the computation burden.

12.5 Conclusion

This chapter introduces the concept of encrypting linear controllers with the ElGamal encryption scheme, which uses multiplicative homomorphic encryption. With controller encryption, the controller calculates the encrypted output (i.e., control input) directly from the encrypted input with the encrypted controller parameters using the modified decryption process. The numerical example showed that the encrypted control system provides feedback and control input signals that look like noise and that are concealed by the encryption scheme without any system's performance being altered.

Future work will involve developing a more efficient and secure encrypted control system with dynamic keys to resolve the real-time computation issues for realizing a practical control system.

Algorithm 12.6 roundG

Require: $u \in \mathbb{R}$, pk, p, d
Ensure: $m \in \mathcal{M}$
1: **if** $u \geq 0$ **then**
2:　　$mid \leftarrow$ round(u)
3:　　**if** $mid \leq 1$ **then**
4:　　　　**return** $m \leftarrow 1$
5:　　**else if** $q \leq mid$ **then**
6:　　　　$upper\_limits \leftarrow q$, $lower\_limits \leftarrow 1$, $upper \leftarrow q$, $lower \leftarrow q$
7:　　　　**return** CheckLower($upper$, $lower$, $upper\_limits$, $lower\_limits$, pk)
8:　　**else**
9:　　　　$upper\_limits \leftarrow q$, $lower\_limits \leftarrow 1$, $upper \leftarrow mid + 1$, $lower \leftarrow mid$
10:　　　　**return** CheckLower($upper$, $lower$, $upper\_limits$, $lower\_limits$, pk)
11:　　**end if**
12: **else**
13:　　$mid \leftarrow$ round$(-u)$
14:　　**if** $q \leq mid$ **then**
15:　　　　$upper\_limits \leftarrow p$, $lower\_limits \leftarrow q$, $upper \leftarrow q$, $lower \leftarrow q$
16:　　　　**return** CheckUpper($upper$, $lower$, $upper\_limits$, $lower\_limits$, pk)
17:　　**else if** $mid \leq d$ **then**
18:　　　　**if** $d - 2mid \leq 1$ **then**
19:　　　　　　**return** $m \leftarrow p - d$
20:　　　　**else**
21:　　　　　　**return** $m \leftarrow 1$
22:　　　　**end if**
23:　　**end if**
24:　　$tmp \leftarrow 2mid + 1 - $ceil$(mid - u)$, $mid \leftarrow p - mid$, $upper\_limits \leftarrow p - 1$
25:　　**if** $tmp < q$ **then**
26:　　　　$lower\_limits \leftarrow p - tmp$
27:　　**else**
28:　　　　$lower\_limits \leftarrow q + 1$
29:　　**end if**
30:　　$upper \leftarrow mid$, $lower \leftarrow mid - 1$
31:　　**return** CheckUpper($upper$, $lower$, $upper\_limits$, $lower\_limits$, pk)
32: **end if**
33: **return** 0

Acknowledgements The author express his gratitude to Mr. Masahiro Kusaka, a PhD student at the University of Electro-Communications, who developed the software for simulating encrypted control systems and conducted practical verification of the encrypted control methods.

Appendix

This appendix presents the algorithms for rounding real numbers (quantization).

Algorithm 12.7 functions called in the roundG

1: **function** CHECKUPPER(*upper, lower, upper_limits, lower_limits,* pk)
2:    **if** $upper^q \equiv 1 \bmod p$ **then**
3:       **return** $m \leftarrow upper$
4:    **else**
5:       $upper \leftarrow upper + 1$
6:       **if** $lower\_limits \leq lower$ **then**
7:          **return** CheckLower(*upper, lower, upper_limits, lower_limits,* pk)
8:       **else if** $upper \leq upper\_limits$ **then**
9:          **return** CheckUpper(*upper, lower, upper_limits, lower_limits,* pk)
10:       **end if**
11:    **end if**
12:    **return** 0
13: **end function**
14: **function** CHECKLOWER(*upper, lower, upper_limits, lower_limits,* pk)
15:    **if** $lower^q \equiv 1 \bmod p$ **then**
16:       **return** $m \leftarrow lower$
17:    **else**
18:       $lower \leftarrow lower - 1$
19:       **if** $upper \leq upper\_limits$ **then**
20:          **return** CheckUpper(*upper, lower, upper_limits, lower_limits,* pk)
21:       **else if** $lower\_limits \leq lower$ **then return** CheckLower(*upper, lower, upper_limits, lower_limits,* pk)
22:       **else return** $m \leftarrow 1$
23:       **end if**
24:    **end if**
25:    **return** 0
26: **end function**

References

1. Alexandru AB, Morari M, Pappas GJ (2018) Cloud-based MPC with encrypted data. In: IEEE conference on decision and control, pp 5014–5019. Miami Beach, FL
2. Chong MS, Wakaiki M, Hespanha JP (2015) Observability of linear systems under adversarial attacks. In: American control conference, pp 2439–2444
3. Darup MS, Redder A, Quevedo DE (2018) Encrypted cloud-based MPC for linear systems with input constraints. IFAC-PapersOnLine 51(20):535–542
4. Darup MS, Redder A, Quevedo DE (2019) Encrypted cooperative control based on structured feedback. IEEE Control Syst Lett 3(1):37–42
5. Darup MS, Redder A, Shames I, Farokhi F, Quevedo DE (2018) Towards encrypted MPC for linear constrained systems. IEEE Control Syst Lett 2(2):195–200

6. E-ISAC: (2016) Analysis of the Cyber Attack on the Ukrainian Power Grid. SANS Industrial Control Systems
7. ElGamal T (1984) A public key cryptosystem and a signature scheme based on discrete logarithms. In: CRYPTO '84, vol 196, pp 10–18
8. Farokhi F, Shames I, Batterham N (2016) Secure and private cloud-based control using semi-homomorphic encryption. IFAC-PapersOnLine 49(22):163–168
9. Farokhi F, Shames I, Batterham N (2017) Secure and private control using semi-homomorphic encryption. Control Eng Pract 67:13–20
10. Fawzi H, Tabuada P, Diggavi S (2014) Secure estimation and control for cyber-physical systems under adversarial attacks. IEEE Trans Autom Control 59(6):6294–6303
11. Kim J, Lee C, Shim H, Cheon JH, Kim A, Kim M, Song Y (2016) Encrypting controller using fully homomorphic encryption for security of cyber-physical systems. IFAC-PapersOnLine 49(22):175–180
12. Kishida M (2018) Encrypted average consensus with quantized control law. In: IEEE Conference on decision and control, pp 5850–5856. Miami Beach, FL
13. Kishida M (2019) Encrypted control system with quantizer. IET Control Theor Appl 13(1):146–151
14. Kogiso K (2018) Attack detection and prevention for encrypted control systems by application of switching-key management. In: IEEE Conference on decision and control, pp 5032–5037
15. Kogiso K (2018) Upper-bound analysis of performance degradation in encrypted control system. In: American control conference, pp 1250–1255
16. Kogiso K, Fujita T (2015) Cyber-security enhancement of networked control systems using homomorphic encryption. In: IEEE conference on decision and control, pp 6836–6843
17. Langner R (2011) Stuxnet: dissecting a cyberwarfare weapon. IEEE Secur Priv 9(3):49–51
18. Miao F, Pajic M, Pappas GJ (2013) Stochastic game approach for replay attack detection. In: IEEE Conference on decision and control, pp 1854–1859
19. Mo Y, Chabukswar R, Sinopoli B (2014) Detecting integrity attacks on SCADA systems. IEEE Trans Control Syst Technol 22(4):1396–1407
20. Mo Y, Sinopoli B (2009) Secure control against replay attacks. In: Annual Allerton conference on communication, control, and computing, pp 911–918
21. Paillier P (1999) Public-key cryptosystem based on composite degree residuosity classes. In: EUROCRYPTO '99, vol 1592, pp 223–238
22. Rivest RL, Shamir A, Adleman L (1978) A method for obtaining digital signatures and public-key cryptosystem. Commun ACM 21:120–126
23. Sasaki T, Sawada K, Shin S, Hosokawa S (2017) Fallback and recovery control system of industrial control system for cybersecurity. IFAC-PapersOnLine 50(1):15247–15252
24. Sasaki T, Sawada K, Shin S, Hosokawa S (2017) Model based fallback control for networked control system via switched lyapunov function. IEICE Trans Fund Electron Commun Comput Sci E100-A(10), 2086–2094
25. Teixeira A, Pérez D, Sandberg H, Johansson KH (2012) Attack models and scenarios for networked control systems. In: Proceedings of the 1st international conference on high confidence networked systems, pp 55–64
26. Yaseen AA, Bayart M (2016) Attack-tolerant networked control system in presence of the controller hijacking attack. In: International conference on military communications and information systems, pp 1–8